# ONE WEEK LOAN

## UNIVERSITY OF GLAMORGAN
### TREFOREST LEARNING RESOURCES CENTRE
Pontypridd, CF37 1DL
Telephone: (01443) 482626
*Books are to be returned on or before the last date below*

**Renew Items on PHONE-it: 01443 654456**
**Help Desk: 01443 482625 Media Services Reception: 01443 482610**
*Items are to be returned on or before the last date below*

# MODERN METHODS OF VALUATION

## OF

## LAND, HOUSES AND BUILDINGS

(Ninth Edition)

by

### TONY JOHNSON

BSc (Estate Management) (Lond), FRICS, Honorary Fellow of the College
of Estate Management. Consultant, Edwin Hill. Visiting Professor, City
University and University of Portsmouth.

### KEITH DAVIES

MA (Oxon), LLM (Lond), Hon Assoc RICS, of Grays Inn, Barrister-at-Law.
Emeritus Professor of Law, University of Reading.

### ERIC SHAPIRO

BSc (Estate Management) (Lond), FRICS, IRRV, FCIArb. Senior Partner,
Moss Kaye & Roy Frank.

2000

A division of Reed Business Information
ESTATES GAZETTE
151 WARDOUR STREET, LONDON W1V 4BN

NINTH EDITION 2000

| 1943 | FIRST EDITION | by | David M. Lawrance and Harold G. May |
| 1949 and 1949 | SECOND EDITION THIRD EDITION | by | David M. Lawrance, Harold G. May and W.H. Rees |
| 1956 | FOURTH EDITION | by | David M. Lawrance and W.H. Rees |
| 1962 and 1971 | FIFTH EDITION SIXTH EDITION | by | David M. Lawrance, W.H. Rees and W. Britton |
| 1980 and 1989 | SEVENTH EDITION EIGHTH EDITION | by | William Britton, Keith Davies and Tony Johnson |

ISBN 0 7282 0346 4

© T.A. Johnson, K. Davies, and E.F. Shapiro, 2000

Printed in Great Britain by Bell & Bain Ltd., Glasgow

"Valuation is an art and not a science, but it is not astrology. I think the valuers were star struck by what they saw and did not have their feet on the ground"
Jacob J. in *Platform Home Loans Ltd* v *Oyston Shipways Ltd* [1996] 2 EGLR 110

# Preface to the Ninth Edition

It is eleven years since the last edition and, unsurprisingly, much has changed. This is reflected in the changes and additions to, and deletions from that Eighth Edition.

One significant change is that Bill Britton felt unable to continue as co- author, having been involved with four editions starting with the Fifth Edition in 1962. Happily, he has contributed the section on Depreciated Replacement Cost in Chapter 20 to prolong his welcome involvement.

A problem with a book that covers a wide range of topics is that it tends to become too long as fresh material needs to be added. We have therefore excised some topics. There is no longer a chapter on agricultural valuations. However, the Third Edition of *Agricultural Valuations, A Practical Guide* by R.G. Williams published by the Estates Gazette in 1998 more than compensates for this deletion.

Similarly, we now concentrate on mainstream properties, leaving the reader interested in specialist properties to refer to other works such as *Valuation: Principles into Practice* edited by W.H. Rees, published by the Estates Gazette.

We also decided to omit the chapter on Life Interests, but we feel that the chapter in the Eighth Edition provides the valuation aspects that the reader might seek, as there have been no developments since then.

We were made aware of the proposed campaign of The Investment Property Forum to persuade valuers to switch from the rental payment assumption of annually in arrear to quarterly in advance and were able to discuss it with them. We have decided to retain annually in arrear as we believe that this reflects the general practice throughout the profession. However, we would not be surprised if we had to switch to quarterly in advance for the Tenth Edition.

As we have kept to mainstream properties we have not needed to seek as much assistance as for earlier editions. Indeed, apart from Bill Britton, we have sought a contribution from only one specialist, Keith Murray FRICS, a partner with Edwin Hill, who revised the chapter on rating and to whom we express our appreciation.

A large part of the book contains new material as distinct from revised text from the previous edition. This created a huge task of preparing the text on both disk and printed copy and we are indebted to Joanne Wilkinson who produced the final material after many rewrites and revisions with great skill and imagination.

London, 2000
Tony Johnson
Keith Davies
Eric Shapiro

# Contents

# Table of Cases

# Table of Statutes

# Table of Statutory Instruments

*Chapter 1*

# Principles of Value

## 1. The Valuer's Role

The services of a valuer may be sought by anyone with an interest in, or contemplating a transaction involving land and buildings. For example, a valuer may be required to advise a vendor on the price he should ask for his property, a prospective tenant on the annual rent he should pay, a mortgagee on the value of the security, and a person dispossessed under compulsory powers on the compensation he can claim.

Knowledge of the purpose for which his valuation is required is usually vital to the valuer, for the advice of the valuer in respect of a particular interest in landed property is not necessarily the same for all purposes. In many cases the valuer will also require to know details of the circumstances of the person for whom the valuation is being prepared because the subjective value of a particular interest may be different for different individuals according to, for example, their liability to income tax.

In the majority of cases, however, the primary concern of the valuer is to estimate the market value; that is the capital sum or the annual rental which at a particular time, on specified terms and subject to legislation, should be asked or paid for a particular interest in property.

It might reasonably be asked next, what are the special characteristics of landed property which make the services of a person with special knowledge desirable, or in many cases essential, in dealing with it? There are three primary reasons:

(i)   Imperfections in the property market.
(ii)  The heterogeneity of landed property and the interests which can exist therein.
(iii) Legal factors.

The need for special knowledge of the law relating to landed property will become apparent in the later chapters of this book; the varied interests which can exist in landed property are considered in Chapter 3.

## 2. The Property Market

The nature of landed property, the method of conducting transactions in it and the lack of information generally available for transactions all contribute to the imperfection of competition in the landed property market. Apart from structural differences in any building, each piece of landed property is unique by reason of location. The majority of transactions in the property market are conducted privately and even if the results of the transactions were available they would not necessarily be particularly helpful in the absence of detailed information on such matters as the extent and state of the buildings and the tenure. The degree of imperfection does, however, differ in different parts of the market. First-class shop investments, for example, are relatively homogeneous and this fact will decrease the imperfection of competition in the market for them. This comes about because shops in prime locations attract the same tenants.

It is, perhaps, misleading to talk of the property market as though it were a single entity. In fact, there are a number of markets, some local, some national and some international. For example, residential properties required for occupation are normally dealt with locally. A person looking for a house in which to live is rarely indifferent to its location because it must be conveniently situated, usually in relation to his place of work and perhaps that of his partner, and to educational facilities for his children. The market for first-class investments on the other hand is national. This type of property appeals to larger institutions such as insurance companies who are largely indifferent to the situation of the property. For example, they will tend to own commercial investments such as shops, offices and factories in all the major urban centres. The market for development properties can be local, national or international. Small, bare sites suitable for the erection of a few houses are of interest to small local developers, whereas large sites for residential or commercial development are sought by larger organisations operating in a much wider area. At the extreme, such organisations will be dealing with sites in several countries to take advantage of the market conditions in an international context.

## 3. A Definition of Value

Much paper can be and, indeed, has been consumed in discussing and defining value. It is not appropriate here to contribute to this particular store of literature but merely to provide a definition which will serve the present purpose. Thus, attention will be concentrated on value in a market situation, value in exchange or market value. Concepts of social value, aesthetic value or other values are not appropriate to this work, but it should be remembered that value can be considered from these points of view.

The market value or market price of a particular interest in landed property may be defined as the amount of money which can be obtained for the interest at a particular time from persons able and willing to purchase it. Value is not intrinsic but results from estimates, made subjectively by able and willing purchasers, of the benefit or satisfaction they will derive from ownership of the interest. The valuer must, therefore, in order to value an interest, be able to assess the probable estimate of benefit to potential purchasers. It must be explained that what is valued is not the physical land or buildings but the property interest which gives legal rights of use or enjoyment of the land or buildings. More precise definitions will be given later in this work dealing with particular aspects of valuations.

A potential purchaser is a person who proposes to tie up a certain amount of capital in land or in land and buildings and there are three main angles from which he may view the transaction. First, if he wishes to occupy the property he will be concerned with the benefits, commercial or social, which he anticipates he will derive from that occupation. Secondly, he may regard the property as an investment capable of yielding an annual return in the form of income. Thirdly, his motive may be profit; that is buying at one price now, in the hope that he may sell at a higher price at some time in the future, perhaps having injected further capital, and thus making a gain. These motives are not, however, mutually exclusive and a transaction may be entered into with more than one motive in mind. In any case the price which the purchaser will be prepared to pay at any given time will be influenced largely by the supply of that particular type of property and the extent of the demand for it. Supply and demand are discussed at length later in this chapter but it might be noted here that demand must be effective; that is that it must be backed up by money so that the desire to possess should be translatable into the action of purchasing.

## 4. Value and Valuation

Although the aim of the valuer is to provide an estimate of market value, it should not be assumed that each valuer's estimate of value and the market price or market value will always be the same. Different valuers could well place different values on a particular interest at a particular time because they are making estimates and there is normally room, within certain limits, for differences of opinion. In the majority of cases this is the most serious difference that should arise between competent valuers in times of stable market conditions because market prices result from estimates of value made by vendors and purchasers on the basis of prices previously paid for other similar interests; but in times when market conditions are not stable, more serious differences may arise. In such unstable times the valuer's estimate of value will be based on prices previously paid, but he must adjust his basis to allow for the changes since the previous transaction took place. The accuracy of his estimate will, therefore, depend on his knowledge of the changes and his skill in quantifying their effect.

Unstable times bring into focus the fact that in any market at any time there may be many buyers and sellers, all of whom will have their own personal views, desires and judgments on what the commodity in question is worth; or there may be a dearth of buyers or sellers. Thus, a market constitutes an amalgamation of individuals who will strike individual bargains, so that a "market price" can only represent an average view of all these factors. Where the commodities are all different, as in the case of property, then the problems of arriving at this average are multiplied. Where the valuer is making statutory valuations in hypothetical market conditions it is inevitable that differences of opinion as to value will become even more marked.

## 5. Demand, Supply and Price

It is axiomatic that quantity bought is equal to quantity sold regardless of price. This does not mean, however, that price is unimportant. Demand and supply are in fact equated at any particular time by some level of price. Other things being equal, at any given time, any increase in demand or decrease in supply will cause price to rise; conversely, any decrease in demand or increase in supply will cause price to fall. Therefore, whatever the given demand and supply, in a freely operating market, price will ration the supply and match it to the demand.

Price has, however, a wider role. If, over a long period, price remains persistently high in relation to the cost of producing more of the thing in question, it will call forth increased supplies which will gradually have the effect of pulling down price to a long-run normal or equilibrium level. Price, therefore, acts also as an indicator and an incentive.

The extent of price changes caused by changes in demand or supply will depend upon the price elasticity of supply or demand. Thus, if supply is highly elastic, a change in demand can be matched quickly and smoothly by an expansion of supply and the price change, if any, will be small. Conversely, if supply is highly inelastic, a change in demand cannot be matched quickly and smoothly by an expansion in supply, thus the price change will be more marked. If demand is elastic, a change in supply will quickly generate an appropriate contraction or expansion in demand and the price change, if any, will be small; but if demand is inelastic this means that any change in supply is likely to generate a contraction or expansion in demand less quickly and the price change will be more marked.

This is a simple statement of economic laws which are applicable to all types of commodity or service, but to consider their application to a particular type of commodity or service it is necessary to look at the special characteristics of the commodity or service in question. In the case of property, a separate discipline of land economics has emerged which reflects the unique status of property.

It is proposed, therefore, to look next at some of the principal factors which affect the demand for, and the supply of, landed property.

## 6. Demand for and Supply of Landed Property

The term landed property covers such a wide range of types of land and buildings and interests therein that generalisation is difficult, but it is correct to say that, generally, the demand for and the supply of landed property are relatively inelastic with respect to price changes.

The most important single factor restricting the supply of land and buildings at the present time is Town and Country Planning legislation. There are, of course, natural limitations on the supply of land. First, the overall supply of land is fixed and, secondly, the supply of land suitable for particular purposes is limited, but these natural limitations are overshadowed by planning limitations.

The powers of the planner are considered in more detail in Part 2 but the following are the most important in this particular context:

(1)   The power to allocate land for particular uses, for example, agricultural use or residential use. This allocation is shown in the development plans for the area. Thus, it would not normally be possible to erect a factory on a bare site in an area allocated for residential use.
(2)   The power to restrict changes in the use of buildings. For example, before the use of a house can be changed to offices, planning permission must be obtained and this permission will not usually be forthcoming unless the change is in accordance with the provisions of the planning policy for the area.
(3)   The power to restrict the intensity of use of land. Thus, the owner of a bare site in an area allocated for residential use is not free to choose between erecting on it, for example, a four-storey or a 15-storey block of flats. Any permission is subject to the further restrictions of building regulations which may negate or limit that which the planners would allow.

To illustrate the operation of these powers in practice, assume that in a particular area in which all available office space is already taken up, there is an increase in demand for office space. This increased demand might be met by an increased supply in three ways. Suitable buildings at present used for other purposes might be converted for use as offices; suitable bare sites might be developed by the erection of office blocks; and buildings which are not fully utilising their sites might be demolished so that larger office blocks could be erected. However, planning may affect all three possibilities.

In the extreme case it might be that: although suitable buildings are available for conversion, planning permission for the change of use is not forthcoming; although bare sites are available, none is in an area allocated for office use; and although existing buildings do not fully utilise their site, they provide the maximum floor space the planner will permit or the buildings have preservation orders on them which prevent their demolition. In such a case, in both the short and the long run, the effect of the increase in demand will be to increase the price of the existing office space to the point at which the high price so reduces demand as to equate it once again with the fixed supply.

In a less extreme case, where it would be possible to increase the supply by one or more of the three methods, some, perhaps considerable, time would elapse before any increased supply was available. In the case of conversion of existing buildings, it would be necessary to obtain planning permission, dispossess any tenants and carry out any necessary works of conversion. The time taken would depend on the administrative speed of the planning authority, the security of tenure of the tenants and the extent of the works. With the other methods, planning permission would again have to be obtained and with the last method, any tenants of the existing buildings would need to be dispossessed and the buildings demolished. The erection of the new buildings could take perhaps one or two years. Thus, in this less extreme case the effect of the increase in demand would be to increase the price of the existing space in the short run. In the long run, as the new space becomes available, the price would tend to stabilise or fall. If the increased supply was sufficient to satisfy the whole of the increased demand the price might return to its former level, but if, as is more likely in practice, the increased supply was insufficient, or, if sufficient was of a higher quality, the new equilibrium price would be higher than the old.

A reduction in the supply of a given type of space is difficult to achieve; the existing stock can only be reduced by demolition or change of use. The loss involved in failing to complete buildings already started would usually be higher than the loss resulting from a fall in price but their use might be changed during construction.

It has already been observed that the demand for land and buildings is also generally inelastic, but the degree of inelasticity will depend to some extent on the purpose for which land and buildings are held and by economic conditions at the time.

Elasticity of demand for any commodity or service depends on whether the commodity or service is regarded as a necessity and on the existence of satisfactory substitutes.

Take, for example, the demand for landed property for residential occupation. This, viewed as a whole, can generally be regarded as a necessity and no satisfactory substitutes exist. Thus, an increase in the price of living accommodation within a limited range would usually not result in any marked contraction in demand because a particular standard of such accommodation is regarded as a necessity and a caravan or a tent would not, to the majority of people, offer a satisfactory substitute. This would not, however, be true if the increase were outside an acceptable range or times were less prosperous when an increase in price would probably, subject to legal restrictions on sub-division and over-crowding, lead to economies in the use of living space as was shown during the recession of the early 1990s. The effect of price changes on demand will depend on whether the demand is "need related" or "standards related".

The importance of the difference between the desire of a prospective purchaser to possess a property and his ability to translate that desire into the actuality of purchasing was mentioned earlier, but further elaboration is desirable at this stage.

The majority of purchases of landed property are not effected wholly by the use of the purchaser's own capital. To take houses as an example, the purchaser will normally pay only a small proportion of the purchase price out of his own capital and the remainder he borrows by way of mortgage from a financial institution such as a building society, an insurance company or a bank. In some cases this will be done because the purchaser prefers to invest the remainder of his capital elsewhere but, in most cases, the purchaser has insufficient capital to pay the whole of the purchase price although his income is sufficient to pay interest on the mortgage loan and to repay the capital gradually over a long period of years.

Therefore, the ability of the majority of prospective purchasers of houses to make their demand effective depends on obtaining the necessary loan from a financial institution. Over the years these financial institutions have experienced two extremes in relation to money resources for lending which illustrate well the impact of effective demand on value. In the period 1970 to 1973, the lending institutions had plentiful sums available for lending, and borrowers were able to obtain mortgage sums for house purchase representing a large proportion of the purchase price and at low rates of interest. The effect of this lending situation was a rapid increase in house prices as borrowers competed fiercely for the houses available for purchase. At the end of 1973, the situation changed dramatically as lending institutions experienced a rapid fall in the sums they had available to lend and the government introduced a limit on the amount of mortgage attracting tax relief (£25,000).

In the succeeding three years mortgages became difficult to obtain, certainly for any substantial proportion of purchase price, and then at high rates of interest, so that house prices actually steadied or even declined before returning to a slow rate of increase. This was particularly marked in the upper price ranges, which suffered most from the mortgage tax relief limit, where in many cases they failed by 1976 to reach the levels of 1973 as a result of three years of mortgage restriction. In 1976 the situation began to ease, and the situation thereafter returned to that seen in the early 1970s.

This rapid increase came to a sudden end following the introduction of high interest rates in 1979 and the market virtually stagnated until the end of 1982 when again a combination of low interest charges and aggressive lending policies led to a sharp increase in prices which continued, with only one or two short pauses for breath, until the end of 1988; during this period prices rose by an average of 125%. This increase was brought to a halt by a change of government policy in respect of the tax deductibility of mortgages and changes in government fiscal policies. Multiple income relief for mortgages of up to £30,000 was withdrawn with effect from August 1 1988 (but announced in the preceding March), after which a combination of high interest rates and increasing unemployment caused a significant fall in demand and an actual fall in prices so that by the middle of 1993 prices had fallen back to their levels of the early 1980s. Prices continued to fall until 1995 when a recovery in prices began to occur (the market had "bottomed out") and by the end of 1996 had reached the levels of 1993. The policy of financial institutions such as building societies and banks has, therefore, an important bearing on the demand for certain types of landed property.

Legislation often affects, either directly or indirectly, the supply and demand for landed property. The various statutes affecting different types of landed property are considered in detail in Part 2 but, as an example at this stage, the supply of smaller or poorer types of residential accommodation for letting was directly affected by the Rent Acts which both restrict the income which the owner may derive from his property and also limit his powers to obtain possession. The removal of this regulatory system for new lettings in 1989 led to a sharp increase in supply.

The other principal factors affecting the demand for landed property are mainly long term so that their effects are felt only gradually over a long period of years.

Some of the effects of planning in relation to the supply of land and buildings have been noted, but planning may also have important effects on the demand side. The creation of new towns, the extension of existing towns to accommodate overspill from the large conurbations and the redevelopment of central shopping areas are examples of planning schemes which may increase demand in and around the areas concerned, although it may reduce demand in other areas. Thus, the new shopping centre may draw demand to adjacent properties but, in shifting the demand pattern, will diminish the demand for other nearby centres which decline in popularity.

Changes in the overall size, the location and the composition of the population will affect the demand for landed property. An overall increase in population, particularly if accompanied by an increase in prosperity, will increase the demand for most types of landed property. The increased population must be housed and its increased demands for necessities and luxuries will have to be met through the medium of such properties as shops, factories, offices, hospitals, schools and playing fields. The movement of population from one part of the country to another will have the dual effect of increasing demand in the reception area and reducing it in the area of origin. Changes in the composition of the population also have important effects. For example, a reduction in family size will increase the demand for separate dwelling units, or an increasing proportion of retired people will increase the demand for bungalows in the favoured retirement areas such as the South Coast or for flats specially developed for such people.

Improvements in transport facilities have encouraged people who work in towns to commute, sometimes over very long distances. As the number of commuters increases, the transport facilities may become overburdened and the process may be reversed, particularly when coupled with above-average increases in fares.

Technological developments have their impact on supply and demand. For example, changes in handling freight can render low-height and multi-storey warehouses obsolescent. Similarly, offices built in the early post war-years may be unable to accommodate the requirements for a computer-led environment.

## 7. Landed Property as an Investment

Any purchase of an interest in landed property can be regarded as an investment. There is, first, the straightforward case of the purchase of an interest in a property which is to be let to someone in order to provide the benefit of an income. But purchase for occupation is also an investment, the benefit in this case being the annual value of the occupation. A businessman, for example, may make a conscious choice between investing his capital in purchasing a property for occupation for his business or alternatively renting a similar property, owned by someone else, so that he can invest his capital elsewhere. The point can be further illustrated by considering a case where a businessman has at some time in the past made the choice of investing part of his business capital by purchasing business premises for occupation. If he now wishes to realise the capital tied up in the property and invest it elsewhere in the business he might enter into a lease-back or a mortgage transaction. In the lease-back transaction he will sell his interest in the property but take the property back on lease paying the annual value by way of rent. In the mortgage transaction he obtains a loan on the security of his interest in the property paying annual interest on the loan.

*Chapter 2*

# Methods of Valuation

## 1. Variety of Methods

A valuer is called upon to give his opinion as to the value of many differing types of interest in many differing types of property for many differing purposes. Given such a multiplicity of situations the approach to the determination of value in one case may well be inappropriate to another and there have consequently evolved over the years several distinct approaches which constitute separate methods of valuation. The purpose of this book is to examine each of these and to explain their application to individual circumstances. At this stage it is useful to consider the methods in broad terms so as to provide the framework for the following chapters.

## 2. Comparisons

Although, as will be seen, the methods of valuation differ significantly, underlying each of them is the need to make comparisons, since this is the essential ingredient in arriving at a market view. In any market the vendors compete with other vendors in attracting the purchaser to them and the purchaser is thus faced with a choice. In reaching his decision the purchaser will compare what is available and at what price and he will purchase the good which in his opinion gives the best return for the price paid; he seeks value for money. This will be true in all but the most monopolistic or monopsonistic market.

The valuer, in arriving at his opinion of value, must try and judge what prices vendors generally would seek and do obtain and what choice purchasers would make. He must therefore assess what is now or has recently been available in the market place and make comparisons between them. In this he acts no differently from any other person making a valuation; for example, a person valuing a motor car would assess the number and types of cars available and the prices currently being obtained and so arrive at the price which in his opinion would attract a purchaser for the motor car in question. The key to the accuracy of the opinion is the knowledge of what prices have been obtained recently for similar goods with which comparison can be made. This evidence of comparable transactions in the property market is termed "comparables" and, as will be seen, the availability and the nature of comparables provide the basis of whatever method of valuation is adopted and, indeed, the choice of the method itself.

## 3. Principal Methods of Valuation

### 1. *Direct Capital Comparison*

The simplest and most direct approach in arriving at a value is to compare the object to be valued with the prices obtained for other similar objects. The method works best if the comparable objects are identical. For example, if ordinary shares in a company are each selling at 108p then the value of further shares is likely to be around 108p. Special factors may alter this view; if a large block of shares were offered which would give control of the company then a purchaser might be prepared to offer more than 108p per share. Thus, even in the case of identical goods some judgment is required in arriving at a value. Property can never be absolutely identical, so that the use of this method is limited to the simplest cases. The application of the method is shown in Chapter 4.

### 2. *Investment Approach*

A large part of the property market comprises properties where ownership and occupation are separated. The purpose of property is to provide an appropriate environment for different requirements – houses for living in, factories for making goods, shops for selling goods, and so on. For each property there is an occupier. However, in many instances properties are occupied under a contract whereby the occupier pays to the owner a sum of money, normally termed rent, in return for the right to occupy. Thus, the owner surrenders the right of occupation in return for money. This is attractive to those who wish to invest capital and obtain a return thereon and the property market is a major source of such investment opportunities. The ability to make such investments is closely allied to the various forms of interests in property which may exist, and which are set out in Chapter 3.

Thus, the valuer is frequently called upon to make a valuation of an interest in property where the value is the amount of capital required to purchase the interest, and this value is clearly dependent on the amount of rent which an occupier would be prepared to pay for the right to occupy and the level of return which an investor would require on his capital. Hence, there are several fundamental elements in the valuation, each of which requires different types of comparables.

The basic principle is that an investor wishes to invest capital to obtain an annual return thereon in the form of a net income which represents an acceptable rate of return.

For example, A may wish to invest £100,000 in shop premises and he requires a return of 6% on his money. What net income will be required?

$$6\% \text{ of } 100,000 = 100,000 \times \frac{6}{100} = £6,000 \text{ net income}$$

However, in practice the position is generally reversed. The valuer will determine the net income which a property will produce. He knows that an investor will require a specific return on his capital. He can therefore calculate the capital sum which the investor will pay for the property.

For example, suppose that shop premises produce a net income of £8,000 p.a. If the appropriate rate of return is 8%, what capital sum will an investor pay?

Let the capital sum be C, then

$$C \times 8\% = C \times \frac{8}{100} = £8,000 \text{ p.a.}$$

$$C = 8,000 \text{ p.a.} \times \frac{100}{8}$$

$$C = 8,000 \times 12.5$$

$$C = \underline{£100,000}$$

As can be seen, the capital sum required is found by multiplying the net income by 100 times the reciprocal of the rate of return (the net income is "capitalised"). This reciprocal figure is termed a years' purchase or YP, a traditional expression in use for very many years, and might be said to represent the number of years to get back the capital invested, although this is a misleading concept as will become clear. It merely expresses the ratio of income to capital. The YP for any rate of interest can easily be found, e.g.

$$\text{YP at } 5\% \quad = \frac{100}{5} = 20 \text{ YP}$$

$$\text{YP at } 7\% \quad = \frac{100}{7} = 14.286 \text{ YP}$$

$$\text{YP at } 10\% \quad = \frac{100}{10} = 10 \text{ YP}$$

It is clear that the lower the rate of interest the higher the YP. The determination of net income is explained fully in Chapters 5 and 6, and the determination of a YP and the capitalising of incomes in various situations in Chapters 7, 8 and 9.


## 3. Residual Approach

So far, the methods considered have been related to property which exists. Property is constantly being destroyed and created under the inevitable process of development or demolition and re-development which is required to meet the changing demands of society. Thus, the valuer often needs to give a valuation of land or buildings which are to be developed or re-developed. He may be able to arrive at such a value by direct comparison with the sale of other similar property which is to be developed in a similar manner. This can be done by analysing the comparables as being so much per unit of proposed development. However, this may be impracticable because of the unique nature of the property in question and the development proposed.

For example, it may be intended to develop some land with new warehouses in an area where no warehouses have been built in recent years. It will be possible to predict the rent which will be obtained from an occupier, and thus the capital sum which an investor would pay for the warehouses when built by adopting the investment approach. Further, it is possible to predict the cost of building the warehouses and the profits which someone would require to carry out such an operation. Given the value of the finished product and the cost of producing it including required profits, it is clear that any difference represents the sum which can be paid for the land. For example, if the value of the finished warehouses is £800,000, the cost of producing them is £500,000 and the required profit is £70,000, then someone can afford to spend £800,000 - (£500,000 + 70,000) = £230,000 on the land. Thus the land value is £230,000. This sum represents the residue available and this approach is termed the residual approach.

This example represents a simple illustration of a method which has been developed in practice into an often complex calculation. The method and its application are explained in detail in Chapter 11.


## 4. Profits Approach

Many types of property depend for their value on various factors which combine to produce a potential level of business. In some instances the factors are so unique that comparison with other similar properties is impracticable and the value must therefore be determined by looking at the actual level of business achieved in the property or achievable from it.

A typical example of this is a petrol filling station. The designs of such premises are relatively similar, and they tend to be located in prominent positions on busy traffic routes. Nonetheless, a comparison of one with another is difficult since each site is susceptible to unique factors which may have a dramatic effect on the sales achieved. Two stations may appear to enjoy almost identical qualities yet one will sell 2 million litres a year and the other 3.5 million litres. The level of sales clearly determines the level of profits, and the profits determine the price someone will pay for the property and the opportunity to obtain the profits.

It follows that the value of the property can be determined from a knowledge of the profits. For example, if a property is producing profits of £50,000 p.a. and a purchaser will invest capital at 6 times the level of profits, the value of the property is £300,000 (50,000 × 6). Alternatively, if an occupier is prepared to pay a rent of £20,000 per annum out of the profits to be able to earn the £50,000 and a purchaser will invest capital at 15 times the rent, then again the value is £300,000 (20,000 × 15). Even here there is a need to obtain comparable evidence, in this instance the ratio of profits or rent to capital. This is a very simplified version of what is termed the profits approach. The method is explained more fully in Chapter 21, when it will be seen that knowledge of the type of business and the ability to interpret accounts and to analyse the profits are needed to arrive at the value.

## 5. *Cost of Replacement Approach*

Within the wide range of properties which exist there are some which are designed and used for a special purpose to meet specific requirements and which are outside the general range of commercial and residential properties. Typical examples are churches, town halls, schools, police stations and other similar properties which perform non-profitable community functions.

In nearly all cases, such properties are built by the authority or organisation responsible for the provision of the special service or use and commonly there is no alternative body which requires the property. In such cases there are no sales in the market and thus no comparables on which to base a valuation. Indeed, such properties are rarely sold and, when they are, they generally need to be replaced by alternative premises which have to be newly built since alternatives rarely exist. As a result, the minimum price required by a body owning such properties is the cost of providing equivalent alternative accommodation. The price for the site will be based on the value of comparable sites whilst the cost of erection is derived from prevailing building costs.

Thus, the valuation of such properties is derived from the value of alternative sites plus cost of building. This cost of replacement approach is sometimes referred to as the contractor's approach or contractor's test or depreciated replacement cost. Its application is restricted to special cases, as described in Chapters 20, 21 and 26. This valuation approach should not be used for property for which there is a general demand and for which comparables exist. Its use is limited in terms of providing a valuation, as opposed to an estimate of replacement cost, to those situations where there is a proven demand, for the specific use or uses for which the property is designed, in this area.

This summary of the principal methods of valuation adopted by valuers indicates the need to take account of the special characteristics of each case. It may be possible to approach a valuation by adopting more than one method so as to check one against the other. It is clear that the accuracy of any method depends to a great extent on the comparable evidence which can be obtained by the valuer from his knowledge of the market. In the following Chapters the examples will make assumptions as to the nature of the evidence. It is stressed that any assumptions are solely for illustrative purposes and are not intended to indicate either levels for any specified factor or the relationship between factors currently pertaining in the market.

*Chapter 3*

# Interests in Land

## 1. Property Rights in General

The two principal interests or rights in land, or in land and buildings, with which the valuer is concerned are known respectively as freehold and leasehold. Both must be distinguished from mere permission to enter upon land (even permission by virtue of a contract, e.g. for payment) which is known in law as a "licence", since permission can normally be withdrawn unilaterally on reasonable notice by the licensor (i.e. the landowner) to the licensee. A licence therefore has a negligible value in itself at common law; but sometimes an "equitable" right is grafted on to it which does have some value or at least reduces the value of the licensor's legal title.

An example of this is the decision of the Court of Appeal in *Crabb* v *Arun District Council*,[1] in which the plaintiff sold land with a legal right of access and retained a landlocked parcel of land in reliance on a promise by the defendants that he would be granted a right of access over a lane on their adjoining property. He could not enforce a right of way of necessity against them, and they executed no deed of grant of a right of way for his benefit. They then withdrew permission and refused him legal access unless he paid for it. It was held on appeal that no common law right of access existed, but in equity the defendants were "estopped" from going back on their promise. He therefore enjoyed by "equitable estoppel" a right of access to his landlocked land. This case is one example of many in modern law in which "equitable" property rights have been recognised, even though "legal" rights have not been created.

To the valuer the term "freehold" implies a property which the owner holds absolutely (i.e. unconditionally) and in perpetuity, and of which he is either in physical possession or in receipt of rents arising from leases or tenancies which have been created out of the freehold interest.

---

[1] [1976] Ch 179.

In law, the term "freehold" has a wider general meaning and includes lesser interests such as entails and life interests. The legal term for the valuer's conception of freehold is the "fee simple absolute in possession". The Law of Property Act 1925 (Section 1) states that this and the leasehold (or "term of years absolute") are now the only forms of ownership capable of being a "legal estate" in land and thus existing at common law. Other interests are merely "equitable". A "legal estate" is "good against the whole world", but an equitable interest is enforceable against some persons and not others. A "fee simple absolute in possession" exists in all land in England and Wales; whereas leaseholds and equitable interests, where they exist, are superimposed on (or "carved out" of) such ownership, and subsequently terminated, depending on circumstances.

In the following sections it will be convenient to consider first the position of the freeholder who is in physical possession of his property, and then note the nature and incidence of the various types of leasehold interest which may be created out of the freehold. Later it will be explained how successive interests in property – such as "entailed interests", "life interests", and "reversions and remainders", which are "equitable interests" only – may sometimes arise under wills or settlements.

Physical ownership of "land" extends indefinitely above and below the surface of the land, within reason, and thus in a normal case includes both air space and subsurface strata of minerals, but not air or water in their natural state nor chattels belonging to some other identifiable owner. Boundaries should be unambiguously specified. Ancient hidden gold and silver artefacts did belong to the Crown as "treasure trove"; but the Treasure Act 1996 replaced this by a more modern system.

## 2. Freeholds

Under the Crown the legal freeholder is inherently the absolute owner of the property, and is sometimes said to be able to do what he likes with his land, subject only to the general law of the land (notably planning control) and to the lawful rights of others. He may thus in a normal case develop it, transfer it, or create lesser interests in it, without the consent of any other private person. His ownership of the property is perpetual even though he is mortal. Buildings may become worn out, the use and character of the property may change; but the land remains the permanent property of the freeholder and his successors, whether the latter obtain it from him by gift, sale or disposition on death.

The legal freeholder, as absolute owner of his property, usually holds it without payment in the nature of rent, though in strict theory he is always nominally the tenant of the Crown. But in some parts of the country freeholds will be found to be subject to annual payments known as "rent charges", "fee farm rents" or "chief rents". These usually arise through a vendor of freehold property agreeing to accept a rent charge in perpetuity or for a limited term in lieu of the whole, or sometimes part, of the full purchase price. Although the rent charge owner has the right, in the event of non-payment, to enter and take the rents and profits of land until arrears are satisfied, a freehold subject to a charge of this kind still ranks in law as a fee simple absolute in possession.[2] In many deeds creating such rent charges there is a power of absolute re-entry where the rent has been in arrears for two years. The Rent Charges Act 1977, however, in general terms prohibits the creation of new rent charges and makes provision for the gradual extinguishment of existing ones over the next 60 years.

A rent charge or fee farm rent will be treated as an outgoing in arriving at the net income of the freehold for valuation purposes. As for the interest of the rent charge owner (which is technically a form of legal estate in the land), although the rent charge is commonly small in amount compared with the net rental of the buildings and land on which it is secured, and thus almost certain to be received in practice, it depreciates in real terms as does any fixed income in periods of inflation.

During a state of national emergency, statute may give the Crown drastic rights of user and possession over the property of its subjects. Even in normal times there are numerous Acts whereby an owner can be forced to sell his land to government departments, local authorities or other statutory bodies, and even some private companies for various purposes. Such compulsory purchases are to be paid for at market value; but this is determined statutorily under the Land Compensation Acts 1961 and 1973. To develop land an owner must obtain planning permission under the Town and Country Planning Act 1990 (as amended). Legislation governing

---

[2] Law of Property (Amendment) Act 1926: Schedule.

environmental health, housing and other uses of land, including Local Acts and Building Regulations, restricts the erection or reconstruction of buildings or compels the owner to keep them in a state of good repair and sanitation. Other legislation gives various leaseholders of residential property, including public sector tenants, the right to require their landlords to sell them the freehold reversion or grant long leases.

The owner's freedom to do as he likes with his property is also limited by the fact that he must not interfere with the natural rights of others as, for instance, by depriving his neighbour's land of support or polluting or diminishing the flow of a stream. Often his enjoyment is lessened by the fact that some other owner has acquired an easement over his property or is entitled to the benefit of covenants restricting the use of it; these rights are discussed below.

Being a "legal estate", a freehold can normally be transferred (by a living owner) only on the execution of a deed, that is to say a document expressly stating that it is "a deed" (Law of Property Act 1925, Section 52, and Law of Property (Miscellaneous Provisions) Act 1989, Section 1). Under the Land Registration Acts 1925-97 legal freeholds (and leaseholds for over 21 years) must, when transferred (or created), be entered in the Land Registry within two months, and the title deed, or deed of "conveyance", will in such cases take the form of a "land certificate". Registration provides a state guarantee of the legal ownership or "title".

Occupation of land counts as legal "possession" if it rests on genuine "title", whether registered or unregistered. The common law remedy for an owner wrongly excluded from possession is a writ or order of possession obtainable from the court (and enforceable by the sheriff's officers); and damages for financial loss are also obtainable against a wrongdoer. But if an owner is excluded from possession and acquiesces in "adverse possession" by a trespasser who treats the land as his own, the Limitation Act 1980 provides that an action for possession against the latter must be brought within 12 years; if not, the owner's rights are extinguished and in their place the trespasser (or "squatter") acquires a lawful title himself.

It commonly happens that the freeholder, instead of occupying the property himself, grants or "lets" the exclusive possession of the premises to another for a certain period, usually in consideration of the payment of rent, under a "lease" or "tenancy".

A letting may be for quite a short period – e.g. yearly, quarterly, monthly or weekly – thus producing a "periodic" tenancy automatically renewing itself until terminated by notice to quit from either side; or it may be for any definite fixed period (see below). A "building lease" of 99 or 999 years is often met with; but occupation leases for terms such as 5, 10, 15, 25 or 50 years are also commonly found in practice.

Like a freehold conveyance, a lease or tenancy taking effect as a legal estate must be granted by deed; but the Law of Property Act 1925, Section 54, provides that no formalities whatever are needed for the grant of a tenancy at a full market rent for not more than three years, taking effect immediately.

Grants of tenancies (but not the obligations contained in them) may be back-dated; or they may be made to begin at a future date ("reversionary leases") but not more than 21 years after the date of the grant.

When a freeholder grants a short-term lease or tenancy the lessee seldom has any valuable interest in the property. This is not only owing to the shortness of his term, but also to the fact that he is usually paying the full rental value which the premises are worth, as distinct from a lump-sum purchase price. The valuation of leasehold interests therefore tends to be significant only in those cases where property is leased for an appreciable term of years and where the rent reserved under the lease is less than the rental value and not subject to frequent adjustment under rent review clauses, so that the lessee is enjoying a "profit rent" from his possession of the property. Normally in such a case the rent is low because the lessee has also paid a lump-sum purchase price to the lessor. This low rent is often loosely called a "ground rent". It should be noted that a lease can be transferred – "assigned" – by gift or sale, or on death, just as a freehold can, unless the terms of the lease provide otherwise.

A freeholder who grants leases or tenancies retains the right at common law to regain physical possession of his property when these leases or tenancies come to an end. For this reason the landlord's interest in the land is known as his "reversion".

The common law right of a freeholder to regain physical possession of his property at the end of a lease granted by him (including, as a normal common law rule, any buildings or other "fixtures" added to the land by the tenant which are not removable by him as "tenant's" fixtures) is now much restricted by legislation.

The most important statutes are the Rent Act 1977, the Housing Act 1988, the Landlord and Tenant Act 1954, the Agricultural Tenancies Act 1995, the Leasehold Reform Act 1967, the Leasehold Reform, Housing and Urban Development Act 1993, and the Housing Act 1996. The provisions contained in these Acts (extending to control of rents as well as security of tenure) are considered in greater detail in Chapters 17 and 18.

## 3.  Leaseholds

A leasehold interest in strict legal parlance is a "term of years". This expression covers leaseholds of all kinds no matter how long or short the period for which they are "demised" (i.e. granted). Any leasehold can be renewed (unless it is agreed otherwise) but if granted for the term of a year or less it will usually be a "periodic tenancy" which renews itself automatically and so appears, superficially but incorrectly, to be of indefinite length. A periodic tenancy can only be terminated by a "notice to quit" taking effect at the end of one of the periods in question (which are most commonly a year, a quarter, a month or a week) and must have a minimum length of one of the periods of the tenancy – with certain exceptions. Leaseholds which are not periodic tenancies end automatically (subject to any renewal or extension agreed by the parties or imposed by statute) by "effluxion of time". In some circumstances, carefully regulated in law, a lease can be cut short by forfeiture or by a "break clause".

A lease is normally subject to the payment of an annual rent and to the observance of covenants contained in the lease. Commercial leases often include provision for "rent reviews" to take account of general changes (assumed normally to be increases) in property values. The grantor of the lease is called the "lessor" and his interest while the lease lasts is the "reversion". The grantee is the "lessee" or "leaseholder". An alternative expression commonly used for lessor is landlord, and tenant for lessee.

If the rent the lessee has covenanted to pay is less than the true rental value of the premises, the lessee will be in receipt of a net income from the property, the capitalised value of which, for the balance of his term, will represent the market value of his leasehold interest. Even if the lessee is paying the full rental value of the property, this interest may have some value in that it confers a right on him to occupy the premises for the unexpired term of his lease and may enable him to claim a new lease or to continue in possession at the end of his term under the legislation mentioned above. If he is paying a rent above the full value, the lease will have a negative value owing to the liability which it creates.

A lessee is very much more restricted in his dealings with the property than a freeholder. In many cases he will be under express covenants to keep the premises in good repair and redecorate internally and externally at stated intervals. Usually he cannot carry out alterations or structural improvements to the property without first consulting his lessor, and possibly covenanting to reinstate the premises in their original condition at the end of the term. Often he may not sublease or assign his interest or part with possession without first obtaining his landlord's consent, although that consent cannot be unreasonably withheld.

If the lessee subleases, even though the sublease contains precisely the same covenants as his own lease, there is still the risk of his incurring liability to his lessor if the sublessee fails to keep his part of the bargain. Similarly, if he assigns he may still be held liable on his express covenants under common law or (in leases granted after 1995) by virtue of a guarantee given as part of the terms for assignment under the provisions of the Landlord and Tenant (Covenants) Act 1995 – such as the covenant to repair – should his assignee prove to be a "man of straw". Assignment to the landlord himself is "surrender".

For these and other reasons, leaseholds are less attractive to the investor than freeholds and the income from a leasehold interest in a particular class of property will usually be capitalised at a higher rate per cent than would be used for a freehold interest in the same property. A lease is normally both a right in land and a contract.

The following are the principal types of lease met with in practice.

## (i) Building Leases

These are leases of land ripe for building – i.e. development – under which the lessee undertakes to pay a yearly ground rent for the land, as such, to erect suitable buildings upon the land, and to keep those buildings in repair and pay all outgoings in connection with them throughout the term. The ground rent represents the rental value of the bare site at the time the lease is granted, and the difference between this and the net rent obtainable from the buildings when erected will constitute the lessee's profit rent or net income. It is common practice for the lessee to pay a capital sum (a premium) to the lessor at the start of the lease, coupled with a reduction in the ground rent payable. As a result, nominal rents or even no rent at all are then payable. For legal reasons where no rent is payable the lease will reserve the right to some payment, traditionally a peppercorn. Hence the reference to a "peppercorn rent" means that the rent is nil.

Building leases are usually granted for terms of 99 years, although 120, 125 and 150 year terms have become common in recent years, particularly for large-scale commercial development. Longer terms of 999 years are found occasionally. Lesser terms than 99 years are met with, but it is obvious that no would-be leaseholder will incur the expense of erecting buildings unless he is granted a reasonably long term in which to enjoy the use of them, since they revert to the landlord with the land at the end of the lease. Leases of residential property exceeding 21 years (but not flats or maisonettes) may, under the Leasehold Reform Act 1967, be extended by 50 years at a modern ground rent; alternatively, the lessee may buy the freehold reversion. The freehold ownership of leases of flats and maisonettes exceeding 21 years may be collectively enfranchised by the tenants (through a nominee company in most cases) under the Leasehold Reform, Housing and Urban Development Act 1993, Part I; or individual tenants may claim a 90-year extension to their leases. When a freehold is enfranchised by tenants it means that the owner of the freehold is compelled to sell the freehold interest to the tenants.

### (ii) Occupation Leases

This term is applied to a lease of land and buildings for occupation in their present state by the lessee. In practice the length of the lease varies according to the type of property. Dwelling-houses and older types of commercial and industrial premises are often leased for fairly short terms, such as one, three or five years or on periodic tenancies. In the case of modern commercial property, before 1990 the tendency was to grant leases for 25 years but since then, because of the slump in the property market, leases for shorter periods, such as three or five years, have become common, or clauses have been included allowing a party to terminate the lease after a short period of years ("break clauses").

In the case of medium- and long-term leases of commercial or industrial premises, it has become standard practice to incorporate a "rent revision" clause which enables the rent to be revised, usually upwards, at fixed periods of time. The tendency in recent years, due largely to the effects of inflation, has been for review periods to become shorter, with five years being a fairly standard period. Thus leases are normally granted for multiples of the review period, i.e. 10, 15, 20 or longer periods of years. Before this, multipliers of seven years were often met.

Ground rents are now normally subject to review. The review periods are often the same as for the duration of occupation leases, although sometimes they are less frequent.

Where the rent reserved under an occupation lease represents the full rental value of land and buildings, or near it, it is called a market rent or rack rent. Where it is less than the full rental value, although obviously not a ground rent, it is called a head rent. It is on the difference between the head rent and the net rack rental value of the premises that the capital value (if any) of the leaseholder's interest normally depends in the case of an occupation lease.

The term head rent is also used to distinguish the rent paid by the head tenant to the freeholder from any rent paid in respect of the same property by a subtenant to that tenant as the sublessor.

## (iii) Subleases

Subject to the terms of his headlease, a leaseholder may sublease the property for any shorter term than he himself holds, either at the same rent or at any other figure he may be able to obtain, although leases normally prohibit subletting at less than the rent payable under the lease. In this way the head lessee himself becomes entitled to a "reversion" on the falling in of the sublease. Often this reversion will be a purely nominal one of a day or a few days only. Thus, a head lessee having 14 years of his lease still unexpired might sublease the property at an improved rent for a term of 14 years less (say) 10 days. The reservation of at least a day's reversion is necessary to preserve the nature of the sublease. An attempt to lease for the full term remaining would have the effect of an outright assignment, not a sublease.

A sublessee in occupation will often have the benefit of statutory security of tenure and control of rents already referred to.

Where a lessee has been granted a lease of building land at a ground rent, he may sublease the land, as a whole or in smaller plots, either at the same rent as he himself pays, in which case the new rent is known as a leasehold ground rent, or, as is usually the case, at a higher rent than that reserved in his own lease, in which case the rent is known as an improved ground rent.

### (iv) Leases for Life

It was possible until 1925 to grant leases for the duration of the life of the lessee or some other person. By the Law of Property Act 1925, these "leases for life" now take effect as leases for 90 years determinable on the death of the party in question by one month's notice, given either by the lessor or the personal representatives of the deceased, expiring on one of the usual quarter days. This is because a grant for life is an uncertain period and the aim of the 1925 Act is to restrict leaseholds to ascertainable periods.

This provision has little effect on the duration of such a lease because if the premises are held at less than their true rental value the lessor will take the first opportunity of terminating the lease on the lessee's death, and if the lessee was paying a higher rent than the premises were worth his personal representatives will be equally anxious to determine.

From a valuation standpoint, therefore, the period of the lease may still be safely regarded as that of the lessee's life.

### 4. Restrictive Covenants and Easements

Land, whether leased or not, may be subject to restrictions as to user. Some previous conveyance of the freehold may have imposed easements or restrictive covenants, for example, to prevent land being used for business purposes or to prohibit the erection of houses costing less than a stipulated figure, or to preserve a right of way over land for the transferor's benefit.

Such restrictions are often inserted in leases. For instance, the lessee of a house may covenant not to use it except as a private dwelling-house, or the lessee of a shop may covenant not to use it for any offensive trade, or to use it only for one particular type of trade.

A "covenant" is a contractual obligation in a deed. Normally at common law contracts are only enforceable against the individual party who has agreed to be bound (or assignees, in the case of normal leasehold covenants). But the rules of equity enable a covenant which "touches and concerns" land to "run" automatically with that land to burden all subsequent owners who have "notice" of it provided that it is negative ("restrictive") in nature, for the benefit of specified neighbouring landowners.

This converts the covenant into an equitable interest in the burdened land. Unless it is an obligation in a lease it is registrable as a "minor interest" in registered land (Land Registration Act 1925) or a "land charge" in unregistered land (Land Charges Act 1972). Such registration acts as "notice to the whole world" – indeed, actual notice then becomes irrelevant. Failure thus to register such a covenant renders it void against a purchaser. But the burden of a restrictive covenant cannot "run with the land" unless the benefit of it is annexed to nearby land capable of benefiting (often in the form of a scheme of development) or else is assigned with such land. Enforcement is principally by means of an injunction against infringement, though a court may also award damages.

Restrictions on the user of land are sometimes imposed solely for the benefit of neighbouring land which the vendor is retaining. In other cases they are intended to safeguard the general development of property in an area. For example, where a building estate is sold in freehold plots to various purchasers, each purchaser usually enters into the same covenants as regards the use of his land. Normally such covenants are mutually enforceable between the various purchasers and their successors in title, so that the owner of any one plot may compel the observance of the restrictions by the owner of any other plot. Yet although restrictive covenants which are reasonable in character may help to maintain values over a particular area, such as a building estate, in some cases – particularly when imposed a considerable number of years ago – they may restrict the normal development of a property and detract from its value.

The right to enforce restrictive covenants may be lost. This may happen if the party entitled to enforce them has for many years acquiesced in open breaches of the covenant or has consented to a breach of covenant on one occasion in such a way as to suggest that other breaches will be disregarded, or if the character of the neighbourhood has so entirely changed that it would be inequitable to enforce the covenant.

There are also statutory means by which a modification or removal of restrictive covenants may be obtained. Thus, the Housing Act 1985, Section 610, enables any interested persons or the local authority to apply to the County Court for permission to disregard leasehold or freehold covenants which would prevent houses being converted into maisonettes, if satisfied that changes in the neighbourhood render it reasonable to do so, and that, while the house is unlettable as a single house, it would readily let if converted into two or more tenements. More generally, under the Law of Property Act 1925,[3] the Lands Tribunal may wholly or partially modify or discharge restrictive covenants on freeholds and leaseholds, if satisfied that by reason of changes in the character of the property or the neighbourhood, or other material circumstances, the restrictions ought to be deemed obsolete or that their continuance would obstruct the reasonable use of the land for public or private purposes without securing practical benefits to other persons. Current planning policies of local planning authorities should be taken into account. These powers also extend to restrictive covenants on leaseholds where the original term was for over 40 years, of which at least 25 years have elapsed.[4]

The Lands Tribunal may make such order as it deems proper for the payment of compensation to any person entitled to the benefit of the restrictions who is likely to suffer loss in consequence of the removal or modification of the restrictions. In practice, these payments are of small amounts (usually less than £1,000) as any larger award would automatically mean that the covenant is of substantial value.

The 1925 Act also gives power to the court to declare whether land is, or is not, affected by restrictions and whether restrictions are enforceable, and if so, by whom. This is a valuable provision which enables many difficulties to be removed where the existence of restrictions or the right to enforce them is uncertain.

Unlike restrictive covenants which only take effect as property rights in equity, easements are common law property rights and thus enforceable against the whole world regardless of notice or registration (though easements can also exist in equity, and these are registrable in the same way as restrictive covenants). An

---

[3] Section 84 as amended by the Lands Tribunal Act 1949, Section 1(14) and Law of Property Act 1969, Section 28.
[4] Landlord and Tenant Act 1954, Section 52.

easement burdens particular land, termed a servient tenement, for the benefit of other land nearby, termed a dominant tenement, and its presence or absence obviously affects the value of each. Typical easements are private (not public) rights of way, and also rights of support, light and ventilation. Many other varieties of easement (for example, rights to install and maintain pipes or cables) have been upheld from time to time. A right to support of land as distinct from buildings is, however, a natural right enjoyed by all owners, and not an easement.

Easements must, to be "legal estates", be granted expressly in a deed. But rules of law have been evolved whereby (for reasons of necessity or otherwise) the grant of easements is often implied into conveyances of land which in fact make no express mention of them. Also, it is possible to have a presumed grant of an easement by "prescription", that is to say long user over the servient land provided it has been exercised "as of right" and not by permission of that land's owner. The Prescription Act 1832 provides that 20 years is the normal minimum period required. Easements can be terminated by express discharge or by abandoning them.[5] Similar to easements are "profits à prendre"; these involve taking such things as minerals, plants, game or fish from the servient land. A dominant tenement is not essential for a profit à prendre. Easements and profits are enforced, like restrictive covenants, by damages or injunctions.

Similar in some ways to a restrictive covenant or easement is the equitable right known as "equitable (or proprietary) estoppel". This has been discussed above, at the beginning of the chapter.

## 5. Successive Interests in Property

Provisions are frequently made in settlements by deed or by will whereby, so far as the law permits, the future succession to the enjoyment of the property is controlled.

A once common example (now extremely rare in consequence of taxation) is the creation of an "entailed interest" in freehold or long leasehold property whereby the property is settled on a certain

---

[5] More than mere non-user is required to prove "abandonment": See *Benn v Hardinge* (1992) 66 P&CR 246 (175 years of non-user is not "abandonment").

person for life and after his death on his eldest son and the "heirs of his body". In this way the property will descend from generation to generation unless the entail is barred or unless heirs fail, in which latter case the land will revert to the original grantor or his successors or else pass to any other person entitled under the settlement on termination of the entail.

Under this and other forms of settlement certain persons may be entitled, either alone or jointly with others, to successive life interests or entails in a property, the ownership finally passing to some other person in fee simple. The first owner enjoys his interest "in possession"; but each of the others has at the outset an interest which is only a future right and holds it "in remainder". Such a person is called a remainderman. Sometimes a party may be entitled to succeed to a property only subject to certain conditions – for example, that he reaches a prescribed age. In such a case, the interest is spoken of as "contingent" and if the contingency is too remote in the future, the interest will infringe what in law is called the "perpetuity rule" and be void. The details of this are too complex to be given here; however, it may be noted that a grantor is allowed to specify a period not exceeding 80 years in which prescribed contingencies (e.g. birth of grandchildren) may operate (Perpetuities and Accumulations Act 1964).

Life interests and future interests in property rank as equitable interests not legal estates. But from a valuation point of view the distinction is of little consequence. To a valuer the problem they present is that of finding the capital value of an income receivable for the duration of a life or lives, or the capital value of an income for a term of years or in perpetuity receivable after the lapse of a previous life interest. Readers interested in the valuation of such interests should refer to earlier editions of this book.

A person enjoying the interest of tenant for life has vested in him, and may sell, the whole freehold or leasehold interest (the legal estate) in the settled property, the capital money being paid to trustees and settled to the same uses (i.e. beneficial interests) as was the land itself. A valuer advising on a sale of this description is concerned not with the value of the party's life interest, but with the market value of the freehold or leasehold property which is the subject of the settlement; and in such a case the equitable (beneficial) interests in the settlement can be ignored for valuation purposes because the land will be sold free of them for its full unencumbered market value. They are said to be "overreached", that is to say preserved not against the land but against the capital of the settlement in its new form in money, investments, etc. There have been two main kinds of trust of land: strict settlements and trusts for sale. The Trusts of Land and Appointment of Trustees Act 1996 replaces these by a single "trust of land", referred to below under "Trusts in General".

### 6. Co-ownership

Land may be concurrently owned by two or more persons, either by way of joint tenancy or tenancy in common. "Tenancy" here simply means "ownership".

In a joint tenancy the parties have equal rights to the benefits of the property, but when one joint tenant dies his share passes to the surviving joint tenants by "survivorship".

In a tenancy in common shares are "undivided" (i.e. there is no physical partition of the property) but are not necessarily equal. Thus, one tenant in common might hold a quarter share in the property, another a half share and so on. Moreover, there is no survivorship, as in joint tenancies, for if one tenant in common dies his share passes under his will or intestacy. Thus the capital value and not merely the enjoyment of the property is shared between the co-owners.

Under the Law of Property Act 1925, the holding of "undivided shares" in the legal estate in land is abolished. This means that tenancies in common can only be equitable rights in the land. Where a tenancy in common exists, the legal estate of the property as a whole will be vested in trustees for sale as joint tenants; the owners of the shares (who may be the same persons as the trustees) are entitled to the benefit in equity, though not at common law, to their respective proportions of the proceeds of sale and of net income from the land pending sale. Joint tenancies in equity are permissible, but may be "severed" into tenancies in common if their holders individually so desire and indicate.

From a valuation standpoint although a tenant in common or owner of a share in property no longer has any legal estate but only an equitable interest in the land, he is entitled to all the benefits of ownership, such as participation in the rents and profits and a share of the capital sum realised by the property on sale. The best procedure is therefore to value the property as a whole and assign an appropriate fraction of the total to the value of the party's share.

It is generally recognised, however, that a share in property is less attractive to prospective investors than absolute ownership, and a percentage deduction varying according to the size of share and the degree of control it confers is usually made to allow for this.

As for joint tenancies, the survivorship rule prevents the rights of each of the joint tenants in equity from having any market value. But the ultimate survivor will hold an absolute freehold (or in some cases leasehold) which should be valued in the ordinary way. However, where joint tenants are holders of the legal title as trustees for one or more beneficial owners in equity it is normally permissible to appoint new trustees, and this counteracts the working of the survivorship rule. Even if new trustees are not appointed the trust continues to exist and the court will if necessary take charge of it.

## 7. Trusts in General

Wherever an equitable interest is expressly or impliedly created in property separately from the legal estate there is a "trust" relationship. The legal owner is a "trustee", the equitable owner is a "beneficiary". Co-ownership, for this reason, will involve a trust. Trusts of land created intentionally or under statute have the effect that "beneficial interests" in equity are "overreached" on a sale of the legal title (see above) and that title has a market value in consequence. There can be trusts in other circumstances, usually termed "bare trusts", as when X holds the legal title of property as nominee for Y, and if so it is normally the equitable interests in such a trust which are of value, not the legal estate of the trustee or trustees.[6] A vendor is an implied trustee for the purchaser in the interval between the contract of sale and the conveyance ("completion"). All these trusts of whatever kind will become simply "trusts of land" under the Trusts of Land and Appointment of Trustees Act 1996, referred to above.

---

[6] Any such equitable interest cannot, however, be enforced against a commercial purchaser of a legal title to the propery unless, on the facts, the purchaser ought to have been aware of the existence of that interest.

## 8. Transfers of Interests in Land and Security in Land

The rules of the law of property in land have evolved chiefly in connection with transfers of ownership. A life interest is a right which cannot be transferred at the life owner's death because that is the time when it comes to an end, but it can be transferred for so long as he remains alive (e.g. for mortgage purposes). The legal freehold or "fee simple", on the other hand, is perpetual in its duration, and so can be transferred by a current owner during his lifetime (*inter vivos*) or at his death. The same is true in principle of leaseholds, except, of course, that the term of a leasehold may end while its holder is still alive, in which case no property right remains to him in respect of it (though, of course, he may be able to secure a renewal).

The valuer needs to take proper account of these various characteristics because what he is valuing is not "land" in the purely physical sense, but property rights in land. Death should perhaps be considered first. Transfer on death is governed by the deceased's (valid) will, if any. Failing that, it goes by operation of law, under the Administration of Estates Act 1925 (as amended), to the deceased's next-of-kin, in accordance with various elaborate rules governing "intestacy". Frequently, in practice this simply involves a sale by his personal representatives; in which case what the next-of-kin actually receive is a distribution of the money proceeds of the sale. But, where necessary, the Inheritance (Provision for Family and Dependants) Act 1975 and Law Reform (Succession) Act 1995 empower the courts to order payments to be made out of a deceased person's estate to any surviving spouses, co-habitees or other dependants who would otherwise in effect be left destitute because no proper provision has been made for them.

Bankruptcy (or liquidation in the case of a company) is treated in a similar way under the Insolvency Act 1986, the property in question being sold not by personal representatives but by a "trustee in bankruptcy" or a liquidator and the proceeds of sale (normally inadequate) going not to next-of-kin but to unsecured creditors. This is, in a sense, a sort of financial death, except that a bankrupt individual will probably qualify for return to life in due course by receiving a discharge from bankruptcy. An insolvent owner's land is thus dealt with in much the same way as an intestate owner's land.

Transfer *inter vivos* occurs either by gift or by sale, that is to say either non-commercially or commercially. The latter can for this purpose be taken to cover transfers in return for a financial consideration in the form of a purchase price or a series of regular payments or both (long leases being often granted in return for both a lump sum purchase price and a "ground rent") or in the form of an exchange. Grants of new leases, assignments of existing leases and transfers of freeholds, are all varieties of transfer *inter vivos* whether as gifts or as commercial transactions.

Transfers of land may also operate to create security for loans and similar financial obligations, because the creditor is granted (expressly or impliedly) a "demise" (lease) or "legal charge" by way of "mortgage" pending repayment in full. This is dealt with in a later chapter.

A gift of land, on death or *inter vivos*, requires the carrying out of the appropriate legal conveyancing formality, which is the execution of a deed of "conveyance" (or a written "assent" by personal representatives) of the legal freehold or leasehold as the case may be. The same is true of commercial transfers. An equitable property right, however, need only be in writing as a rule; indeed, there are many circumstances where owners acquire equitable rights merely by implication of law. Thus, if land is "conveyed" by legal deed of grant to a nominee, in equity the benefit goes by implication not to him but to the real purchaser. Legal leasehold tenancies, if at a market rent and taking immediate effect, do not, however, require to be granted by any formality at all, provided that they are created for a period not exceeding three years (thus oral short-term tenancies of this kind are still "legal estates" in land).[7]

But commercial transfers, unlike gifts, involve another stage in addition to (and prior to) the "conveyance", namely the contract or bargain constituting the sale, lease or mortgage. Section 2 of the Law of Property (Miscellaneous Provisions) Act 1989, provides that a land contract "can only be made in writing". This must be done "by incorporating all the terms which the parties have expressly agreed in one document or, where contracts are exchanged, in each". But these terms can "be incorporated in a document either by being set out in it or by reference to some other document". The contract "must be signed by or on behalf of each party to the contract".

---

[7] Law of Property Act 1925, Section 54(2).

If in correspondence the phrase "subject to contract", or similar wording is used, there is no contract at all at that stage, merely negotiation[8] from which either party can back out. But once the contract itself has been made, in accordance with the 1989 Act it is binding and enforceable by an order of "specific performance" whereby the Court compels a defaulting party to proceed to the completion stage. Damages, if appropriate, may be awarded, and a defaulting vendor will normally be ordered to return a deposit.

If a definite date is prescribed for completion it is binding at common law, but not in equity unless on reasonable notice one of the parties expresses an intention to make a specified date "of the essence of the contract". The appropriate legal remedies for breach of the contract, whether specific performance (or the converse, i.e. rescission of the contract), or damages, will be available accordingly.[9] The availability as a general rule of specific performance to vendor and purchaser alike (or lessor and lessee) means that in equity the property, as a capital asset, passes to the purchaser at the contract stage and is termed an "estate contract" (protected by registration as a minor interest or land charge in registered and unregistered land respectively as the case may be) even though the *legal* title stays with the vendor until completion, which amounts to an implied trust (mentioned above), in the intervening period.

---

[8] *Tiverton Estates Ltd* v *Wearwell* [1974] 1 All ER 209.
[9] See the House of Lords decision in *Stickney* v *Keeble* [1915] AC 386.

# Chapter 4

# Direct Value Comparison

## 1. Generally

As was described in Chapter 2, the simplest and most direct method of valuation is direct comparison. The method is based on comparing the property to be valued with similar properties and the prices achieved for them and allowing for differences between them, so determining the price likely to be achieved for the property in question.

The method is based on a comparison of like with like. Properties may be similar but each property is unique so that they can never be totally alike. As they move away from the ideal situation of absolute similarity so the method becomes more unreliable. The reasons for dissimilarities are mainly:

(a) *Location*

The position of any property is clearly unique. The actual position is an important factor in the value of a property and for some, such as shops, the major factor. For example, on an estate built to the same design at the same time, some houses will be nearer to a busy traffic road than others; some will be on higher ground with good views whilst others will be surrounded by other houses; some will back south, others back north; some will have larger plots than others. Even in a parade of shops, some will be nearer a street crossing than others or be adjacent to a pedestrian exit from the car park or nearer to a popular large store.

(b) *Physical State*

The physical condition of a property depends on the amount of attention which has been given to maintenance and repair and decoration. Two otherwise physically identical properties can be in sharp contrast if one is well maintained and the other in poor order. Repair apart, occupiers of properties typically make alterations, varying from a minor change to major improvements. This is perhaps most apparent with residential properties where occupiers may install or change the form of central heating, renovate kitchens and bathrooms, change windows, remove fireplaces, add extensions and so on, to the extent that houses which from external appearances are identical are seen to be totally different on closer inspection.

(c) *Tenure*

As was shown in Chapter 3 there are various interests in property. Apart from a freehold interest with no subsidiary leasehold interests, the likelihood of interests being similar is remote in view of the almost endless variations that may exist between them. Even on an estate with estate leases which allow for each tenant to hold on a similarly drafted lease, the terms are likely to vary as to expiry date or rent payable or permitted use. In the market at large the differences will multiply considerably and the method is strained to compare, say, a leasehold shop with 45 years to run at a rent of £100 p.a. with the tenant responsible for all outgoings apart from external repairs and no limitations as to use, with the leasehold interest in a physically identical shop with eight years to run at a rent of £8,000 p.a. with rent review in three years, tenant responsible for all outgoings and the use limited to the sale of childrens' clothes. If the first lease sells for £100,000 it is impossible to derive the capital value of the second lease directly from that figure – it is certainly less, but how much so? This contrast between the types and nature of interests is the principal barrier to the use of the direct comparison method for the valuation of interests other than freehold in possession.

(d) *Purpose*

The purpose of a valuation is an important element in deciding upon the method to be adopted. Where the valuation is, for example, for investment purposes, the direct comparison method would be inappropriate. Other methods give a more realistic answer.

(e) *Time*

Transactions take place in the market regularly. The market is not static so that the price obtained for a property at one time will not necessarily be achieved if the same property is sold again. Thus the reliability of evidence of prices diminishes as time elapses since the transaction took place. In a volatile market, only a short time will pass before the evidence becomes unreliable.

Given the special characteristics of property and the drawbacks these create in adopting the direct comparison method, it is clear that it is of limited application to establish the value directly. The comparison method will have wider application to establish constituent parts of more complicated valuation methods. In practice, only three types of property lend themselves readily to the use of the method and normally only when the freehold interest in the property, free from any leasehold estate, is being valued. Two types are considered below. The third is agricultural property.

## 2. Residential Property

The great majority of all property transactions comprise the sale of houses and flats with vacant possession. As the direct comparison method is adopted for the valuation of such properties this is the method of valuation most commonly used. It is true that flats are generally held as leasehold interests, but since these tend to be for a long period of years at a relatively small rent, they often have the general characteristics of a freehold so allowing direct comparison.

*Example 4–1*

A owns the freehold interest in a house which is to be sold with vacant possession. The house was built in 1934. It is semi-detached with three bedrooms, bathroom and separate WC on the first floor and a through lounge and kitchen on the ground floor. There is a garage lying to the rear of the house with an access shared by the adjoining houses. Apart from taking down the wall between the former front and rear rooms to form a single lounge, and the installation of central heating, no improvements have been carried out. It is considered that both the bathroom and kitchen need to be modernised. The property is in a fair state of repair.

Several nearby similar properties have sold recently for prices in the range £76,000 to £88,000. The most recent include a house in the next pair to A's house, built at the same time and with identical accommodation but without central heating. However, the bathroom has had a suite installed with matching tiling, and new kitchen units were installed some four years ago. It is in good repair. The house sold for £83,000. Another house in the same street, also built at the same time and with the same basic accommodation but with modernised kitchen and bathroom, central heating and ground floor cloakroom, and also having an independent drive to the garage, in good repair, sold recently for £88,000.

*Valuation*

It appears that a house such as A's with the basic amenities and modernised and in good repair is worth around £83,000 without central heating.

Assume that with central heating the value would be £84,500. This appears to be realistic since a similar house but with ground floor cloakroom and independent drive to garage sold for £88,000 (an allowance of £3,500 for these further factors seems reasonable).

Allowing £4,000 for the cost and trouble of modernising the bathroom and kitchen, it appears that the value of A's house is £80,500 in good repair.

Allowing for the lower standard of repair, the value of A's house is in the region of £79,500.

It is impossible to be absolutely precise in a valuation of this kind because different purchasers will make different allowances for modernisation costs and the costs of repairs. Some purchasers would replace the bathroom suite even if it were almost new if they did not like its colour and therefore would pay more for this house than £79,500 because they would have been willing to spend £84,500 and still incur expenditure.

It is clear from this example that it is necessary to make various allowances for the differences in quality between the houses. The level of allowance is subjective and depends on the experience and knowledge of the valuer. This example illustrates that the method is simple in its general approach, but is dependent on considerable valuation judgment for its application.

The valuation of residential property is considered in detail in Chapter 17.

## 3. Development Land

In many instances property will be sold which is suitable for development or redevelopment and this is referred to as development land or building land. The types of development land range widely, reflecting the many forms of development which are possible and the variation in size and location of such land. There is one type, however, which tends to have many qualities common to each: development land for residential development. For this reason, residential building land may be valued using the direct comparison method.

The valuer, in analysing the comparables will need to allow for such factors as the type of development contemplated, e.g. from luxury houses to high-density low-cost accommodation; the site conditions, e.g. level or sloping, well drained or wet; and location in relation to general facilities, such as transport, schools, hospitals. The comparable sites will vary in size and the valuer therefore reduces the evidence to a common yardstick. This may be value per hectare (acre), value per plot or per unit, or value per habitable room. For example, if a site of 4 hectares (10 acres) sold for £2,500,000 with permission to build either 90 4-room houses or 120 3-room flats the price may be expressed as:

£625,000 per hectare (£250,000 per acre) – £27,778 per plot – £20,833 per unit – £6,944 per habitable room.

The valuer will use whichever yardstick is appropriate to the type of sites he is valuing; the main essential is to apply the same yardstick to all cases.

*Example 4–2*

Value the freehold interest in 2 hectares (5 acres) of residential building land with consent to build 42 houses. Recent comparable land sales include:

(a) 3 hectares (7.5 acres) with consent to build 60 houses, sold for £1,800,000.
(b) 5 hectares (12.5 acres) with consent to build 125 houses, sold for £3,125,000.
(c) 0.4 hectares (1 acre) with consent to build 6 houses, sold for £300,000.

The analysis (p.44) indicates that building land is worth between £600,000 and £750,000 per hectare (£240,000 and £300,000 per acre) or between £25,000 and £50,000 per house. Where more houses are to be built on a given area (the "density") it is assumed that the houses will be smaller and less valuable, so that the value per house falls, and also the value of the land on which it stands, although conversely the value per acre rises since the builder will have higher output and presumably higher profits overall. Site (c) has values outside the general range since it is a small development which would attract a different type of buyer, and it also has a low density so that large and high value houses are to be built. The higher value of (c) supports the general value evidence of (a) or (b).

*Analysis of Comparables*

| Site | £/hectacre | (£/ acre) | £/House | Density (houses/hectacre) |
|------|------------|-----------|---------|---------------------------|
| (a)  | 600,000    | (240,000) | 30,000  | 20                        |
| (b)  | 625,000    | (250,000) | 25,000  | 25                        |
| (c)  | 750,000    | (300,000) | 50,000  | 15                        |

Hence the site to be valued, which has a density of 21 houses per hectare (8.4 houses per acre), appears to lie within the general band of values.

*Valuation*

| | | |
|---|---|---|
| 2 hectares at say £610,000 per hectare | = | £1,220,000 |
| or 42 houses at say, £29,000 per house plot | = | £1,218,000 |
| *Value* say | | £1,220,000 |

Although residential building land may be valued by direct comparison, it is usual to value the interest using the residual method of valuation, which is the typical method for valuing development land, as was stated in Chapter 2, and to cross-check the valuation by valuing using direct comparison. The valuation of development land is considered fully in Chapters 12 and 16.

# Chapter 5

# Rental Value

## 1. Generally

In Chapter 2 the basis of valuation using the investment method was described. It involves the calculation of the capital value from a consideration of the future income which was derived from the property. There are thus two constituent parts of the valuation, namely the net receivable income and the yield.

This chapter and Chapter 6 are concerned with the factors affecting the income and Chapter 7 deals with the factors determining yield.

## 2. Rental Value and Net Income

Property can be let on the basis that the tenant will bear all of the costs and outgoings associated with the property including repairs, insurance, rates etc. These lettings are normally known as full repairing and insuring lettings or lettings on FR and I terms or FRI Leases.

Alternatively property can be let on the basis that the landlord bears some or all of the property outgoings and in this case the net income receivable by the landlord will be arrived at by deducting the outgoings from the rent payable.

It is only the net rental income which is capitalised to arrive at the capital value. Therefore, where reference is made to "rent" it is to be understood that "net rental income" is intended. The usual outgoings connected with property are outlined in Chapter 6 and in the chapters dealing with particular types of property.

Rent may be defined as "an annual or periodic payment for the use of land or of land and buildings". In fixing the rental value of a property a valuer is largely influenced in practice by the evidence he can find of rents actually paid for comparable properties in the same district. But it is essential that he should appreciate the economic factors which govern those rents. Knowledge of economic factors may not be necessary for the immediate purpose of his valuation, but it is vital in understanding fluctuations in rental value, and in advising on the reasonableness or otherwise of existing rents and market prices.

## 3. Economic Factors Affecting Rent

### (a) Supply and Demand

In Chapter 1 it was observed that the general laws of supply and demand govern capital values and that value is not an intrinsic characteristic.

Variations in capital value may be due either to changes in rental value or to changes in the return that may be expected from a particular type of investment. Either or both of these factors may operate at any time.

Up until recent times the general trend was for rents to rise over the years but there is no law of economics which dictates that this must happen. Over-supply and an economic recession led to a sharp fall in rental values in London and the South-East from 1989 to 1993 and increasing obsolescence of design and location could have the same result even in normal economic conditions. Movements in interest rates follow general market movements but not necessarily to the same degree. It follows that at times rental values rise and returns fall, so combining to produce a sharp rise in capital values, whilst at other times the changes have opposing effects on capital values.

For example, shops in an improving position may increase in rental value over a period of years. At the same time, there may be changes in the investment market which cause that particular type of security to be less well regarded; a higher yield will be expected and consequently a lower multiplier (or Years' Purchase) will have to be applied to the net rent to arrive at the capital value. The result may be that, whilst the rental value has increased, the capital value remains unchanged.

Both changes in rental value and changes in the yield required from property investments are the result of the operation of the laws of supply and demand. For rental value the influence of supply and demand is on the income which the property can produce; as for yield it is on the Years' Purchase which investors will give for the right to receive that income. In either case any variation has a direct bearing on the capital value of the property.

In this chapter we are concerned with the influence of supply and demand on rental value.

When we say that value depends on supply and demand, it must be remembered that supply means the effective supply at a given price, not the total of a particular type of property in existence. In a particular shopping centre there may be perhaps 30 shops of a very similar type, but 20 may not be available to let because they are already let or owner occupiers are trading from them and they are not on the market. The effective supply consists of the other 10 shops whose owners are willing to let.

Similarly, demand means the effective demand at a given price, i.e. a desire backed up by money and ability to trade. In the shopping centre example, demand is derived from potential tenants who have weighed the advantages and disadvantages of the shops offered against other properties elsewhere.

The prevailing level of rents will be determined by the interaction of supply and demand, what is sometimes referred to as "the higgling of the market". If there is a large or increasing demand for a certain type of property, rents are likely to increase; if the demand is a falling one, rents will diminish.

As pointed out in Chapter 1, landed property can be distinguished in two important respects from other types of commodity in that the supply of land is to a large extent a fixed one and that, in the case of buildings, supply may respond slowly to changes in demand as supply is usually fixed in the short term.

*Example 5–1*

| Rental Value today | £ 5,000 | Rental value 5 years hence | £ 6,250 |
|---|---|---|---|
| Years' Purchase (YP) to show 8% yield | 12.5 | Years' Purchase (YP) to show 10% yield | 10 |
| Capital Value | £62,500 | Capital Value | £62,500 |

*(b) Demand Factors*

It is obvious that the prime factor in fixing rent is demand.

The demand for land and buildings is a basic one in human society, as it derives from three essential needs of the community:

(i)    the need for living accommodation;
(ii)   the need for land, or land and buildings, for industrial and commercial enterprise – including the agricultural industry;
(iii)  the need for land, or land and buildings, to satisfy social demands such as schools and hospitals and sports grounds.

The demand for any particular property, or for a particular class of property, will be influenced by a number of factors, which it will be convenient to consider under appropriate headings.

*General Level of Prosperity*
This is perhaps the most important of the factors governing demand.

When times are good and business is thriving, values will tend to increase. There will be a demand for additional accommodation for new and expanding enterprises. There will be a corresponding increase in the money available for new and improved housing accommodation and for additional associated amenities.

The increase in rental values will not necessarily be exactly comparable to the increase in the general standard of living. It may not apply to all types of property and its extent may vary in different parts of the country or in different parts of the same area owing to the influence of other factors such as changing tastes.

Experience in this and other countries has shown, however, that rental values generally show a steady upward tendency in times of increasing prosperity.

*Population Changes*
Times of increasing prosperity have in the past been associated with the factor of a growing population.

The supply of land being limited, it is obvious that an increase in demand due to increased prosperity or increase of population will tend to increase values. On the other hand, when these causes cease to operate or diminish in intensity, values will tend to fall.

Increase in population will, in the first place, influence the demand for housing, but indirectly will affect many other types of property the need for which is dependent upon the population in the locality – as with, for instance, retail shops. Increases in population within a locality may occur in various ways. The birthrate may exceed the number of deaths so that there is a natural increase in the size of the existing population. Alternatively, population may rise locally due to migration of people from other areas to the locality. A further change in population may arise for a temporary period; this is typical of holiday centres, where the numbers rise significantly for certain periods of the year. The impact of tourists on demand can be very significant.

### Changes in Character of Demand

Demand, in addition to being variable in quantity, may vary in quality. Associated with an increase in the standard of living, there is a change in the character of the demand for many types of property.

For example, in houses and offices, many amenities regarded today as basic requirements would have been regarded as refinements only to be expected in the highest class of property two decades ago. Similar factors operate in relation to industrial and commercial properties.

The increased standards imposed by legislation or technical improvements in the layout, design, and equipment of buildings tend to cause many buildings to become obsolete and so diminish their rental value.

### Rent as a Proportion of Personal Income

In relation to residential property, the rent paid may be expressed as a proportion of the income of the family occupying the property. It is obvious that, in a free market, the general level of rents for a particular type of living accommodation will be related to the general level of income of the type of person likely to occupy that accommodation. There will, of course, be substantial variations in individual cases owing to different views on the relative importance of living accommodation, motor cars, television sets, holidays and other goods and services. However, in the United Kingdom, there was not, for many years, a free market in living accommodation to rent. The slump in the housing market after 1988, coupled with changes in legislation, led to a revival in the free market with lettings at market rents. The cause was a reluctance to buy and take on mortgage commitments against a background of job insecurity.

There is a free market in residential properties for sale and a useful parallel today in relation to capital, not rental, values is that building societies, in determining an appropriate mortgage loan, assume that an individual should not pay more than a maximum proportion of his income on mortgage repayment plus rates, typically one quarter to one third.

## *Rent as a Proportion of the Profit Margin*

Both commercial and industrial properties and land used for farming are occupied as a rule by tenants who expect to make a profit out of their occupation, and that expectation of profit will determine the rent that such a tenant is prepared to pay.

Out of his gross earnings, he has first to meet his expenses; from the gross profit that is left, he will require remuneration for his own labour, and interest on the capital that he has to provide; the balance that is left is the margin that the tenant is prepared to pay in rent. Obviously, no hard and fast rule can be laid down as to the proportion that rent will form of profit, but the expectation of profit is the primary cause of the tenant's demand and the total amount of rent he will be prepared to pay will be influenced by this estimate of the future trend of his receipts and expenses and the profit he requires for himself.

## *Competitive Demand*

Certain types of premises may be adapted for use by more than one trade or for more than one purpose, and there will be potential demand from a larger number of possible tenants. For instance, a block of property in a convenient position in a large town may be suitable for use either for warehouse or for light industrial purposes. The rent that can consequently be expected to be paid for either use will reflect the relative demands; that having the greater demand produces the greater rent.

In a free market it would be expected that the enterprise yielding the largest margin out of profits for the payment of rent would obtain the use of the premises, for where there are competing demands for various purposes, properties will tend to be put to the most profitable use. In practice, the operation of this rule will be affected by the fact that the use to which the property may be put will be restricted by town planning legislation.

## (c) Supply Factors

It has already been stated that land differs from other commodities in being to a large extent fixed in amount. This limit on the amount of land available is the most important of those factors which govern supply and thus the supply of land and building is regarded as being relatively static.

### Limitation of Supply

At any one time there will be in the country a certain quantity of land suitable for agricultural purposes and a certain quantity of accommodation suitable for industrial, commercial and residential purposes.

If these quantities were entirely static, the rents would be affected only by changes in demand. In fact, they are not static, but respond, although slowly, to changes in demand. There is, ultimately, only a certain quantity of land available for all purposes, but increase in demand will cause changes in the use to which land is put.

In agriculture, conditions of increasing prosperity, leading to higher prices, will bring into cultivation additional land which previously it was not profitable to work. Conversely, in bad times land will go out of cultivation and revert to waste land. Similarly, increasing demand will render it profitable for additional accommodation to be built for commercial and residential uses and the supply of accommodation will be increased by new buildings until the demand is satisfied.

These statements assume that the market is working freely. In practice, there is considerable intervention by government, either to support and stimulate some uses or to dampen them down. For example, agriculture has generally enjoyed a strong measure of support for the past 40 years from all governments, but this has now led within the European Community to substantial over-production. To reduce production to the level of demand within the EC, a number of restrictive measures, such as milk quotas, have been introduced and a substantial amount of land has been deliberately taken out of agricultural production. Office development has often been curbed too in parts of the country, despite a strong demand for new accommodation, in the largely mistaken belief that this will encourage similar development in other parts of the country.

The increase in rents that would occur if the supply remained unchanged will be modified by additions to the supply. But, as was pointed out in Chapter 1, adjustments in the supply of land and buildings are necessarily made slowly in comparison with other commodities.

Where there is a tendency for the demand to fall, the supply will not adjust itself very quickly; building operations may well continue beyond the peak of the demand period and some time may elapse before expansion in building work is checked by a fall in demand.

It would enlarge the scope of this book unduly to discuss the various causes of inelasticity in supply. But the principal ones are probably:

(i)   the fact that building development is a long-term enterprise, i.e. one where the period between the initiation of a scheme and completion of the product is comparatively lengthy; and

(ii)  the difficulty involved in endeavouring to forecast demand in any locality.

The above considerations introduce the subject of what economists call "marginal land". In the instance given of land being brought into cultivation for agricultural purposes, there will come a point where land is only just worth the trouble of cultivation. Its situation, fertility or other factors will render it just possible for the profit margin available as a result of cultivation to recompense the tenant for his enterprise and for the use of his capital. There will, however, be no margin for rent. This will be "no-rent" land or marginal land.

It is obvious that land better adapted for its purpose will be used first and yield a rent, but that also, with increasing demand, the margin will be pushed farther out and land previously not worth cultivation will come into use.

Similar considerations will apply to other types of property. For instance, there will be marginal land in connection with building development. A piece of land may be so situated that, if a factory is built on it, the rent likely to be obtained will just cover interest on the cost of construction, but leave no balance for the land. With growth of demand, the margin will spread outwards, the rent likely to be obtained for the factory will increase, and the land will yield a rent.

Land may have reached its optimum value for one purpose and be marginal land for another purpose. For example, land in the vicinity of a town may have a high rental value for market garden purposes and have reached its maximum utility for agricultural purposes. For building purposes it may be unripe for development – in other words, "sub-marginal" – or it may be capable of development, but incapable of yielding a building rent – in other words, "marginal".

From the practical point of view, it is unlikely that land will be developed immediately it appears to have improved a little beyond its marginal point. Some margin of error in the forecast of demand will have to be allowed for, and no prudent developer would be likely to embark on development unless there is a reasonable prospect of profit. In many actual cases planning control will prevent development where land has improved well beyond its marginal point. As an example the pressure to develop open land around cities is contained by green belt planning policies.

*Relation of Cost to Supply*
Another factor which may influence supply is the question of cost.

If, at any time, values as determined by market conditions are less than cost, the provision of new buildings will be checked, or may even cease, and building will not recommence until the disparity is removed.

The disparity which may exist at any time between cost and value may be subsequently removed either by a reduction in cost, or by an increase in demand or a reduction in supply due to demolition of marginal property. As an example, offices and flats built in the 1950s and 1960s are being demolished. Governments may intervene by providing grants or other incentives so as to reduce the effective cost or to stimulate demand. It might be further modified by the non-replacement of obsolete buildings, which might in time bring about an excess of demand over supply and a consequent increase in value.

## 4. Determination of Rental Value

### (a) Generally

It has already been pointed out that in estimating the capital value of a property from an investment point of view, the valuer must usually first determine its rental value.

In doing so, he will have regard to the trend of values in the locality and to those general factors affecting rent discussed above. These will form, as it were, the general background of his valuation, although he may not be concerned to investigate in detail all the points referred to when dealing with an individual property.

The two factors most likely to influence his judgment are:

(i)   the rents paid for other properties; and
(ii)  the rent, if any, at which the property itself is let.

In most cases when preparing an investment valuation it is assumed that the rental value at the time when the valuation is made will continue unchanged in the future. This assumption is reasonable as normal changes, either upwards or downwards, are reflected in the rate per cent which an investment is expected to yield. The rent payable at the date of the valuation must be compared with current open market rental values and regard had to the length of the lease, the incidence of rent reviews, and the terms of the lease as regard matters such as repair and user. Where the property is over-rented, i.e. the rent payable exceeds the rental value, this must be allowed for in the calculation. The logic of the assumption is considered in greater detail in Chapter 9.

In general, it may be said that if rental value is likely to increase, a higher Years' Purchase will be applied; while if there is a doubt as to the present level being maintained, the Years' Purchase will be reduced to cover the risk.

Future variations in rental value may sometimes arise from the fact that the premises producing the income are old, so that it may be anticipated that their useful life is limited. Together with the likelihood of a fall in rental value there may be a possibility that, in the future, considerable expense may be incurred either in rebuilding or in modernisation to maintain the rental value. In such cases the rate per cent which the investment will be expected to yield will be increased and the Years' Purchase to be applied to the net income correspondingly reduced. This particular aspect, commonly termed obsolescence, will be considered in greater detail in Chapter 9.

Where a future change in rental value is reasonably certain it should, of course, be taken into account by a variation in the net rent to which the Years' Purchase is to be applied. For example, a valuation may be required of a shop which is offered for sale freehold subject to a lease being granted to an intending lessee on terms which have been agreed. These terms provide for a specified increase in rent every five years. It has been ascertained that many similar shops in the immediate vicinity are let on leases for 10, 15 and 20 years with a provision for similar increases in the rents at the end of each five-year period. Assuming that the position is an improving one and that the valuer is satisfied that such increasing rents are justified, he will be correct in assuming that the shop with which he is dealing can be let on similar terms, and he will take future increases in rent into account in preparing his valuation.

The method of dealing with such varying incomes from property will be considered in detail in Chapter 9.

### (b) Basis of the Rent Actually Paid

Where premises are let at a rent and the letting is a recent one, the rent actually paid is usually the best possible evidence of rental value.

But the valuer should always check this rent with the prevailing rents for similar properties in the vicinity or, where comparison with neighbouring properties is for some reason difficult, with the rents paid for similar properties in comparable positions elsewhere.

There are many reasons why the rent paid for a property may be less than the rental value. For instance, a premium may have been paid for the lease, or the lease may have been granted in consideration of the surrender of the unexpired term of a previous lease. The lessee may have covenanted to carry out improvements to the premises or to forego compensation due to him from the landlord. In many cases the rental value may have increased since the existing rent was agreed. Differences between rent paid and rental value may also be accounted for by the relationship between lessor and lessee, e.g. father to son, parent company to a subsidiary.

In some instances a rent from a recent letting may be above rental value. For example, an owner may sell an asset to realise capital and take back a lease from the purchaser (a sale and leaseback). If the owner is a strong covenant the purchaser may agree to pay above market value and accept a higher rent, relying on the strength of the covenant for security of the income.

If, as a result of the valuer's investigations, he is satisfied that the actual rent paid is a fair one, he will adopt it as the basis of his valuation, will ascertain what outgoings, if any, are borne by the landlord and by making a deduction in respect thereof will arrive at net rental value.

If, in his opinion, the true rental value exceeds the rent paid under the existing lease, his valuation will be made in two stages. The first stage will be the capitalisation of the present net income for the remainder of the term. The second stage will be the capitalisation of the full rental value after the end of the term.

If, on the other hand, he considers that the rent fixed by the present lease or tenancy is in excess of the true rental value, he will have to allow for the fact that the tenant, at the first opportunity, may refuse to continue the tenancy at the present rent, and also for the possible risk, in the case of business premises, that the tenant may fail and that the premises will remain vacant until a new tenant can be found.

As a general rule, any excess of actual rent over estimated fair rental value may be regarded as indicating a certain lack of security in the income from the property. But it must be borne in mind that rent is secured not only by the value of the premises but also by the tenant's covenant to pay, and a valuer may sometimes be justified in regarding the tenant's covenant as adequate security for a rent in excess of true rental value.

An example of such cases is a sale and leaseback deal of the type referred to above. Another is where a shop is let on long lease to some substantial concern such as one of the large retail companies with many shops – a "multiple". Here, the value of the goodwill the tenants have created may make them reluctant to terminate the tenancy, even if they have the right to do so by a break in the lease; they may, however, threaten to do so unless the rent is reduced. Where there is no such break, they will in any case be bound by their covenant to pay. Since rent is a first charge, ranking even before debenture interest, and since such a concern will have ample financial resources behind it, the security of the income is reasonably assured, but the valuer should be aware of the true financial state of the tenant company before making the assumption. Sometimes the lease is held by a subsidiary of the major company; in the absence of a guarantee from the parent, the covenant is only as good as the accounts of the subsidiary justify.

## (c) Comparison of Rents

It has been suggested that the actual rent paid should be checked by comparison with the general level of values in the district. Not only is this desirable where the rent is considered to be *prima facie* a fair one, but it is obviously essential where premises are vacant or let on old leases at rents considerably below true rental value.

In such circumstances the valuer has to rely upon the evidence provided by the actual letting of other similar properties. His skill and judgment come into play in estimating the rental value of the premises under consideration in the light of such evidence. He must have regard not only to the rents of other properties that are let, but also to the dates when the rents were fixed, to the age, condition and location of the buildings as compared with that with which he is concerned, and to the amount of similar accommodation in the vicinity to be let or sold.

In many cases, as for instance similar shops in a parade, it may be a fairly simple matter to compare an unlet property with several that are let and to assume, for example, that since the latter command a rent of £20,000 a year on lease it is reasonable to assume the same rental value for the shop under consideration which is similar to them in all respects.

But where the vacant premises are more extensive and there are differences of planning and accommodation to be taken into account in comparing them with other similar properties, it is necessary to reduce rents to the basis of some convenient "unit of comparison" which will vary according to the type of property under consideration.

For example, it may be desired to ascertain the rental value of a factory having a total floor space of 1,000 m². Analysis of the rents of other similar factories in the neighbourhood reveals that factories with areas of 500 m² or thereabouts are let at rents equivalent to £50 per m² of floor space, whereas other factories with areas of 4,000 m² or thereabouts are let at rents of £40 per m² of floor space. This indicates that size is a factor in determining the unit rental value. After giving consideration to the situation, the building, and other relevant factors, it may be reasonable to assume that the rental value of the factory in question should be calculated on a basis of £47.50 per m² of floor space.

The unit chosen for comparison depends on the practice adopted by valuers in that area or for the type of property. Notwithstanding the transition to metric units, over past years valuers have shown a marked reluctance to abandon imperial units. Thus the most commonly met units of measure are still values per square foot or per acre and most, if not quite all, comparative statistics are based on Imperial measures but this is likely to change over the life of this edition. Metric units are commonly adopted for agricultural purposes and in valuations for rating purposes. The student must be prepared to accept and work with either method. The practitioner will follow his normal practice but he should be aware that the Royal Institution of Chartered Surveyors advocates the use of metric units, either with or without the imperial equivalent. There is, therefore, a compelling reason for surveyors to adopt metric units.

The method of measuring also varies according to the type of property concerned. In some cases the areas occupied by lavatories, corridors, etc. are excluded and in other cases the valuer works on the gross internal area. In an attempt to bring some uniformity, the professional societies have produced recommended approaches. Thus, the Royal Institution of Chartered Surveyors and the Incorporated Society of Valuers and Auctioneers jointly published a Code of Measuring Practice (4th Edition, 1993). This Code both identifies various measuring practices, such as gross internal area and net internal area, and also recommends the occasions when such practices would be appropriate, for example gross internal area for industrial valuation and net internal area for office valuation. Though not bound to follow the Code by the rules of conduct of the professional bodies, it is irrational not to do so since comparable evidence and calculation of rental value are usually on the same basis when obtained from other surveyors.

When quoting rental figures the amount per annum should always be qualified by reference to such terms of the lease as the liability for rates, repairs and insurances, and by reference to the length of the term. The date of the valuation should also be stated.

### (d) Rental Value and Lease Terms

The amount of rent which a tenant will be prepared to offer will be influenced by other terms of the lease. Rent is but one factor in the overall contract and if additional burdens are placed on the tenant in one respect he will require relief in another to keep the balance. For example, if a landlord wishes to make a tenant responsible for all outgoings in respect of a property, he cannot expect the rent offered by the tenant to be the same as where the landlord bears responsibility for some of them.

Similarly, if a landlord offers a lease for a short term with regular and frequent upward revisions of the rent, he cannot expect the same initial rent as would be offered if a longer term were offered with less frequent reviews, particularly if his terms are untypical of the general market arrangements. Again, if a landlord seeks to limit severely the way in which premises may be used, the rent will be less than where a range of uses will be allowed.

It is clear, therefore, that rental value will reflect the other terms of the lease and cannot be considered in isolation. Indeed, a rental value cannot be determined until the other terms of the lease are known.

In making comparisons, it must be remembered that the terms of the lettings of different classes of property vary considerably. It is convenient, therefore, to compare similar premises on the basis of rents on the terms usually applicable to that type of property. Good shop property and industrial and warehouse premises are usually let on terms where the tenant is liable for all outgoings including rates, repairs and insurance. Thus, net rents are compared on this type of property, assuming a full repairing and insuring lease will be agreed. With blocks of offices let in suites, however, the tenant is usually liable for internal repairs to the suite and rates and the landlord for external repairs, repairs to common parts and insurance. The landlord provides services such as lifts, central heating and porters, but these are dealt with separately by means of a service charge on the tenants (and which may also include the landlord's liability to repair and insure). Thus, rents used for comparing different suites of offices would be inclusive of external repairs, repairs to common parts and insurance but exclusive of the service charge. Having arrived at the rental value of the office block on this basis, the net income would be arrived at by deducting the outgoings which are included in the rents and for which the landlord is liable. In cases where landlords include within the service charge their liability to repair and insure and also the supervisory management fees, the rental value will be the net income. This latter form of lease is known as a "clear lease" as the landlord recovers all of his costs from the tenants and thus the rent is usually comparable with the rental on an FRI lease.

*Example 5–2*

A prospective tenant of a suite of offices on the third floor of a modern building with a total floor space of 250 m$^2$ requires advice on the rent which should be paid on a five-year lease. The tenant will be responsible for rates and internal repairs to the suite. The landlord will be responsible for external repairs, repairs to common parts and insurance. Adequate services are provided, for which a reasonable additional charge will be made. The following information is available of other recent lettings on the same lease terms in similar buildings, all of suites of rooms:

| Floor | Area in m$^2$ | Rent £ | Rent £ per m$^2$ |
|-------|--------------|--------|------------------|
| Ground | 200 | 36,000 | 180 |
|        | 300 | 48,000 | 160 |
| 1st | 150 | 18,000 | 120 |
| 2nd | 160 | 20,000 | 125 |
| 4th | 400 | 52,000 | 130 |
| 5th | 500 | 67,500 | 135 |
| 6th | 200 | 26,000 | 130 |

*Estimate of Rent Payable*

As these other lettings are on the same terms as that proposed, it is unnecessary for purposes of comparison to consider the outgoings borne by the landlord or to arrive at net rental values.

The broad picture which emerges from the evidence is that the highest rents are paid on the ground floor but thereafter the rents are approximately the same. This is what would be expected in modern buildings with adequate high-speed lifts. In older buildings with inadequate lifts or no lifts, the higher the floor the lower the rent. The inconsistencies in the rents may be due merely to market imperfections or to differences in light or noise.

Subject to an inspection of the offices under consideration so that due weight can be given to any differences in quality and amenities, a reasonable rent would appear to be, say, £130 per m$^2$ or £32,500 per annum.

The different methods used in practice for fixing rent by comparison are referred to in more detail in later chapters dealing with various classes of property.

## 5. Effect of Capital Improvements on Rental Value

Since, in most cases, an owner is not likely to spend capital on his property unless he anticipates a fair return by way of increased rent, it is usually reasonable to assume that money spent by a landlord on improvements or additions to a property will increase the rental value by an amount approximating to simple interest at a reasonable rate on the sum expended. The rate per cent at which the increase is calculated will naturally depend upon the type of property.

Since, however, the real test of rental value is not solely what the owner expects to obtain for his premises, but more importantly what tenants are prepared to pay for the accommodation offered, in practice any estimate of increased rental value should be carefully checked by comparison with the rents obtained for other premises to which similar additions or improvements have been made.

Capital expenditure is only likely to increase rental value where it is judicious and suited to the type of property in question. For example, a man may make additions or alterations to his house which are quite out of keeping with the general character of the neighbourhood and are designed solely to satisfy some personal whim or hobby. So long as he continues to occupy the premises he may consider he is getting an adequate return on his money in the shape of personal enjoyment, but there will be no increase in the rental value of the premises since the work is unlikely to appeal to the needs or tastes of prospective tenants.

Again, expenditure may be made with the sole object of benefiting the occupier's trade, as, for instance, where a manufacturer spends a considerable sum in adapting premises to the needs of his particular business. In this case, the occupier may expect to see an adequate return on his capital in the form of increased profits, but it is unlikely that the expenditure will affect the rental value unless, of course, it is of a type which would appeal to any tenant of the class likely to occupy the premises – as, for instance, where an occupier of offices installs air-conditioning which significantly improves the working conditions.

On the other hand, the increase in rental value due to capital expenditure may considerably exceed the normal rate of simple interest on the sum expended. This will be so in those cases where the site has not hitherto been fully developed or where it is to some extent encumbered by obsolescent or unsuitable buildings. For example, a site in the business quarter of a flourishing town may be covered by old, ill-planned and inconvenient premises. If, by a wise expenditure of capital, the owner can improve these premises so as to make them worthy of their situation, he should secure an increase in rental value which will not only include a reasonable return on the capital sum expended, but also a certain amount of rental value which has been latent in the site and which will have been released by the development.

*Chapter 6*

# Outgoings

## 1. Generally

A major difference between property and most other forms of investment is that considerable sums of money need to be expended regularly to maintain, insure and let it. The duties of ensuring that sums are paid when due and work done when needed demand the attention of managers and this in itself is a cost. All such expenses are referred to as outgoings.

Where the property is occupied by the owner he will be responsible for meeting all these costs. In the case of property which has been let, the terms of the lease will determine whether the landlord or the tenant is liable for any particular expense (subject to various statutes which impose duties on particular classes of persons, notwithstanding their contractual arrangements).

It is common in practice for the landlord to seek to impose the duty of meeting all of the various expenditures on property on the tenant by means of a full repairing and insuring lease (an FRI lease) so that the rent payable to him is free from any deduction. This is particularly so in the case of non-residential property let to a single tenant. In the case of property in multi-occupation, where services and other activities can only be met through a central or common approach, for example, the cost of maintaining a lift serving various tenants on different floors, landlords tend to attempt to recoup the expenditure by means of a charge additional to rent known as a service charge so as to separate rent from outgoings.

The reason for landlords' concern over this aspect is that rents tend to be fixed for a given period, whereas some outgoings are subject to unpredictable and substantial changes due to their nature, and are also affected by inflation. Where a rent is fixed for a period and the landlord has to meet outgoings from this income, if the cost of the outgoings rises, it follows that the net return to him will fall. A variable income is unsatisfactory to an investor at all times and during a time of inflation it could result in a negative income.

From a tenant's point of view, the burden of outgoings is a cost of occupying the premises which is additional to the rent required and which, in aggregate, represents the cost to him of occupation. It is the aggregate cost which he will consider in deciding on whether or not to take a lease of property and what rent he will offer. Hence, if the burden and cost of outgoings are increased then he will have a smaller sum available out of his aggregate budget figure which he will be prepared to offer as rent. This is known as the "total cost concept".

For example, if two factories are offered of equal quality, but in one case the tenant must meet the cost of all outgoings, costing £6,000 p.a., whereas, in the other, the landlord will bear half, the tenant will have to spend £3,000 p.a. less in the second case and would be prepared to pay around £3,000 p.a. extra in rent in consequence. This aspect was referred to in Chapter 5 on rental value.

In the case of properties of similar facilities but different burdens, the tenant will pay a higher rent for the building with cheaper outgoings. For example, a prospective tenant may be offered two similar factory premises which, on other considerations, are of equal value; one of them he finds to be of indifferent construction and to have a considerable extent of external painting, with a consequent heavier liability to repair. In such a case he will be willing to offer a higher rent for the premises where the cost of repairs is likely to be less.

It is clear, therefore, that the level of outgoings has a significant effect on property dealings. It follows that it is important to determine the cost of meeting the various types of outgoings, and these are now considered.

## 2. Repairs

### (a) Generally

The state of repair of a property is an important factor to be taken into account when arriving at the capital value. Likewise the burden of keeping a property in repair is an important factor in determining rental value.

Hence, regard must be had to the need for immediate repairs, if any; to the probable annual cost; and to the possibility of extensive works of repair, rebuilding or modernisation in the future.

The age, nature and construction of the buildings will affect the allowance to be made for repairs. Thus, a modern house, substantially built, with a minimum of external paintwork, will cost less to keep in repair than a house of some age, of indifferent construction and with extensive woodwork.

### (b) Immediate Repairs

If such repairs are necessary, the usual practice is to calculate the capital value of the premises in good condition and then to deduct the estimated cost of putting them into that condition. This is known as making an "end allowance".

### (c) Annual Repairs Allowance

It is obvious that the cost of repair of premises will vary from year to year.

For the purpose of arriving at the net income of a property where the cost of repairs has to be deducted, it is necessary to reduce the periodic and variable cost to an average annual equivalent. This may be done by reference to past records, if available; by an estimate based on experience, possibly, in some cases, expressed as a percentage of the rental value; by an estimate based on records of costs incurred on similar buildings, expressed in terms of units of floor space; or by examining in detail the cost of the various items of expenditure and their periods of recurrence.

Where reference is made to past records, it is essential to check the average cost thus shown by an independent estimate of what the same item of work would cost at the present day.

One of the dangers of using a percentage on rental value can be illustrated by the following example. Two shops are very similar physically, but one of them is so situated as to command a rent considerably in excess of the other; the cost of repair is likely to be similar, but different percentages would have to be taken to arrive at the correct answer. For example, one shop may be well situated and have a rental value of £50,000 p.a. Another similar shop may be in a secondary position with a rental value of £7,000 p.a. If the annual costs of repair for each are £900 p.a., in the first case this is 1.8% of rental value whilst in the second it is 12.9%. This discrepancy will remain even if the standard of repair of the better placed property is higher than the other. It is clear that the repair costs of the better shop would need to be £6,450 p.a. if 12.9% were to be the general percentage of rent spent on repairs. The percentage basis is therefore reliable only where the properties are similar physically and attract a similar rental value. While the percentage basis may be of some use, it is suggested that the only wise course is to judge each case on its merits and to make an estimate of the cost of the work required to put and keep the premises in good condition.

An estimate may be based on costs incurred on similar buildings. If the costs of keeping in repair several office buildings of similar quality and age are known, an analysis may show that they tend to cost a similar amount expressed in terms of a unit of floor space. For example, if the analysis showed that around £15 per square metre was being spent, it would be reasonable to adopt this sum in valuing a further office building in the area of similar qualities and age.

The fourth method of estimating the probable average annual cost of repairs is illustrated in the following example.

*Example 6–1*

A small and well-built house of recent construction, comprising three bedrooms, two reception rooms, kitchen and bathroom, is let on an annual tenancy, the landlord being responsible for all repairs, at a Fair Rental of £2,000 p.a. From inspection it is found that the house is brick built, in good structural condition, with a minimum of external paintwork. It is required to estimate the average annual cost of repairs.

(a)  Experience might suggest an allowance of 40% of the rent.
(b)  This estimate might be checked by considering the various items of expenditure and how often they are likely to be incurred as follows:

| | | |
|---|---|---|
| (1) | External decorations every 4th year, £1,200 | £300 p.a. |
| (2) | Internal decorations every 7th year, £1,400 | £200 p.a. |
| (3) | Pointing, every 25th year, £3,000 | £120 p.a. |
| (4) | Sundry repairs, say | £250 p.a. |
| | | £870 p.a. |

The method used in the example can be equally well applied to other types of property, regard being had in each case to the mode of construction, the age of the premises and all other relevant factors which will affect the annual cost.

## (d) Conditions of Tenancy

Regard must be had to the conditions of the tenancy in estimating the annual cost. Where premises are let on lease, the tenant is usually liable for all repairs and no deduction has to be made (a full repairing lease). In other cases the liability may be limited to internal repairs (internal repairing lease), particularly where a building is in multi-occupation; in valuing the landlord's interest, a deduction for external repairs will therefore be needed. This will be so even if the landlord has not covenanted to carry out external repairs and he is not therefore contractually liable for such repairs (unless statutory provisions override the contractual position) since the rent paid is assumed to exclude any liability on the tenant for this element. In practice, most landlords would regard it as prudent estate management to keep the property in repair. On the other hand, it is common practice for landlords to impose a service charge on tenants which will include the cost of external repairs. In such cases no deduction from the rent will be made.

Before the valuation is carried out the solicitor acting for the client should be contacted, or the lease(s) should be examined, to verify the exact liability to repair of the parties. In the case of short lettings, it must also be remembered that, on the conclusion of the tenancy, the premises will need redecoration before re-letting or sale even if in a fair state of repair at the date of valuation. The same may be true where a longer lease is coming to an end, depending on the extent of liability to repair imposed upon the tenant. A tenant under a general covenant to repair is liable to leave the premises in repair on termination of the lease and any works required at that time to bring the premises into proper condition are termed "dilapidations". Nonetheless, a valuer would need to decide if some allowance would be appropriate in valuing the landlord's interest, particularly if the tenant is a poor covenant and there is the risk of repairs not being done.

If the cost of repairs is expressed as a percentage of the rental value it must not be forgotten that the landlord may also have undertaken to pay other outgoings besides repairs. The percentage to be adopted in such cases where applied to the inclusive rent will be somewhat more than where only repairs are included within the gross rent.

## (e) Future Repairs

In considering the value of property it may be necessary to make allowance for the possibility of extensive works of repair, improvement, or even ultimate rebuilding in the future. Where the date for such works is long deferred it is not usual in practice to make a specific allowance, but, where such works are possible in the near future (say up to 10 years or so), the present cost of the works deferred for the estimated period is deducted from the capital value. In the case of older properties the works will tend to reflect a considerable element of improvement to allow for the changes in standards that have occurred since the property was erected. In fact, the rate of improvement has tended to increase over recent years as changes in standards accelerate, and a generous allowance for repairs as an annual outgoing may be prudent to reflect the cost of these works which are often cosmetic rather than actual repair. Alternatively a provision for a capital investment in the property on the termination of the lease might be more appropriate. This would be so even where a tenant will contribute to dilapidations since these only cover putting into repair what was there, whereas the landlord might wish to carry out more extensive works of improvement or refurbishment.

## 3. Sinking Funds

It is sometimes suggested that provision should be made, in addition to the cost of repairs, for a sinking fund for the reinstatement of buildings over the period of their useful life. In some countries it is standard practice to do so.

In the case of properties where the buildings are estimated to be nearing the end of their useful life this provision may have to be made, but in most valuations the risk of diminution in the value of buildings in the future is reflected in the percentage yield to be derived from the property. This aspect is considered further in Chapters 9 and 19.

## 4. Business Rates, Water Rates and Council Taxes

Rates and Council Taxes are normally the personal liability of the tenant, but in some cases, such as lettings on weekly tenancies, the landlord may undertake to pay rates or Council Taxes.

Business rates are levied on the rateable value at a poundage which is consistent across the Country (the Uniform Business Rate or UBR). Council Taxes vary from local authority to local authority. Water rates are calculated in a number of different ways including metering, on the number of employees in occupation or as a percentage of the historic rateable value. This subject is considered further in Chapter 21.

To arrive at a fair amount to be deducted, regard must be had not only to the actual amount paid at the present time but also to possible changes in rate poundage and assessment in the future. This is particularly important if the rent cannot be increased on account of the increase in charges. However, in most cases increases in charges will be borne by the tenant, either because he bears them directly or because the landlord is entitled to recover such increases from him.

As regards changes of assessment, the present rateable value of a property must not necessarily be taken to be correct. All assessments are capable of revision to take account of changes to the property or its use, or under a general revaluation. If an assessment appears to be too high, regard should be had to the possibility of its amendment at the earliest possible moment. If too low, the possibility of an upward revision should be anticipated.

It would obviously be incorrect, where no provision exists to adjust the rent paid in respect of alterations in rates, to base a valuation on the assumption that rates will continue at their present level, when there is every possibility of an increase in assessment.

With regard to possible variation in rate poundage, unless there is some clear indication of a higher than average change in the future, it is usual to assume continuance at the present amount if the rental income can be varied in the short term. Where the rent is fixed for the medium or long term then regard must be had to future changes in these outgoings. This can be done by changing the net income from year to year or by deducting a sum above the present level of payment or by adopting a higher yield to reflect the prospect of a falling net income.

It should be borne in mind that any alterations in rates due to changes in assessment and in rate poundage are passed on to the tenant in the case of residential tenancies under the Rent Acts since Registered Rents are expressed on an exclusive basis. This may not be the case for other kinds of tenancies. In these cases the provisions of the lease or tenancy agreement must be examined to see if a similar provision is made. Where such "excess rates" clauses are present no account need be taken of possible changes in the future for the term of the lease or agreement.

## 5. Income Tax

The income from landed property, equally with that from other sources, is subject to income or corporation tax, which is assessed in accordance with the rules of the Income and Corporation Taxes Act 1988.

It is customary to disregard tax as an outgoing, since most types of income are equally liable. This aspect of taxation is further considered in Chapter 9.

Allowance may have to be made for tax, however, when valuing leaseholds and other assets of a wasting nature, such as sand and gravel pits. The reason for this is that tax is levied on the full estimated net annual value of the property without regard to the necessity, in such cases, of a substantial proportion of the income having to be set aside as a sinking fund to replace capital. This aspect of taxation is further considered in Chapters 9 and 12.

## 6. Insurances

### (a) Fire Insurance

This outgoing is sometimes borne by the landlord and will be deducted in finding the net income. A growing practice, when premises are let on lease, is that the lessee usually undertakes to pay the premium or to reimburse the premium paid by the landlord. Where the tenant's occupation is for some hazardous purpose, such as a wood-working factory or chemicals store, the tenant may, in addition, have to pay the whole or a portion of the premium payable in respect of other premises owned by the same landlord where the insurance cost is increased by virtue of the adjoining hazardous use.

The cost of fire insurance is often small in relation to the rental value, particularly for modern, high-value premises. In the case of older premises of substantial construction, where the cost of remedying damage will be expensive, the premium may be significant, particularly as such property tends to command relatively lower rental values. The use of a percentage figure of rental value can therefore be misleading, and it is advisable to establish the current premium payable and to check the level of cover under the policy against current rebuilding costs. The premium is calculated on the replacement cost of the building and not on its value; cost assessments for fire insurance purposes are considered in Chapter 20.

### (b) Other Insurances

In many cases of inclusive lettings, e.g. where blocks of flats or offices are let in suites to numerous tenants, allowance must be made for special insurance, including insurance of lifts, employers' liability, third-party insurance and national insurance contributions. In the case of shops, the insurance for replacement of the plate glass shop windows is normally a duty placed on the tenant.

### 7. Management

Agency charges on lettings and management must be allowed for as a separate outgoing in certain cases. An allowance should be made even where the investor will manage the property himself, as his efforts must have an opportunity cost which should be reflected. In properties where there is a service charge, this charge often includes the cost of management, but the valuer should check the service charge provisions in order to decide whether or not an allowance needs to be made. The existence of a service charge does not automatically mean that all costs incurred by the landlord are recoverable.

In cases where the amount of management is minimal, as with ground rents and property let on full repairing lease, the allowance is normally ignored.

In yet other cases, such as agricultural lettings and houses let on yearly tenancies or short agreements, the item is often included in a general percentage allowance for "repairs, insurance and management".

Where it is appropriate to make a separate deduction for "management" it can usually be estimated as being from 5% to 10% of the gross rents. Management charges on service costs are usually higher and can range up to 20% or more of the cost of the services.

It may be mentioned in this connection that the allowance to be made for this type of outgoing, and for voids and losses of rent referred to below, cannot be entirely dissociated from the Years' Purchase to be used. In making an analysis of sales of properties let at inclusive rents, the outgoings have to be deducted before a comparison can be made between net income and sale price to arrive at a Years' Purchase. If management is allowed for in the analysis, the net income is reduced and a higher Years' Purchase is shown to have been paid than would have been the case if management was not allowed for.

Accordingly when, in making a valuation, a figure of Years' Purchase is used derived from analysis of previous sales, it is important that similar allowances are made for outgoings as were made in the cases analysed.

## 8. Voids

In some areas and for some properties, it is customary to make an allowance for those periods when a property will be unlet and non-revenue producing. For example, a poor quality block of offices in multiple occupation may expect to have some part of the accommodation empty at any time. These periods, termed voids, may be allowed for by deducting an appropriate proportion of the annual rental as an outgoing.

In addition, a landlord will be liable for rates, insurance and service charges during such periods, so there may be a need to allow for these outgoings in respect of void parts in a property where such costs are otherwise recoverable from the tenants.

Where a property is unlet but it is anticipated that an early letting can be achieved, an allowance for voids may still be required if the normal market terms include a long initial rent-free period. However, no deduction for outgoings would be needed.

## 9. Service Charges

Where buildings are in multiple occupation, the responsibility for maintenance of external and common parts is inevitably excluded from the individual leases. As an example, the maintenance of the roof benefits not only the tenant immediately below it but all the tenants of the building. It would thus be inequitable for one tenant to be responsible for the repair of the roof.

The responsibility may be retained by the landlord, although, in some cases, particularly blocks of flats which are sold on long lease, the responsibility may be passed to a management company under the control and ownership of the tenants.

Where a landlord does retain the responsibility, he will commonly seek to recoup the costs incurred from all of the tenants by means of a service charge. The goal is to recoup the full costs so that the rent payable under the various leases is net of all outgoings. Such a situation typically arises with office blocks let in suites, but is also found in shopping centres, industrial estates and other cases of multiple tenancies.

Typical items covered by a service charge are repairs to the structure; repair and maintenance of common parts including halls, staircases, lifts and shared toilets; cleaning, lighting and heating of common parts; employment of staff such as a receptionist, caretaker, maintenance worker and security staff; insurance; and management costs of operating the services. In addition to these types of expenditure which relate to the day-to-day functioning of the building, service charges are increasingly extended to provide for the replacement of plant and machinery such as lifts and heating equipment.

It is desirable for costs to be spread as evenly as possible over time so that an annual charge is normally levied, with surpluses representing funds which build up to meet those costs which arise on an irregular basis such as major repairs or replacements of plant. These latter funds are known as "sinking funds" or "reserve funds".

Various methods of allocating the costs to individual tenants are adopted, including proportion of total floor area occupied by each tenant, and proportion of Rateable Value of each part to the aggregate Rateable Values. Where some tenants do not benefit from services as much as others, then special adjustments are needed. For example, where there is an office building but with lock-up shops on the ground floor, it would clearly be unfair to seek to recover costs relating to the entrance hall, staircases and lifts from the tenants of the shops who make no use of and receive no benefit from these facilities.

It is clear that the existence of extensive service charges should allow the valuer to assume that the lease rents received are net of further outgoings, once he is satisfied that the charges are set at a realistic level, that the lease terms permit the landlord to levy such a charge and to include all the items which are included, and that statutory requirements which apply, for example, to blocks of flats, are met. However, if there are any voids or costs not recoverable, the landlord will become responsible for the charges relating thereto. If the valuer feels that voids are likely to occur fairly frequently, and past evidence may support this view, he may need to consider making some deduction on this account. For example, if the space is extensive it may be reasonable to assume that 4% or 5% will be empty at any one time and so allow, as an outgoing for the landlord, 4% or 5% of the predicted service charges.[1]

## 10. Value Added Tax

In the case of some commercial property, value added tax (VAT) is chargeable on rents and service charges and all VAT expenditure is recoverable by the landlord. However, in many cases the property will not be "vatable". In such cases, any outgoings on which VAT is charged are deducted gross of VAT. This is considered in further detail in Chapter 22.

## 11. Outgoings and Rent

At the start of this chapter it was emphasised that the level of outgoings will have a considerable influence on a tenant's offer of rent since he will be concerned with his overall budget. This can now be illustrated by the following chart.

| Terms of Lease | External Repairs | Internal Repairs | Insurance | Management | Rent | Budget Figure |
|---|---|---|---|---|---|---|
| Full Repairing & Insuring Lease (FRI) | 1,300 | 1,200 | 160 | 140 | 12,000 | 14,800 |
| Internal Repairing & Insuring Lease | – | 1,200 | 160 | 140 | 13,300 | 14,800 |
| Fully Inclusive Terms | – | – | – | – | 14,800 | 14,800 |

[1] For further reading on service charges and VAT on outgoings, see *Service Charges – Law & Practice* (1997) by P. Freedman, E. Shapiro and B. Slater (Jordans).

It is assumed that a prospective tenant is offered a property on three different bases, and the costs of meeting the various liabilities are as indicated.

Although in practice the effects on rent will not be quite so clear cut, the chart does demonstrate how rents are affected by lease liabilities. It also shows that rents are at their lowest when a lease is on FRI terms. The valuer, when comparing rents on different terms of leases, will need to adjust by adding back or subtracting allowances for outgoings as appropriate.

*Chapter 7*

# Yield

## 1. Generally

In Chapter 2 it was shown that, in order to use the investment method of valuation, the valuer must determine the yield, which is the rate of interest appropriate to the particular interest in property being valued. He will normally do this by an analysis of other market transactions. The valuer is not, however, merely an analyst. He must have a clear idea not only of what the market is doing but also why the market is doing it and, if he is to advise adequately on the quality of the investment, what the market is likely to do in the future.

The valuer should have some knowledge of the levels of the interest rates on most types of investment and of the principal factors which influence them. This is because property is only one form of investment and it must compete for funds with all other forms. In some cases, the characteristics of a property investment will be similar to stock market quoted investments and thus the yields will be related to each other.

This chapter is concerned with these points and in it will be considered the principles governing interest rates generally and yields on the main types of landed property.

Before considering these matters, however, it is necessary to dispose of two preliminary points which may be a source of confusion to some.

## 2. Nominal and Actual Rates of Interest

The nominal rate of interest, or dividend, from an investment in stocks or shares is the annual return to the investor in respect of every £100 face value of the stock. Where stock is selling at its face value, that is at par, the nominal rate of interest and actual rate of interest, or yield, are the same. Thus, to take a Government Security as an example, the nominal rate of interest on 2½% Consolidated Stock ("Consols") is fixed at 2.5%; that is £2.50 interest will be received each year for each £100 face value of the stock held. If the stock is selling at £100 for each £100 face value, an investor will receive £2.50 interest each year for every £100 invested which gives a yield of 2.5%. But assume now that £100 face value of 2.5% Consols is selling at £35, or below par. Each £35 of capital actually invested would be earning £2.50 interest annually

$$\therefore \text{Yield} = \frac{2.5}{35} \times \frac{100}{1} = 7.14 \text{ or } 7.14\%$$

Thus, the actual rate of interest is 7.14% whilst the nominal rate of interest remains 2.5%.

If a large industrial concern declares a dividend of, say, 25% on its ordinary shares, this means that the company will pay 25% of the nominal value of each share as the dividend. Hence, if the shares are £1 shares, the dividend per share will be 25% of £1 = 25p per £1 share. But if the price of each £1 share on the market is £4 then:

$$\text{Yield} = \frac{25p}{400p} \times \frac{100}{1} = 6.25\%$$

If the price of these shares in the following year had risen to £4.50 and the same dividend of 25% had been declared, then:

$$\text{Yield} = \frac{25p}{450p} \times \frac{100}{1} = 5.56\%$$

From these examples two important points can be seen:

(i)   That a comparison of income receivable from various types of investment can only be made on the basis of yields and that nominal rates of interest derived from face values are of no significance for this purpose.

(ii)  That a rise or fall in the price of a security involves a change in the yield of that security.

### 3. Timing of Payments and Yields

A yield is expressed as the interest accruing to capital in a year. Hence, if an investment is made of £1,000, and the investor receives £100 in interest payments in each year, the yield is said to be:

$$\frac{100}{1000} \times \frac{100}{1} = 10\%$$

This assumes that the interest payment of £100 is made at the end of the year. However, if the payment of interest is made in instalments, then that produces a different result.

Suppose, for example, that the investor receives £50 after six months and a further £50 at the end of the year. He is receiving:

$$\frac{50}{1000} \times \frac{100}{1} = 5\%$$

per half year. The payment received after 6 months can, in turn, be invested. If it is assumed that it is invested in a similar investment, then further interest of 5% for the remaining half year will be earned. The total interest payments at the end of the year are thus:

£50 (after 6 months) + [(5% of £50) + £50] (end of year payment)

In this case the total interest for the year is £102.50, representing:

$$\frac{102.50}{1000} \times \frac{100}{1} = 10.25\%$$

As the payment patterns change, such as quarterly in arrears or quarterly in advance, so will interest for the year change. This phenomenon is recognised in everyday life by the adoption of APR (annual percentage rate) figures which are quoted in respect of loan rates for borrowers or interest payments for credit card borrowers. The APR reflects the timing of the interest charged on the loans (which will rarely be interest added at the end of the year).

When a reference is made to a yield, it is the yield as determined by the total annual interest expressed as a return on capital ignoring the timing of the payments. Thus, in both of the foregoing examples, the notional yield is 10% whereas the true yield in the first example is 10% but in the second example it is 10.25%. It is the simplistic approach which is adopted by valuers and property investors generally, although the final yield adopted will reflect the timing of the payments. Thus, if a property produces a rent of £1,000 per quarter payable in advance, and if it is offered for sale at £40,000, the yield will be calculated by property investors as being 10% although the APR is in fact a little over 10.38%. This will be examined in more detail in Chapters 9 and 12, and see Preface.

## 4. Principles Governing Yields from Investments

The precise nature of interest and the relationship between, and the level of, long-term, medium-term and short-term rates of interest are subjects for the economist and market analyst and are not dealt with in this book. The valuer is interested primarily in long-term securities and the relative yields from them. He is interested in why the investor requires investment A to yield 6%, investment B to yield 3% and investment C to yield 12%.

A reasonably simple explanation of the many complex matters that the investor must take into account in determining the yield he requires from an investment can be derived from the creation, first, of an imaginary situation. For this purpose the following assumptions are made:

(i)   the real value of money is being maintained over a reasonable period of years – that is, £1 will purchase the same quantity of goods in say, 10 years' time as it will now; and, either
(ii)  there is no taxation or the rates of tax are so moderate as not to influence the investor significantly; or
(iii) the system of taxation is such that taxes bear as heavily on capital as on income.

In these circumstances the yield required by the investor would depend on:

(a)  the security and regularity of the income;
(b)  the security of the capital;
(c)  the liquidity of the capital;
(d)  the costs of transfer, i.e. the costs of putting in the capital in the first instance and of taking it out subsequently.

Thus, the greater the security of capital and income, the greater the certainty of the income being received regularly, the greater the ease with which the investor can turn his investment into cash and the lower the costs of transfer, the lower will be the yield he requires.

Therefore, if investment D offered a guaranteed income payable at regular intervals, no possibility of loss of capital and the loan repayable in cash immediately on request at no cost to the investor, the investor would require the minimum yield necessary to induce him to allow the borrower to use his capital. If investment E offered the same terms as investment D except that six months' notice was required before the capital could be withdrawn, the investor would require a yield sufficiently higher than the minimum to offset this difference. If investment F offered the same terms as investment E except that there was some cost of recovering his capital, the investor would require a still higher yield.

Now this imaginary situation must be adjusted to make it accord more closely with actual conditions in recent years.

With regard to the first assumption, that of money maintaining its real value, this does not happen during periods of inflation. Thus, where an investor investing during a period of inflation is guaranteed a secure income of, say, £100 a year, if the value of the pound is halved in 10 years, his "secure" income will in 10 years' time have a real value of only £50 a year. So also if "security" of capital means merely that if he invests £1,000 now he can withdraw £1,000 when he wishes, the real value of his capital would be halved over a period of 10 years. If, in these circumstances, 10%, that is an income of £100 a year from an investment of £1,000, is a reasonable yield, the investor should be prepared to accept a lower rate of interest from an investment which will protect his capital and income from the erosion of inflation. Thus, he might be prepared to accept a yield of 5% now on an investment of £1,000 if there is a reasonable chance that:

(i)   the income of £50 a year now will have doubled to £100 a year in 10 years' time thus maintaining its real value, and

(ii)  the capital of £1,000 will increase to £2,000 in 10 years' time thus maintaining its real value.

An investment which offers the investor the opportunity of maintaining the real value of capital and income in this way is described as a "hedge against inflation" and is said to be "inflation proof".

The second and third assumptions for the imaginary situation related to the level and incidence of taxation. Over recent years the occasions when a tax charge arises and the rates of tax charged have changed frequently. Tax tends to divide between tax on capital and tax on income.

In the case of capital, until 1962 there was no tax on capital apart from estate duty. This led to proceeds arising from the sale of a capital asset being free of tax. In 1962 a capital gains tax imposed on short-term gains was introduced, followed in 1965 by the establishment of capital gains tax on all gains. This tax remains today, though much changed, and the rate of tax was always 30% of the gain realised until 1987. In 1987 the capital gains tax charged on companies became the same as the corporation tax charged on income. In 1988 individuals were put in the same position as companies, with the capital gains tax rate being the same as the income tax rate (see Chapter 22). On top of this, several other capital taxes have come and gone, all of which sought to impose a separate, and higher rate of tax on any capital gain arising from development value. These were betterment levy from 1967 to 1970, development gains tax from 1974 to 1976, and development land tax from 1976 to 1985.

Estate duty was a tax imposed on the assets of a person on his death, but careful tax planning made it possible to keep this down to modest levels or even to avoid it all together: it was regarded as an avoidable tax. Estate duty was replaced by capital transfer tax in 1975. This was a tax charged not only on assets at death but also on the value of gifts made during a person's lifetime. In 1986 capital transfer tax was replaced by inheritance tax which is more akin to estate duty. The rates of tax assessed on the value of the assets in the estate varied from 30% to 60%, but since 1988 a single rate of 40% has been imposed above an annually determined base on which no tax is levied.

As to tax on income, this has seen some significant changes in the rates of tax over the same period but the taxes imposed (corporation tax on companies and income tax on individuals) have existed in some form for many years. The basic rules are set out in the Income and Corporation Taxes Act 1988, though with many changes in detail over the years. At the time of writing, corporation tax has a basic rate of 30% whilst that for income tax is 23%. This contrasts with tax rates in the past when corporation tax had a basic rate of 52% for many years and income tax on investment income reached 98%.

It can be seen from this brief summary that tax is a significant factor in respect of investments, and that, in general, the tax on income has tended to be more penal than the tax on capital, particularly on capital gains. It also indicates that tax is prone to frequent change and fluctuation.

The four principles governing yields enumerated above, adapted to meet conditions of inflation and levels of income and capital taxation have, in fact, governed yields in the investment market in this country in recent years. This is apparent on examination of yields from different types of security during this period. British government securities, which in times of stable prices and moderate taxation have been described as "ideal security", offered the minimum yield because they are practically riskless. Around 1955 the yield on government securities rose above that on ordinary shares in sound companies for the first time and this position has prevailed since then.

There are many reasons for the fall in price, and consequent increase in yield, of government securities, but one of the most important is that neither the capital invested in them nor the income issuing from them is secure in real terms and they are thus "inflation prone". The lower yields on ordinary shares in, for example, many industrial concerns and retailing companies, can be accounted for by the security in real terms which they offer and, in many cases, the probability of substantial capital appreciation. By contrast, the yields from shares in mining companies are often well in excess of those from government securities, reflecting the insecurity of this type of company due to the uncertainty of production from uncertain reserves together with high costs of extraction.

## 5. Yields from Landed Property Investments

A prospective investor in landed property, i.e. from land and buildings, will be aware of the other forms of investment available and of the yields he can expect from them. He will, therefore, judge the yield he requires from a landed property investment by comparison with the yields from other types of investment such as insurance, building societies and stocks and shares. The principles governing yields discussed above are, therefore, applicable to landed property in the same way as to other forms of investments. However, landed property has certain special features which will be considered before looking at the yields from the main types of landed property.

First, there is the question of management. The investor has no management problems with, for example, government securities as he will receive his income by cheque every six months. With most types of landed property, however, some management is involved. The actual cost of management of landed property is allowed for in computing the net income as discussed in Chapter 6. Apart from this factor a landowner will incur costs such as the agreeing of new rental levels and may incur legal and surveyors' fees in disputes with tenants over various matters affecting the property. He will, therefore, require the yield to compensate for this cost and risk. Where, in addition to being costly, management is troublesome, the investor will require a higher yield to compensate for this.

The second special feature of landed property relates to liquidity of capital and costs of transfer. Ordinary shares, for example, can normally be bought or sold through a stock exchange very rapidly and the transfer costs are a small percentage of the capital involved. A transaction in landed property, however, is normally a fairly lengthy process and the costs of transfer, mainly legal and agents' fees and stamp duty, are somewhat higher. The effects of such costs on the yield, and in particular whether the yield is gross or net of such costs, is referred to in Chapter 9.

The last of the special features is legislation. Legislation does, of course, impinge on many types of investment but its effects, direct or indirect, on landed property are frequently of major significance. The Rent Acts and the Town and Country Planning Acts provide excellent examples. The former limit the amount of rent which can be charged for certain categories of dwelling-house and the latter severely restrict the uses to which a property can be put.

As was seen in Chapter 3, there is more than one type of interest in landed property and different interests in the same property may have different yields. The yields considered below are from freehold interests in the type of property concerned. It is important to note that the yields given merely indicate the appropriate range for the particular type of property at the time of writing. The general level of yields from all types of investment, or from landed property investments only, may change and there are often substantial variations between yields from landed property investments of the same type. The location, age and condition of the buildings and the status of the tenant occupier are also factors which influence the yield. Other things being equal, the older the building and the poorer its condition, and the less substantial the tenant, the higher will be the yield. These comments cannot, therefore, be applied to actual valuations where the yield must be determined in accordance with market information.

*Shops*
A normal range of yields is 4.5% to 15%, the lower figure being applicable to shops in first-class positions occupied by national retail organisations and the higher to shops in secondary or tertiary positions occupied by small traders. Position and type of tenant are vital factors in judging a shop as an investment and these matters are dealt with in detail in Chapter 20. Retail trade suffered from the

recession after 1989 with consequential increases in target yields as rents fell and shops became empty, even in first-class centres. The recovery which began in 1995 should restore yields to former levels if it is maintained. The wide range of yields reflects the great variations in types of retail investment.

### Offices

A normal range of yields is between 6% and 12%. The lowest yield would be expected from a modern block let to a single tenant who shoulders the management burden. The same block let in suites where the owner is responsible for a considerable amount of management, including the provision of such services as lifts and central heating, would yield a slightly higher rate. The highest yield would be expected from an older block, possibly let in suites and lacking modern amenities. In areas where the demand for offices is strong, rents have been increasing at a sufficient rate to provide a secure investment in real terms with a substantial appreciation. However, the recession after 1989 caused a re-appraisal of office investments. The most important factors are the length of leases and the quality of the tenant. Where less than 10 years remain on the occupation lease the income is seen as insecure as the building may become and remain vacant before the capital is recouped. The factors which therefore affect yield are location, quality of construction, supply of modern facilities (such as raised access flooring), layout of accommodation, economy of occupation, quality of tenant and length of letting.

### Factories and Warehouses

Until recent years factories and warehouses had not been a popular investment and the range of yields, 8% to 15%, to some extent reflects past unpopularity. However, significant changes have occurred over recent years which have blurred the distinction between factories and warehouses and other types of properties. In the case of factories, a strong demand has arisen for space to be occupied by companies in the computer and electronic fields where the requirements are for standards closer to those of offices than the traditional, more basic, factory. These are commonly termed "high-tech" buildings. Similarly, alongside the traditional warehouse where the occupier houses his wares, there has grown up the development of retail warehouses, where the form of building is a single-storey building but with extensive car parking and the occupier selling direct to the public. Thus, the building form is similar to a warehouse but the activity is similar to that of a shop. Where the use is closer to office or retail, the yield will move to the yields appropriate to such uses.

Yields will tend to be at the bottom of the range in the case of modern single-storey factories and warehouses in areas of good demand. In the case of factories and warehouses in areas of poor demand, particularly in areas suffering from general industrial decline accompanied by high unemployment levels, yields will be at the top of the range. For older buildings, which are frequently multi-storied and with low heights to eaves, the yield may be well in excess of 15%. Indeed, apart from buildings capable of conversion to small workshop units, they may even cease to be considered as investments and will change to other uses or remain vacant pending redevelopment.

## Residential Properties

This expression covers a very wide range of properties and some sub-classification is required before even the broadest generalisation on yields can be made. Initially it should be stressed that whereas, in the past, residential properties formed a major part of the property investment market, in recent times they have shrunk to making a small contribution. The main impetus for this decline has been the growth in owner occupation coupled with restrictive legislation imposed on the powers of landlords. Even where property is let, it is more likely to be seen by a purchaser as a speculation with the hope of obtaining possession followed by sale with vacant possession rather than as a long-term income-producing investment. Thus, the yields referred to in this section are intended to apply to those instances where the property is likely to remain an investment (see Chapter 17).

With the tenement type of residential property the yield is high, from 12% to 18%. Properties of this type will normally be old, rent restricted, subject to a great deal of legislation imposing onerous obligations on owners relating to repair, cleanliness and other matters, and many tenants may be unreliable in payment of rent. Capital and income are not, therefore, secure in any terms, income may not be regular and the property may be difficult to sell. In recent years such properties have tended to be acquired by public authorities or publicly financed organisations (such as Housing

Associations) so that they are gradually ceasing to be investments found in the private sector.

Although the majority of housing is now either housing association owned, owner-occupied or owned by local authorities, a substantial number of housing units are held for investment purposes.

Blocks of flats, if they are modern, are regarded as a good investment, yielding about 8% to 10%. Frequently they sell at much lower yields but these higher prices, known as "break-up" values, are based on the expectation that a substantial proportion of the flats could be sold to the sitting tenants usually on long leases at prices in excess of investment value, or will be sold with vacant possession at a later date. The valuation approach will often be to calculate a percentage of vacant possession value rather than to apply a yield to the rent income. The percentage is usually in the range of 35% to 50%, but this could increase substantially where the tenant is elderly and early vacant possession is likely.

Houses yield in the range 5% to 15%. As was the case with flats, only more so, such properties when bought outside the public sector are purchased with a view to their ultimate sale so that the usual investment criteria are not applicable and very low yields apparently emerge, but these are not to be interpreted as simple investment yields.

Individual flats are also held as investments in the same way as individual houses. The risk factor is greater because the outings include a service charge but otherwise the same considerations apply.

### Ground Rents

A ground rent is a rent reserved under a building lease in respect of the bare land without buildings. In recent times the practice has been to grant the building lease once the buildings have been built, with the developer holding an agreement to be granted the building lease during the period of development. Building leases are normally granted for a long term, 99 years at one time being fairly common, although modern leases are frequently for around 125 years, particularly in the case of commercial properties.

Secured ground rents (i.e. where buildings have been constructed on the land) where the lease has many years to run, yield in the range 6% to 15%. Where the rent is a fixed amount throughout the term, it is comparable in many ways to government securities and the yield will be similar to, but slightly higher than, that on 2½% Consols and at the top of the range. The investor obtains better security in real terms where the lease provides for upward revision of the rent at reasonable intervals and such an investment would provide a lower yield than where the rent is a fixed amount.

The amount of the rent is also a significant factor. A single rent of a few pounds may be unsaleable unless the occupier is in the market, since the cost of collection will absorb most of the rent received. Usually these small ground rents will sell in "blocks".

Where leases are less than 60 years the value of the leasehold interest is markedly affected by market resistance to wasting assets. Thus, there may be considerable marriage value available to the freeholder. Marriage value is the surplus value that sometimes emerges where two interests in a property are combined, especially the merger of a freehold and a leasehold interest to form a freehold in possession (see Chapter 9). In the case of houses with a rateable value of less then £1,000 in London and £500 elsewhere, the owner occupier lessee may force the sale of the freehold at a price which excludes marriage value (Leasehold Reform Act 1967), but, in all other cases, compulsory sale or lease extension (under the Leasehold Reform, Housing and Urban Development Act 1993) allows the freeholder not less than 50% of the marriage value.

Where the lessee is not an owner occupier, then the value of the freehold interest will reflect the reversion to a house or flat which is let on a form of residential tenancy and the yield cannot then be interpreted as a simple investment yield.

The impact of leasehold reform is considered in Chapter 17.

## 6. Changes in Interest Rates

The landed property investment market normally responds to change less rapidly than other investment markets. The reason for this is probably the length of time taken to transfer ownership and the high costs of transfer. If, for example, circumstances are such that it is felt desirable to hold a greater proportion of assets in cash than hitherto, Stock Exchange securities can be realised immediately even if some loss is incurred. Landed property, however, cannot be realised with anything like the same speed. If, therefore, the change in preference is temporary, the landed property market may remain unmoved while other markets are reacting violently.

The landed property investment market tends to respond to longer term changes. In recent years, the movements in interest rates have been more marked and more rapid than in earlier times. This reflects the closer integration of the property market into the general investment market brought about in the main by the growing involvement of pension funds and insurance companies ("the institutional investors").

Yields for property investments are therefore more stable and are comparable to long-term interest rates available in the economy generally and not to short-term money market rates.

# Chapter 8

# The Mathematics of Valuation Tables

## 1. Generally

It is the valuer's business to make a carefully considered estimate of the value of a property.

In making that estimate he must come to certain conclusions regarding the property – for instance, as to the net income it can produce, as to the likelihood of that income increasing or decreasing in the future, as to the possibility of future liabilities in connection with the property, and as to the rate per cent which a prospective purchaser is likely to require as interest on his capital.

The accuracy of his conclusions on these and other points will depend on the extent of his skill, judgment and practical experience. It is then the function of the valuation tables to enable the valuer, by a simple mathematical process, to express his conclusions as a figure of estimated value.

The object of the valuation tables is thus to save the valuer time and reduce the risk of error involved in elaborate mathematical calculations. Proficiency in their use can never be a substitute for practical experience of the property market or real appreciation of the factors which influence value; but it will assist the valuer very substantially in his work. The tables represent a mathematical tool for use by the valuer and he will gain by having a full understanding of their derivation and application.

Although calculators and computers have substantially changed the situation since tables were first published in 1913, they are still extensively used in practice. Whether printed tables or programmed machines are used, it remains essential to understand the underlying mathematics. This and the immediately following chapters, therefore, remain of fundamental importance to the understanding of the valuation process.

The tables on which this chapter is based are those in the current edition[1] of *Parry's Valuation and Investment Tables*. There are various

---

[1] 11th Edition by A.W. Davidson (Estates Gazette).

sets of valuation tables available, some covering similar ground to *Parry's*, others of more limited application. Further comment on the use and limitations of tables can be found in Chapter 12.

The mathematical construction of the valuation tables and the formulae on which they are based are considered in this chapter, together with the nature of the various tables, what the figures in them represent and how they are commonly used in practice.

It should be emphasised at the start that all the tables which the valuer uses are based on the principle of compound interest.

It is for this reason that the following sections deal first with the "Amount of £1 Table" – the table of compound interest – instead of following the order in which the tables are arranged in *Parry's Valuation Tables*.

For the purposes of this chapter, income is usually assumed to be payable annually in arrears and the affects of inflation are ignored.

## 2. Amount of £1 Table

(pp. 93–110 of *Parry's Valuation Tables*)
The figures in this table are simply figures of compound interest. They represent the amount to which £1, invested at various rates of compound interest, will accumulate over any given number of years on the basis that interest is payable annually in arrears.

Such a calculation could, of course, be made manually as shown by the following example.

*Example 8–1*

To what amount will £1 invested at 5% compound interest accumulate in three years?

*Answer*

|  | Principal | Interest | Total |
|---|---|---|---|
| Amount at end of 1 year (£1 plus interest at 5% on £1) | £1.00 | £0.05 | £1.05 |
| Amount at end of 2 years (£1.05 plus interest at 5% on £1.05) | £1.05 | £0.0525 | £1.1025 |
| Amount at end of 3 years (£1.1025 plus interest at 5% on £1.1025) | £1.1025 | £0.055125 | 1.157625 |

This process is obviously a laborious one and it is a great saving of time to be able to take the appropriate figure direct from the valuation tables.

The importance of the Amount of £1 Table lies mainly in the fact that it has been used as the basis for the construction of the other valuation tables.

In practice, the table is sometimes useful in calculating the loss of interest involved where capital sums are expended on a property which, for the time being, is unproductive of income.

### Example 8–2

A building estate was purchased for £5,000,000. A sum of £1,000,000 was spent at once on roads and other costs of development. During a period of five years no return was received from the property. What was the total cost of this property to the purchaser at the end of the five years assuming interest at 9%?

### Answer

If the owner had not tied up £6,000,000 in the purchase and development of this land, he presumably could have invested that sum in some other investment producing interest at 9% and have allowed capital and income to accumulate at compound interest. The cost of the property to him is therefore the sum to which £6,000,000 might have accumulated in five years at 9% compound interest.

| | | |
|---|---|---|
| Capital sum invested | £6,000,000 | |
| Amount of £1 in 5 years at 9% | 1.5386 | |
| Cost to purchaser | | £9,231,600 |

Alternatively, an owner may have borrowed money to purchase a development property on the basis that interest will be charged but not payable until the development is completed (the interest is said to be "rolled up").

### Example 8–3

A development property was purchased three years ago and a loan of £200,000 was obtained for this purpose at a fixed annual interest rate of 8% rolled up until the development is completed. The development will be completed in one year's time. Calculate the sum due for repayment at that time.

| Capital sum borrowed | £200,000 |
| Amount of £1 in 4 years at 8% | 1.360 |
| Sum due for repayment | |
| (loan plus rolled-up interest) | £272,000 |

The formula for the Table is derived as follows:

To find the amount to which £1 will accumulate at compound interest in a given time.

Let the interest per annum on £1 be i
Then the amount at the end of 1 year will be $(1 + i)$
The amount at the end of 2 years will be $(1 + i) + i (1 + i) = 1 + 2i + i^2 = (1 + i)^2$
The amount at the end of 3 years will be similarly $(1 + i)^3$

By similar reasoning the amount in n years will be $(1 + i)^n$
(The total interest paid on £1 in n years will be $(1 + i)^n - 1$)

### 3. Present Value of £1 Table

(pp. 61–80 of Parry's Valuation Tables)
The figures in this table are the inverse of those in the Amount of £1 Table. Whereas the latter table shows the amount to which £1 will accumulate at compound interest over any given number of years, the Present Value of £1 Table shows the sum which invested now at compound interest will amount to £1 in so many years' time.

The figures in the table are the reciprocals of those in the Amount of £1 Table, i.e. it is possible to obtain any required figure of Present Value of £1 by dividing unity by the corresponding figure of Amount of £1.

*Example 8–4*

What sum invested now will, at 5% compound interest, accumulate to £1 in six years' time?

*Answer*
Let V equal the sum in question
V × Amount of £1 in 6 years at 5% = £1

$$\therefore V = \frac{1}{\text{Amount of £1 in 6 yrs at 5\%}}$$

$$= \quad \frac{1}{1.3400956}$$

$$= \quad 0.7462154$$

It is, however, very much quicker to take the figure direct from the Present Value of £1 Table.

The Present Value of £1 Table is widely used in practice for calculating the value at the present time of sums receivable in the future and in making allowances for future expenditure in connection with property.

The value at the present day of the right to receive a capital sum in the future is governed by the fact that whatever capital is invested in purchasing that right will be unproductive until the right matures. So that if £x is spent in purchasing the right to receive £100 in three years' time, it follows that for those three years the purchaser's capital will be showing no return, whereas if the £x had been invested in some other security it might, during those three years, have been earning compound interest. The price which the purchaser can fairly afford to pay, therefore, is that sum which, with compound interest on it (calculated annually in arrears) at a given rate per cent, will amount to £100 in three years' time.

*Example 8–5*

What is the present value of the right to receive £100 in three years' time assuming that the purchaser will require a 5% annual return on his money?

*Answer*

Let V = sum which the purchaser can afford to pay. He will be losing compound interest on this sum during the three years he will have to wait before he receives the £100.

V must therefore be such a sum as, together with compound interest at 5%, will in three years' time equal £100.

V × Amount of £1 in 3 years at 5% = £100.

$$\therefore V = \quad \frac{£100}{\text{Amount of £1 in 3 yrs at 5\%}}$$

$$= \quad \frac{£100}{1.1576}$$

$$= \quad \underline{£86.38}$$

The above method has been used to show the principles involved. The same result would be obtained direct from the Present Value of £1 Table as follows:

| | |
|---|---:|
| Sum receivable | £100 |
| PV £1 in 3 years at 5% | 0.8638376 |
| *Present Value* | £86.38 |

The process of making allowance for the fact that a sum is not receivable or will not be expended until some time in the future is known as "deferring" or "discounting" that sum. In the above example, £86.38 might be described as the present value of £100 "deferred three years at 5%".

Since, in the case of sums receivable in the future, the valuer is concerned with the temporary loss of interest on capital invested in their purchase, the rate per cent at which their value is deferred should generally correspond to that which an investor might expect from the particular type of security if in immediate possession. In other words, the rate should be a "remunerative" one.

*Example 8–6*

What is the present market value of the reversion to a freehold property let for a term of five years at a peppercorn rent but worth £35,000 per annum? Similar property in possession and let at its full market value has recently changed hands on a 5% basis.

*Answer*
Since no rent is payable for the first five years (a "peppercorn rent" being a legal device for leases where no rent is to be paid), no value arises in respect of this period.

| *Value in five years' time* | |
|---|---:|
| Full rental value | £ 35,000  p.a. |
| YP in perp at 5% | 20.00 |
| | £700,000 |
| | |
| PV £1 in 5 years at 5% | 0.783 |
| *Present Value* | £548,100 |

Where the sum for which allowance has to be made is a future expense in the nature of a liability which cannot be avoided, the valuer is not concerned with the question of loss of interest on capital invested, but rather with the rate per cent at which a fund can be accumulated to meet the expense. A future capital liability can be provided for either by the setting aside of an annual amount in the form of a sinking fund, or by the investment of a lump sum which at compound interest will with certainty accumulate to the required amount in the given period. The first method will be discussed later in this chapter. The second method can be effected through an insurance company by what is known as a single premium policy or some other investment with a guaranteed rate of interest for the whole period.

Due to the need to guarantee the future payment there is an interest risk on the insurance company because of fluctuating market rates of interest. Therefore, the rate of compound interest on it allowed by insurance companies is low. It is therefore sounder, as a rule, to defer sums to meet liabilities in the future at a low "accumulative" rate rather than at the "remunerative" rate at which interest on capital is taken. The actual rate depends on the interest rates prevailing at the time and the financial market's view of future trends in interest rates. Historically such rates have tended to be around 2.5% to 3.5%, net after tax.

*Example 8–7*

A freehold factory was recently let to substantial tenants on a 40 years' lease at a rent of £40,000 p.a. The premises are in good repair, and the tenants are under full repairing covenants, but the owner has covenanted with the lessees that, after two years of the lease have expired, he will rebuild, in fire-resisting construction, a staircase which is now built of wood, at a cost of £20,000; and also that, after a further period of four years, he will rebuild the chimney shaft at a cost of £50,000 and that two years later (i.e. when the lease has 32 years unexpired) he will replace a wooden fence with a brick wall; this will cost £10,000. Current accumulative rates are 2.5%.

Assuming that the works in question are necessary to maintain the present rent, and that no higher rent may be expected when the lease comes to an end, what is the value of the freehold interest?

*Valuation*

| | | | |
|---|---|---|---|
| Net Rental Value (with improvements) | | | £ 40,000 |
| YP perp at 8% | | | 12.5 |
| | | | 500,000 |

Deduct cost of:

| | | | |
|---|---|---|---|
| Fire-resisting staircase | 20,000 | | |
| PV £1 in 2 years at 2.5% | 0.952 | 19,040 | |
| Chimney shaft | 50,000 | | |
| PV £1 in 6 years at 2.5% | 0.862 | 43,100 | |
| New brick wall | 10,000 | | |
| PV £1 in 8 years at 2.5% | 0.821 | 8,210 | 70,350 |
| Value of Freehold Interest | | | £429,650 |
| | | say | £430,000 |

An exception to the above general rule occurs in cases where the future capital expense can be met out of moneys arising from the property itself, in which case there will be no need to provide for it by investment of a single premium at a low rate of interest and the sum can be deferred at the appropriate "remunerative" rate.

Where, as is frequently the case, future expenditure is of a kind which is optional - as distinct from a liability which cannot be avoided – it is probably preferable to allow for it at the higher (remunerative) rate, since in this type of case it is thought that the investor would not set aside a sum to accumulate at a low rate of interest, e.g. in a single premium policy.

The formula for this Table and its determination is as follows:

**To find the present value of £1 receivable at the end of a given time.**
Since £1 will accumulate to $(1 + i)^n$ in n years, the present value of £1 due in n years equals

$$\frac{1}{(1 + i)^n}$$

**If it is desired to take into account the payment of interest and its reinvestment at more frequent intervals i and n must be modified.**
Let the number of payments in 1 year be m. Then the total number of payments is mn. As the annual rate of interest is i, the rate of interest for one period i/m and the amount of £1 in n years equals

$$\left(\frac{1+i}{m}\right)^{mn}$$

The present value of £1 correspondingly will be:

$$\frac{1}{\left(\dfrac{1+i}{m}\right)^{mn}}$$

This assumes that the rate of interest is the nominal rate representing the total interest received over the years.

If it is assumed that interest paid quarterly is reinvested at the same annual rate, the effective annual yield would need to be determined. The effective annual yield represents the total interest which would have accumulated by the end of the year as a result of the re-investment at compound interest on a quarterly basis. The formula to determine the annual effective yield is:

$$\left[\left(\frac{1+i}{m}\right)^{mn}\right]^{1/n} -1$$

Thus, by way of an example, if £1 is invested at 10% per annum payable monthly for five years the effective yield is as follows:

$$\left[\left(\frac{1+0.1}{12}\right)^{60}\right]^{1/5} -1$$

Which $= (1.6453)^{1/5} - 1 = 0.1047$
$= 10.47\%$ p.a.

## 4. Amount of £1 per Annum

(pp111–126 of *Parry's Valuation Tables*)
The figures in this table represent the amount to which a series of deposits of £1 at the end of each year will accumulate in a given period at a given rate of compound interest.

A calculation in that precise form does not often come within the scope of a valuer's practice; but the table may be of use to him in calculating the total expense involved over a period of years where annual outgoings are incurred in connection with a property which is for the time being unproductive.

The following example serves to illustrate this use and also to emphasise the distinction between the nature and use of this table and the Amount of £1 Table.

### Example 8–8

A new plantation of timber trees will reach maturity in 80 years' time. The original cost of planting was £2,000 per hectare. The annual expenses average £200 per hectare. What will be the total cost per hectare by the time the timber matures, ignoring any increase in the value of the land and assuming that interest is required on other outstanding capital at 5%?

### Answer

The £2,000 per hectare spent on planting is in the nature of a lump sum which will remain unproductive over a period of 80 years. Its cost to the owner is represented by the sum to which it might have accumulated if it had been invested during that period at compound interest. This part of the calculation requires use of the Amount of £1 Table.

The expenditure of £200 per hectare on upkeep is an annual payment which will also bring no return during a period of 80 years. Its cost to the owner is represented by the sum to which a series of such payments might have accumulated if placed in some form of investment bearing compound interest. Here it will be necessary to use the Amount of £1 per annum Table.

| | | |
|---|---|---|
| Original capital outlay per hectare | £2,000 | |
| Amount of £1 in 80 years at 5% | 49.56 | £ 99,120 |
| Annual cost per hectare | £ 200 | |
| Amount of £1 per annum in 80 years at 5% | 971.229 | £194,246 |
| *Total cost per hectare for period of maturity* | | £293,366 |

The derivation of the formula for this Table is as follows:

### The amount to which £1 per annum invested at the end of each year will accumulate in a given time.

It is conventional to assume that interest is not paid until the end of the first year, hence the first payment £1 will accumulate for n – 1 years if the period of accumulation is n years.

The second payment will accumulate for n – 2 years and so on year by year.

The amount to which the first payment accumulates will, applying the Amount of £1 Table be $(1 + i)^{n-1}$; the second $(1 + i)^{n-2}$; and so on.

Let the amount of £1 per annum in n years be A.

Then $A = (1 + i)^{n-1} + (1 + i)^{n-2} + (1 + i)^{n-3} \ldots (1 + i)^2 + (1 + i) + 1$; or more conveniently

$$A = 1 + (1 + i) + (1 + i)2 \ldots (1 + i)^{n-1}$$

These terms form a geometrical progression for which the general expression is:

$$S = \frac{a(r^n - 1)}{r - 1}$$

Substituting therein $S = A$; $a = 1$; $r = 1 + i$;

$$\text{Hence } A = \frac{1 (1 + i)^n - 1}{i}$$

## 5. Annual Sinking Fund

(pp. 81–92 of *Parry's Valuation Tables*)

This table is the inverse of the Amount of £1 per annum Table. Instead of showing the sum to which a series of deposits of £1 will accumulate over a given period, it shows the sum which must be deposited annually at compound interest in order to produce £1 in so many years' time. The figures in the one table are the reciprocals of those in the other, just as the figures in the PV of £1 Table are the reciprocals of those in the Amount of £1 Table. Thus, any required figure of Annual Sinking Fund can be found by dividing unity by the corresponding figure of Amount of £1 per annum, or vice versa.

The table is of direct use to the valuer when it is required to know what sum ought to be set aside annually out of income in order to meet some capital expense due in the future, such as a possible claim for dilapidations on the termination of a lease, or a sum likely to be required for the rebuilding or reconstruction of premises.

The provision of a sinking fund to meet future capital liabilities, although often neglected by owners in practice, avoids the embarrassment of having to meet the whole of a considerable expense out of a single year's income and enables the owner to see precisely how much of the annual return from the property he can afford to treat as spendable income. It is, however, met quite frequently in service charge agreements where a sinking fund provision is made to meet, for example, the costs of replacing lifts and plant when they reach the end of their useful lives.

*Example 8–9*

An investor recently purchased for £100,000 a freehold property which it is estimated will yield a net income of £20,000 for the next 15 years. At the end of that time it will be necessary to rebuild at a cost of £150,000 in order to maintain the income. How should the owner provide for this and what will be the result on the percentage yield of his investment?

*Answer*

The owner may provide for the cost of rebuilding by means of an annual sinking fund accumulating at, say, 3% over the next 15 years, as follows:

| | |
|---|---|
| Cost of rebuilding | £150,000 |
| Annual sinking fund to produce £1 in 15 years at 3%. | 0.0538 |
| Sinking fund required | £ 8,070 p.a. |

The owner's true income for the next 15 years will therefore be (£20,000 – £8,070) = £11,930, representing a return of 11.93% on the purchase price instead of the 20% which the investment might appear to be yielding.

The necessary provision may be made by means of a sinking fund policy taken out with an insurance company on terms and at rates of interest similar to those referred to in Section 3 of this chapter for single premium policies, i.e. 2.5% to 3.5%.

If the sum to be set aside annually is considerable, it is possible that an owner may find opportunity for the accumulation of it in his own business, or in some other investment, at a higher rate of interest than that usually allowed by an insurance company. It is probable that in many cases purchasers will be inclined to take this fact into account in considering the investment value of property, when otherwise the cost of allowing for replacement capital on ordinary sinking fund terms becomes prohibitive. However, the insurance policy approach guarantees a certain sum which will be available at a future date, which is rare in any form of investment other than those for a short term of up to around five years. In any event the purchaser will need to allow for any income tax or corporation tax which may be levied on the interest arising from other forms of investment. It is the interest *net of any tax payable* which he will need to adopt in comparing alternative forms of investment with an insurance policy approach.

It is proposed in the examples which follow to adopt 3% net as a reasonable average figure where a sinking fund is concerned.

The Annual Sinking Fund tables on pages 87–92 of *Parry's Valuation Tables* are compiled on a net basis – that is, assuming that the accumulations of interest on the sinking fund are free of income tax.

The allowance for income tax on interest on sinking fund accumulations which is made by using a net rate of interest should not be confused with the entirely separate adjustment for income tax on the sinking fund element of income which is made when using dual rate Years' Purchase. The latter point is dealt with in Chapter 9.

The derivation of the formula for this Table is as follows:

**To find the sum which, if invested at the end of each year, will accumulate at compound interest to £1.**
Let the annual sinking fund be S and the period n years, then the first instalment will accumulate to $S(1 + i)^{n-1}$; the second to $S(1 + i)^{n-2}$; and so on.

Hence $1 = S(1 + i)^{n-1} + S(1 + i)^{n-2} \ldots$; or more conveniently

$$S(1 + i)^2 + S(1 + i) + S;$$
$$1 = S + S(1 + i) + S(1 + i)^2 \ldots S(1 + i)^{n-1}$$

The sum of the series will be:

$$1 = \frac{S\,[(1 + i)^n - 1]}{i} \qquad \text{therefore } S = \frac{i}{(1 + i)^n - 1}$$

It will be observed that S is the reciprocal of Amount of £1 p.a.

It will be noted that investment of the sinking fund takes place at the end of each year. This is convenient when dealing with the income from real property, which is commonly assumed to be receivable yearly at the end of each year. For further discussion of this point see Chapters 9 and 12.

In comparing sinking fund tables with amounts payable under a sinking fund policy, it must be remembered that, under a policy, the premium is payable at the beginning of each year. Thus, by similar reasoning to the above in this case:

$$S \text{ really} = \frac{i}{(1 + i)^{n+1} - 1}$$

However, because rent is assumed to be payable yearly in arrears it is assumed that the earlier formula generally applies.

## 6. Present Value of £1 per annum or Years' Purchase Table

(Dual Rate pp. 1–26, Single Rate pp. 27–40 of *Parry's Valuation Tables*)

The figures in the table show, at varying rates of interest, what sum might reasonably be paid for a series of sums of £1 receivable at the end of each of a given number of successive years. By applying the appropriate figure from the table to the net income of the property the valuer arrives at his estimate of market value.

Where the income is perpetual, the appropriate figure of Years' Purchase can be found by dividing 100 by the rate of interest appropriate to the property, or by dividing unity by the interest on £1 in one year at the appropriate rate per cent. Thus (as shown in Chapter 2), Years' Purchase in perpetuity can be expressed as:

$$\frac{100}{\text{Rate per cent}} \qquad \text{or} \qquad \frac{1}{i}$$

But where the income is receivable for a limited term only, as, for example, with a leasehold interest, the relationship between Years' Purchase and the rate of interest is more complex.

For example, if a certain property producing a perpetual net income of £500 p.a. can fairly be regarded as a 5% investment, a purchaser can afford to pay (£500 × 20 YP) = £10,000 for it. Assume now that the income is receivable for six years only but that 5% is still a reasonable return. In this case a purchaser could not afford to pay £10,000, for if he did so, although the income of £500 would represent 5% on the purchase price throughout the term, at the end of the six years his interest in the property would cease and he would then lose both capital and income.

There are two ways to approach such a problem. The first of these is to consider the income flow for the next six years and take each year's income in isolation. Hence what is to be valued is in fact:

| End of year 1 | £500 |
| End of year 2 | £500 |

and so on to
End of year 6 £500

Now it has been shown that in valuing the right to receive a sum in the future, the present "discounted" value is determined by applying the Present Value of £1 to the actual sum to be received. Hence, taking each in turn, it is clear that:

| | | |
|---|---|---|
| End of year 1, sum receivable | £500 | |
| × Present Value of £1 in 1 yr at 5% | 0.9524 | £476 |
| End of year 2, sum receivable | £500 | |
| × Present Value of £1 in 2 yrs at 5% | 0.9070 | £454 |
| End of year 3, sum receivable | £500 | |
| × Present Value of £1 in 3 yrs at 5% | 0.8638 | £432 |
| End of year 4, sum receivable | £500 | |
| × Present Value of £1 in 4 yrs at 5% | 0.8227 | £411 |
| End of year 5, sum receivable | £500 | |
| × Present Value of £1 in 5 yrs at 5% | 0.7835 | £392 |
| End of year 6, sum receivable | £500 | |
| × Present Value of £1 in 6 yrs at 5% | 0.7462 | £373 |
| | | £2,538 |

Hence it can be seen that the value of the right to receive £500 for each of the next six years is £2,538. This is a laborious approach, and if the period were extended would quickly become cumbersome. A tidier method would be as follows:

| | | |
|---|---|---|
| Sum receivable each year | | £500 |
| Present Value of £1 in 1 year at 5% | 0.9524 | |
| Present Value of £1 in 2 years at 5% | 0.9070 | |
| Present Value of £1 in 3 years at 5% | 0.8638 | |
| Present Value of £1 in 4 years at 5% | 0.8227 | |
| Present Value of £1 in 5 years at 5% | 0.7835 | |
| Present Value of £1 in 6 years at 5% | 0.7462 | 5.0756 |
| Present value of £500 p.a. for next 6 years | | £ 2,538 |

In this way the annual rent is multiplied by the sum of the present values of £1 for each year. In fact, a formula can be derived to represent the sum of the present values of £1 p.a. for any number of years known as Years' Purchase Single Rate (see pp. 27–40 of *Parry's Valuation Tables*).

Hence in the above example:

| | |
|---|---|
| Rent Receivable | £ 500 |
| × Years' Purchase (Single Rate) for 6 years at 5% | 5.0757 |
| Capital Value | £2,538 |

An alternative approach to the problem of valuing a terminable income is to recognise that a lower Years' Purchase will be required to capitalise the terminable income, but that at the same time the rate of interest must not be interfered with. If the rate of interest is merely increased to give the lower Years' Purchase necessary, it is no longer performing only its proper function as an indicator of the relative merits of different investments; it is also being required to function as an indicator of the period during which the income will be received. In order to leave the rate of interest to perform only its proper function, it is necessary to make the income comparable in terms of time. This can be achieved by allowing an amount out of the terminable income to be set aside annually as a sinking fund, sufficient to accumulate during the term to the capital originally invested. If this is done the purchaser, having paid £x for the interest in the first instance, will receive the income during the term and set aside out of that income a sinking fund, so that at the end of the term the sinking fund will have accumulated to the original capital of £x. In this way, the terminable income has been perpetuated and is, therefore, directly comparable with the perpetual income.

Thus, the formula for finding the Present Value of £1 per annum for a terminable income is:

$$\frac{1}{i + s}$$

i being the interest on £1 in one year at the appropriate rate per cent and s being the sinking fund to replace £1 at the end of the term.

The next question is at what rate of interest should the annual sinking fund be assumed to accumulate? Again, the proper function of the rate of interest, that of indicating the relative merits of different investments, is the prime consideration. It was explained above that sinking fund arrangements can be made by means of a sinking fund policy with an insurance company. The interest on such a policy is low because the investment is riskless and as near trouble-free as possible. If it is assumed that the sinking fund is arranged in this way, all of the risks attached to the actual investment in the property will be reflected, as they should be, in the rate of interest the purchaser requires on the capital invested.

These two assumptions, (a) that a sinking fund is set aside and (b) that the rate of interest at which the sinking fund accumulates is the rate appropriate to a riskless and trouble-free investment, are in no way invalidated by the fact that many investors make sinking fund provisions in some way other than through a leasehold redemption policy or that many investors make no sinking fund provision at all.

*Example 8–10*

What is the value of a leasehold property producing a net income of £500 for the next six years, assuming that a purchaser requires a return of 5% on his money and that provision is made for a sinking fund for redemption of capital at 3%?

*Answer*

| | | |
|---|---|---|
| Interest on £1 in 1 year at 5% | = | 0.05 |
| SF to produce £1 in 6 years at 3% | = | 0.1546 |

$$\therefore \text{YP 6 years at } 5\% \text{ \& } 3\% = \frac{1}{i+s}$$

$$= \frac{1}{0.05 + 0.1546}$$

$$= 4.888$$

*Valuation*

| | |
|---|---|
| Net income | £500 |
| PV of £1 p.a. or YP 6 years at 5 and 3% | 4.888 |
| Value | £2,444 |

*Notes*

In practice, the figure of 4.888 could have been obtained direct from the table of PV of £1 p.a. at 5 and 3% in *Parry's Valuation Tables*.

*Proof*

The fact that the estimated purchase price does allow both for interest on capital and also for sinking fund may be shown as follows:

| | |
|---|---|
| Interest on £2,444 at 5% (2,444 × 0.05) | £122 |
| Sinking fund to produce £2,444 in 6 yrs at 3% (£2,444 × 0.1546) | 378 |
| Income from property | £500 |

The Years' Purchase used in the above example is called a dual rate Years' Purchase because two different rates of interest are used. The first rate (5% in the example) is called the "remunerative" rate, and the second (3%) is called the "accumulative" rate.

As is shown below, the sum of the Present Values of £1 p.a. has the same formula but the rate of interest for the sinking fund is the same as the remunerative rate. Hence there is a single rate adopted throughout and so the result is a Years' Purchase Single Rate.

It will be noted that the terminable income of £500 p.a., when valued on a YP Single rate produced £2,538 whereas on YP Dual Rate the answer is £2,444. The choice of the appropriate YP Table is determined by the nature of the income and surrounding factors, particularly the nature of the legal estate, as is illustrated in Chapter 9.

The derivation of the formulae for Years' Purchase Dual Rate and Years' Purchase Single Rate is as follows:

(a)   Years' Purchase Single Rate

To find the value of £1 per annum receivable at the end of each year for a given time allowing compound interest.

The present value of the first instalment of income is

$$\frac{1}{1 + i}$$

that of the second $\dfrac{1}{(1 + i)^2}$ the third $\dfrac{1}{(1 + i)^3}$ and so on.

Let the present value be V and the term n years, then

$$V = \frac{1}{1 + i} + \frac{1}{(1 + i)^2} + \ldots \frac{1}{(1 + i)^{n-1}} + \frac{1}{(1 + i)^n}$$

Hence summing the series and changing the sign in the numerator and denominator:

$$V = \frac{1 - \dfrac{1}{(1 + i)^n}}{i} \qquad \text{OR} \qquad \frac{1}{i + \dfrac{i}{(1 + i)^n - 1}}$$

## (b) Years' Purchase Dual Rate

To find the present value of £1 per annum for a given number of years, allowing simple interest at i per annum on capital and the accumulation of an annual sinking fund at s per annum.

$$V = \cfrac{1}{i + \cfrac{s}{(1 + s)^n - 1}}$$

If this is to be adjusted for tax at t%, the formula becomes:

$$V = \cfrac{1}{i + \left[ \left( \cfrac{s}{(1 + s)^n - 1} \right) \times \cfrac{100}{100 - t} \right]}$$

# Chapter 9

# Investment Method – Application and Use of Valuation Tables

In Chapter 8 the principal Valuation Tables were examined and their functions explained. The valuer employs the Tables in many situations but their main use is in valuations using the investment method. This method is adopted for valuing both freehold and leasehold interests and these will be considered in turn.

## 1. Valuation of Freehold Interests

As has been previously explained, the freehold owner of a property may choose to let the property and accept rent in lieu of occupation. Property is let on a lease and normally the freeholder will seek to obtain full rental value, by which is meant the highest rent obtainable from an acceptable tenant on the best appropriate terms. These terms may vary from a full repairing and insuring lease to a fully inclusive lease, as was set out in Chapter 5.

At the start of the lease, therefore, the tenant is paying full rental value. This level of rent remains payable until the end of the lease or until the rent is reviewed under the terms of the lease. This will be so regardless of any changes in the prevailing level of rental value. Suppose for example that a shop is let on a 15-year lease at an initial rent of £30,000 p.a. which is the rental value. The lease provides for the rent to be reviewed after 5 and 10 years. Assume that the rental value of the shop increases by £3,000 each year. It follows from this that rent and rental value coincide in year 1, but differ for the following four years, coinciding in year 6 as the rent review operates. Hence:

| Year | 1 | 2 | 3 | 4 | 5 | 6 | 7 |
|------|------|------|------|------|------|------|------|
| FRV | 30,000 | 33,000 | 36,000 | 39,000 | 42,000 | 45,000 | 48,000 |
| Rent Payable | 30,000 | 30,000 | 30,000 | 30,000 | 30,000 | 45,000 | 45,000 |

Now if the valuer is required to value the freehold interest in year 1 he will be valuing a freehold interest let at full rental value, whereas in, say, year 3 the rent will be £30,000 p.a. whilst the rental value is £36,000 p.a. so that he has to value a freehold interest let at below full rental value. Each of these situations will be considered separately.

## (a) Freehold Let at Full Rental Value

As was explained in Chapter 2, the principle of the investment method is Net Income × Years' Purchase = Capital Value

Net Income is the rent receivable less any outgoings borne by the landlord, other than income tax. Income tax is ignored as this is a generally applicable impost and, insofar as it varies according to the personal status of the recipient, is a personal factor unrelated to the quality of the investment itself. The implications of valuing incomes net of income tax are considered in Chapter 12.

In the case of a freehold interest, as the interest is perpetual then income from the property will be perpetual. It is true that rent arising from buildings rather than land is unlikely to be perpetual since buildings eventually wear out. However, the life of the building rarely has a predictable end and, unless the building's life is relatively short, say up to 25 years, then the likelihood of rent from the building ending at some distant date is insignificant to the valuation. For example, suppose that a building has a predictable life of 30 years and produces a rent of £10,000 p.a. whilst a similar building produces the same rent but will last for 100 years.

Then the rent from the 30-year building has a value of £10,000 p.a. × YP 30 yrs at say 10% = 10,000 × YP 9.427 = £94,270, whilst the second rent has a value of £10,000 p.a. at say 10% = 10,000 × YP 100 yrs at 10% = 10,000 × YP 9.999 = £99,990. The rent for 30 years has a value equal to 94.27% of the rent for 100 years.

The reason for this is that the value of £1 discounted for a short period is higher than for a long period, and the effect of compound interest is to magnify this effect. Hence the Present Value of £1 in 1 year at 10% is £0.909, 2 years is £0.826, 3 years is £0.751 and so on for each succeeding year, the value falling for each subsequent year at an accelerating rate, so that for 30 years it is £0.057 and for 100 years £0.00007. This is illustrated on the graph on page 114. A similar pattern will apply at any rate of interest.

The rent in the earlier years is therefore more significant to the valuation than for later years. As was seen, YP for 30 years at 10% = 9.427, whereas YP in perpetuity at 10% = 10.00. Thus, in valuing any income to perpetuity at 10%, 94.27/100 = 94.27% of the total value lies in the first 30 years' income. An inspection of the YP Single Rate Tables on pages 27–40 of *Parry's Valuation Tables* will indicate the effect of this mathematical consequence at different rates of interest, and as the graph on page 115 shows.

Quite apart from the mathematical reason, a building is normally demolished because it is at the end of its economic life, in that the rental value of the site for a new development exceeds that for the standing building. For example, the 30-year building will probably be pulled down after 30 years because the freeholder can obtain a rent of £10,000 p.a. or more from the site (or its equivalent capital value) and naturally prefers this to £10,000 p.a. from the old building. Hence, in valuing the rent from a building the income may be regarded as perpetual although it is recognised that the rent will not always arise from the existing building.

It is true that in some special circumstances a building will have a predictably short life, for example, because of physical deficiencies or town planning restrictions, where a lower rent will subsequently arise. This situation is considered later.

The valuer is thus faced with a net income at full rental value which he can regard as perpetual. At present the rent is at the prevailing rental value, but the valuer may feel that the rental value will change in future years. Indeed, he would normally anticipate changes to occur as market forces change and, given that inflation has become an established feature of economic life, he would probably anticipate inflationary increases if nothing else. Additionally, he may feel that rental values will rise faster than inflation because the property is in a position which will improve or that there will be an exceptional growth in demand for that type of property. On the other hand, he may sense that the property is in a declining area and will anticipate a low rise in rents or indeed an actual fall in levels.

It follows that the valuer will not regard the rental value as being at a fixed level. However, he cannot know what changes will occur, no matter how clear his crystal ball. The best that he can do is to predict the general movements in rental values and perhaps estimate average growth, above average growth, or whatever. He can then compare the predicted growth with that predicted for other investments. The comparison can then be translated into value effect by adapting the yield to be adopted, since the yield acts as a measure of comparison as well as a barometer of investment expectations.

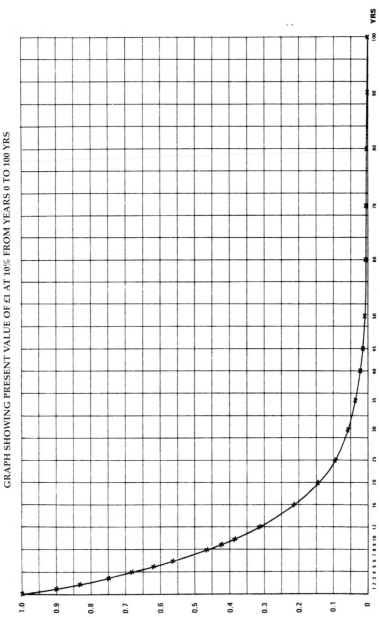

GRAPH SHOWING PRESENT VALUE OF £1 AT 10% FROM YEARS 0 TO 100 YRS

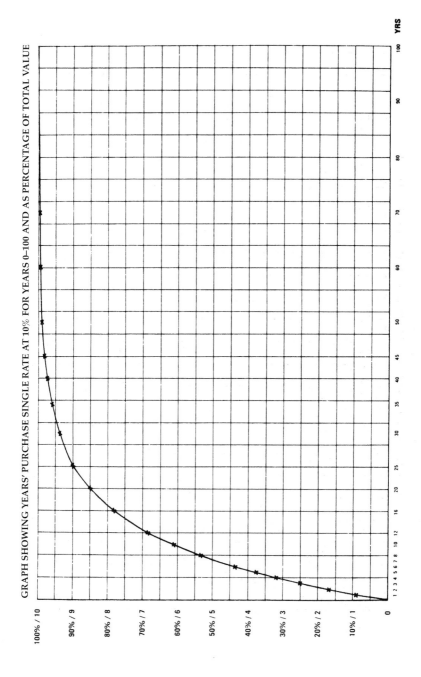

For example, a prime investment where the property is well located, the tenant is sound, and above average growth is predicted may attract a yield of 5%, i.e. investors purchasing such a property would expect a return of 5% on their capital, whereas a similar property but where only average growth is predicted may attract 7%. In effect, the investor would require a higher initial yield to compensate him for the anticipated future shortfall. Thus the choice of yield reflects the future anticipated rental value pattern.

The valuer can therefore take the net income at full rental value and regard it as perpetual and, to allow for future changes in rental, adopt a yield in the valuation appropriate to the predicted rental movement.

*Example 9–1*

Value the freehold interest in shop premises in a prime location.

The premises were recently let to a multiple company at £100,000 p.a. exclusive on full repairing and insuring terms for 35 years with 5-yearly rent reviews. The rent is at full rental value.

*Valuation*

| | |
|---|---:|
| Rent Reserved (and FRV) | £ 100,000  p.a. |
| *Outgoings* (NIL) | Nil |
| Net Income | £ 100,000  p.a. |
| | |
| YP in perp at 4% | 25.00 |
| | £2,500,000 |

*Example 9–2*

Value the freehold interest in shop premises in a secondary location. The premises comprise a shop on the ground floor with flat above which were let recently to a single tenant at £10,000 p.a. exclusive on internal repairing and insuring terms for 15 years with 5-yearly rent reviews. The rent is at full rental value.

*Valuation*

| | | |
|---|---:|---:|
| Rent Reserved (and FRV) | | £10,000  p.a. |
| *less* | | |
| External Repairs say | £1,000 | |
| Management at 5% | 500 | 1,500 |

| | |
|---|---:|
| Net Income (and net FRV) | £8,500 p.a. |
| YP in perp at 9% | 11.11 |
| | £94,435 |
| Value say | £94,000 |

Note that the less attractive investment features of Example 9–2 involve the adoption of a higher yield than in Example 9–1. A further significant factor in determining the yield is that, in Example 9–1, the tenant is tied to the property for a substantial period. If the new lease were for only a short period, say five years, then there is no certainty that the property would be re-let to the same covenant or one of a similar quality and there might be a void period between lettings.

It is worth noting that if a valuer chose to predict the future levels of rent at each review and thereafter and then chose to include them in the valuation, he would need to adopt a higher yield than that which was used. (Which does reflect growth potential) as he would otherwise double-count the growth. The problem then is the choice of a non-growth yield since yields excluding growth are not readily obtainable from comparables. The possibility of incorporating growth rental is discussed, together with comments on equated yields, in Chapter 12 where further aspects of this topic are considered.

It is also worth noting that the valuations determine the sum which an investor can invest to achieve a desired return. They do not, however, represent the whole cost of the investment since costs will arise in relation to the purchase such as professional fees and stamp duty. On all but small investments where no stamp duty is payable, these on-costs tend to average around 3.75% for the purchasers and around 2.5% for the vendor.

It follows, therefore, that the return on the actual sums spent on acquiring an investment will be slightly less than those adopted in the valuation which represents the return on the net purchase price. This in no way invalidates the valuation approach since the returns are based on the net purchase prices of comparable investments. As the on-costs tend to be a constant proportion there will be a fixed relationship between returns to net and gross sums invested.

For example, suppose that an investor pays £100,000 for an income of £10,000 p.a. The return to the net investment is:

$$\frac{10,000}{100,000} \times 100 = 10\%$$

If on-costs are 3.75% of £100,000 = £3,750, the return to the gross investment is:

$$\frac{10,000}{103,750} \times 100 = 9.638\%$$

Thus, an investor seeking a 10% return recognises that he will achieve a 9.638% return on gross sums invested. If any net income is valued at 10% the return to gross investment will be 9.638% in any event. When references are made to the return or yield on an investment this is normally a reference to a net investment. A valuer should of course be sure that this is so, particularly where large investment institutions are involved, since they do tend to talk in terms of returns to gross costs in many instances.

### (b) Freehold Let at Below Full Rental Value

As has been illustrated, a freeholder will frequently be receiving a rent below rental value because the rent was fixed some years ago and rental values have risen subsequently. An alternative reason might be that the lessee paid a premium to the freeholder when he took the lease which entitled him to pay a rent below rental value from the very start.

The valuer faced with such a situation may find that the approach adopted when rent equals rental value is inadequate. He could take the actual rent as net income and use the approach: Net Income × Years' Purchase = Capital Value. However, he has a difficult problem in that, in addition to normal predictions as to underlying rental growth, he has a certain knowledge that the rent will change at a specific time when either the lease terminates or a rent review is to operate, and the rent will rise to a new level. This change is different from normal rental value variations resulting from general market movements. Consequently he will be unable to draw from market transactions the appropriate yield to apply unless he can find similar situations where the rent bears the same proportion to rental value, and the time for review or renewal is the same. Indeed, in practice the only typical situation which tends to follow these requirements is the valuation of freehold ground rents where the rent is relatively nominal in relation to rental value and the time for rent change is distant.

Accordingly, the valuer needs to adapt the general approach for this situation. For example, suppose that a valuation is required of a freehold interest where the rent receivable is £1,000 p.a. for the next three years. The current rental value is £4,000 p.a.

It is clear that for Years 1 to 3 the freeholder will receive £1,000 each year. At the end of Year 3 the rent will be revised to the then prevailing rental value. It is not known what the rental value will be at that time but it is known that the current rental value is £4,000 p.a. The valuation is made in two stages.

*Stage 1. Value current rent passing*

The period for which the rent is fixed is known as the Term. The freeholder will receive £1,000 per year. It may be that it will be payable quarterly, half yearly, or annually, and in advance or in arrears. It is normally assumed for valuation purposes that the rent is payable annually in arrears. This is merely a convention of convenience although it is true that typically leases require rents to be paid in advance, and commonly quarterly in advance. This topic is further considered in Chapter 12.

Hence, following the normal convention, the valuer must value £1,000 receivable for each of the next three years. This he can do as follows:

| | | |
|---|---|---|
| Year 1 | | £1,000 |
| PV £1 in 1 year | = say | £1,000  x |
| *Plus* | | |
| Year 2 | | £1,000 |
| P.V. £1 in 2 years | = say | £1,000  y |
| *Plus* | | |
| Year 3 | | £1,000 |
| PV £1 in 3 years | = say | £1,000  z |
| | | £1,000  (x +y + z) |

However, as was shown in Chapter 8, a simpler method is to use the Years' Purchase Single Rate.

The valuer needs to adopt the appropriate yield to determine the Years' Purchase. He probably has no comparables of sales of similar incomes at one quarter of rental value receivable for three years only. What he does know is the yield adopted when valuing the rental value from similar properties: he can therefore derive the appropriate yield from this.

For example, suppose that if the property were let at its rental value of £4,000 p.a. the yield would be 6%. What yield should be adopted in valuing the lower rent? One view is that the lower rent, being from the same property, has all the investment qualities of the rental value, but in addition it is even more certain of payment since the tenant is enjoying the accommodation at a cheap cost: thus a lower yield is appropriate. On the other hand, if the tenant did fail to pay the rent the freeholder could obtain possession and re-let at the higher rental value, so that this added security might be said to be of no real importance. Further, if the tenant is substantial, the chances of his defaulting at any level are remote if "a good covenant" has any meaning. Meanwhile, a low rent paid by a weak covenant is certainly more certain of being received than where the rent is at its maximum rental value. Yet again, if the low rent is fixed for a period beyond the normal review period then it has the disadvantage of providing a poorer hedge against inflation, a factor which will outweigh any additional certainty of payment: indeed, if the rent is fixed for several years, it takes on the characteristics of a medium fixed interest investment when the yield may be more realistically derived from yields for such investments than from the yield for the property producing the income.

From a purely practical point of view it will be found that, in cases where the term rent is small compared with the reversionary rent and the term is for a short period, the value of the term is only a small part of the whole value. The choice of yield in such cases can only have a limited practical significance since the overall value is hardly affected by whatever yield is chosen. Further it is practically impossible to analyse comparable transactions to demonstrate the yield being used for valuing the term for the same reasons.

The approach therefore is to weigh up the situation and make whatever adjustment appears to be appropriate – the valuer must make a subjective judgement. It is suggested that as a general rule the same yield should be adopted. However, if the reduced rent makes the investment a better quality investment because of this fact or if the additional certainty of payment is specially attractive, for example where the tenants are not financially strong – as is generally the case in poorer properties – then a reduced yield is appropriate. On the other hand, if the rent is fixed for a period significantly longer than the normal review period, it should be increased. In every case, however, the yield is derived from the prevailing yield applicable to the valuation of the property if let at its full rental value.

So, returning to the rent for the term, the valuation is:

| | |
|---|---|
| Rent Reserved | £1,000 p.a. |
| YP 3 yrs at say 6% | 2.673 |
| | £2,673 |

*Stage 2. Value rent for period after the term*

The period following the Term is called the Reversion. The valuer must first determine the rent which will arise at the end of the Term. He has two choices, to adopt a predicted rental value or to adopt the prevailing rental value.

If he takes the former course he has two problems. First, he does not know what the rental value will actually be. Second, if he adopts a predicted future level he is contradicting the general principle established in valuing a freehold interest let at full rental value.

On the other hand, if he adopts current rental value he can have reasonable confidence that at least this rent is likely to be obtainable. In addition the yield adopted in valuing the full rental value reflects the likely trends in rental value. If the property were now let at a rent which was below the current rental value this would anticipate a growth in rent receivable at regular periods in the future which will occur in the property being valued.

Hence the valuer should adopt the current rental value as the rent receivable on reversion, recognising that the yield he applies reflects the possible changes in rental value that will arise in the intervening period.

The valuer can therefore project himself forward to the end of the Term. At that time there will be a freehold interest let at full rental value. The valuation of such a situation has already been established. Hence:

| | |
|---|---|
| Rental Value | £4,000 p.a. |
| YP in perp at 6% | 16.6667 |
| | £66,667 |

However, that is the value in three years' time. What is required is the value now. This can be determined by applying the Present Value of £1 Table. The appropriate rate to use is that adopted for valuing the reversion since it is logical that an investor purchasing a future investment showing 6% invests a sum today to grow at the same rate. Hence:

| | |
|---|---|
| Value in 3 years' time | £66,667 |
| × Present Value of £1 in 3 yrs at 6% | 0.8396 |
| | £55,974 |

The valuation of the Reversion is thus:

| | |
|---|---|
| Full Rental Value | £4,000 |
| × YP in perp at 6% | 16.6667 |
| | £66,667 |
| × Present Value of £1 in 3 yrs at 6% | 0.8396 |
| | £55,974 |

As an alternative approach, it is known that the Years' Purchase Single Rate is the sum of the Present Values of £1. Hence YP in perp = PV £1 in 1 yr + PV £1 in 2 yrs + PV £1 in 3 yrs + PV £1 in 4 yrs and so on. However, the valuation in this case requires the sum of the Present Values of £1 into perpetuity apart from the next 3 years. Hence one can apply YP in perp (which includes PV for 1, 2 and 3 yrs) and deduct PV £1 in 1, 2 and 3 yrs (which equals YP 3 years Single Rate). Thus:

| | | |
|---|---|---|
| Full Rental value | | £4,000 p.a. |
| YP in perp at 6% | 16.6667 | |
| *less* | | |
| YP 3 yrs at 6% | 2.6730 | 13.9937 |
| | | £55,975 |

*Note*

The deduction of YPs does not apply to Years' Purchase Dual Rates as will be shown later.

The valuer can short circuit either of these approaches where he is valuing a reversion to a term which goes on into perpetuity since *Parry's Valuation Tables* contain on pages 42–55 Years' Purchase of a Reversion to a Perpetuity. Hence:

| | |
|---|---|
| Full Rental Value | £ 4,000 p.a. |
| YP in perp deferred 3 yrs at 6% | 13.99367 |
| | £55,975 |

*(c) Term and Reversion*

Thus it is seen that the valuation of a freehold interest let at less than full rental value requires the combined value of the term and reversion. The full valuation of the example becomes:

*Term*

| | | |
|---|---|---|
| Rent Reserved | £ 1,000 p.a. | |
| YP 3 yrs at 6% | 2.673 | |
| | | £ 2,673 |
| | | |
| Reversion to FRV | £4,000 p.a. | |
| YP in perp def'd 3 yrs at 6% | 13.994 | £55,976 |
| | | £58,649 |

An alternative approach to the Term + Reversion is sometimes adopted where the initial rent is valued into perpetuity and the additional rent receivable at reversion is valued into perpetuity but deferred for the period of the term. Such an approach will give the same result as described above so long as a common yield is adopted throughout, but where the yield adopted in valuing the term should be different from the reversion, an artificiality is required for this layer method which makes it less reliable. The *layer hardcore* method is considered in Chapter 12.

The following examples illustrate the valuation of freehold interests let at less than full rental value.

*Example 9–3*

Value the freehold interest in shop premises in a secondary location.

The premises are let at £2,500 p.a. net on lease with four years to run. The current rental value is £6,000 p.a. net. The property is a 10% investment.

*Term*

| | | |
|---|---|---|
| Rent Reserved | £2,500 p.a. | |
| YP 4 yrs at say 9% (rent less | | |
| than FRV and more secure) | 3.240 | £ 8,100 |
| *Reversion* | | |
| to FRV | £6,000 p.a. | |
| YP in perp def'd 4 yrs at 10% | 6.830 | £40,980 |
| | | £49,080 |

*Example 9–4*

Value the freehold interest in well located factory premises let to a good covenant.

The premises are let at £10,000 p.a. net on lease with 12 years to run without review. The current rental value is £16,000 p.a. net. The property is an 8% investment (assuming a lease with 5-yearly rent reviews).

*Term*

| | | |
|---|---|---|
| Rent Reserved | £10,000 p.a. | |
| YP 12 yrs at 10% (rent less than FRV but fixed for a long term) | 6.814 | £68,140 |

*Reversion*

| | | |
|---|---|---|
| to FRV | £16,000 p.a. | |
| YP in perp def'd 12 yrs at 8% | 4.964 | £79,424 |
| | | £147,564 |

*Note*

The reversal of the yields reflects the inflation prone nature of the term rental.


*Example 9–5*

Value the freehold interest in shop premises in a good location. The premises are let on a ground lease with 60 years to run at £1,000 p.a. without provision for review. The current rental value is £50,000 p.a. net. The property is a 7% investment (assuming a lease with 5-yearly rent reviews).

*Term*

| | | |
|---|---|---|
| Rent Reserved | £1,000 p.a. | |
| YP 60 yrs at 14% (the rent is very secure but is fixed for a very long term) | 7.14 | £ 7,140 |

*Reversion*

| | | |
|---|---|---|
| to FRV | £50,000 p.a. | |
| YP perp def'd 60 yrs at 7% | 0.2465 | £12,325 |
| | | £19,465 |

Note that the current rent provides a yield of $1,000/19,465 \times 100 = 5.137\%$ to capital invested. In practice it might be possible to show that similar ground rents sell on the basis of an initial yield of around 5% – or put another way, they sell on a 20 YP basis. This is one case where a direct valuation of the interest may be derived from the present rent, i.e. capital value is £1,000 p.a. × 20 YP =

£20,000. It is not as precise as the detailed approach but it may be considered sufficiently accurate in some circumstances.

## 2. Valuation of Leasehold Interests

Leasehold interests are commonly found in the property market. They arise in two principal ways.

One is where a freeholder of development land offers a ground lease under which the leaseholder will carry out the development and enjoy the benefits of the property he develops for the period of the lease in return for the payment of a ground rent. Such leases are usually for a long term of years, typically 99 years or 125 years, to allow the lessee to recoup his capital investment in the buildings, and the ground rent is low relative to the rental value of the buildings.

The other principal source of leasehold interests is where either a freeholder or leaseholder offers a lease of premises which the tenant will occupy for his own use and enjoyment in return for the payment of a rent which will normally be the rental value of the premises.

As can be seen, a lease may require the payment of a rent at less than full rental value or at full rental value. In addition, as was shown above, even where a lessee takes a lease at full rental value with provision for periodic reviews of rent, if rental values rise there will come a time when the rent paid falls below prevailing rental values.

The importance of the relationship between the rent payable under a lease and the full rental value is that a leasehold interest has no value unless full rental value is greater than the rent payable. The reason for this is that as full rental value represents the maximum rent which a tenant is prepared to pay, he will not be prepared to offer any additional sum to purchase the lease. It must be remembered that when someone purchases a lease he becomes responsible for the observance of all the covenants set out therein, including of course the duty to pay the rent. Hence, if someone purchases a lease for a capital sum and then is required to pay full rental value he is worse off than he would be if he had taken a lease direct of a comparable property.

Thus, as a first principle, a leasehold interest has a value where the full rental value exceeds the rent payable. The difference between these two figures is termed a "Profit Rent". For example, suppose that A holds a lease under which he must pay £800 p.a. when the full rental value is £1,500 p.a. The position is:

| | |
|---|---|
| FRV | £1,500 p.a. |
| *Less* | |
| Rent paid | £800 |
| Profit Rent | £700 p.a. |

Such a lease has a value because if someone purchases the lease he will be entitled to occupy the property under the lease terms and so pay £800 p.a. for something which is worth £1,500 p.a. or alternatively he may sub-let the property at a rent of £1,500 p.a. and so each year receive £1,500, pay £800 and retain the "Profit" of £700 as a profit rent. This benefit will continue until the lease ends or the rent is adjusted under a rent review clause to the higher value. The task of the valuer is to determine the price a purchaser would pay for the benefit – the value of the lease.

The valuation approach normally adopted is the investment method notwithstanding that the purchaser might intend to occupy. This is valid particularly in the case of commercial properties since, although the purchaser does not seek an actual return of rent on his capital investment, he will be receiving an annual rental benefit which will come to him as part of his profits on the capital he invests. That this is so can readily be demonstrated by assuming the facts of the example above, where the lessee will pay £800 p.a. as rent, also assuming in addition that another identical property is available at the full rental value of £1,500 p.a. In the first property he will pay £700 p.a. less than he would in the second property. Since all other conditions are identical, his profits from trading in the first property will be £700 p.a. greater than for the second property. Similarly, when in the first property the lease ends or the rent is reviewed, so that the rent is increased from £800 p.a. to £1,500 p.a., there will be an immediate reduction in the profits of £700.

Thus, when valuing a leasehold interest, the basic principle of the investment method will apply, namely:

$$\text{Net Income} \times \text{Years' Purchase} = \text{Capital Value}$$

## (a) Provision for Loss of Capital

Net Income for this purpose will be the Profit Rent, as has been shown above. However, when an investor purchases a leasehold interest he is faced with one factor which he will not meet in a freehold interest, namely that one day his interest must cease and he will have no further interest in the property: if he obtains a further lease in the property to follow the expired lease that will be a new investment.

The investor knows therefore that one day his interest will cease and he will have no capital asset remaining. His investment will therefore decline from whatever he paid to nothing. Such an investment is know as a wasting asset since the asset wastes away naturally and inevitably.

The investor faced with such a situation must therefore take steps to deal with the loss of his capital. This he can do by investing funds which grow to compensate for the capital he will lose. The sensible approach is to find these funds out of the income produced by the wasting asset so that it is self-compensating. Various methods are open to him but the conventional approach is to assume that he will invest part of his rent each year out of the net income in an annual sinking fund. Since he needs to guarantee the replacement of his capital, he must seek a medium which is as certain as possible that the sums invested will reach a specific and certain figure. The only such medium is normally an insurance policy. Naturally the returns offered by insurance companies must be based on long-term rates of interest which ignore abnormal movements in interest rates and which give a reasonable guarantee that they can meet their obligations. Consequently, the returns offered will be low and an average yield of around 3% has been found to be realistic.

If the investor places some of his net income in such a sinking fund he can be satisfied that his original investment capital will be available at the end of the lease for further investment. In that way he can continue indefinitely so that the return he receives after allowing for sinking fund costs will be perpetual. The effects of inflation on such an arrangement will be considered later.

An alternative approach to setting aside a sinking fund is to write off the investment over its life. This is an accountancy convention often followed where companies such as industrialists employ plant and machinery which by their nature have a limited life. A typical approach would be to adopt straight line depreciation where the capital cost is divided by the predicted years of useful life and this figure deducted as an annual cost of the item in the accounts. Such an approach does not lend itself readily to a valuation and fails to reflect the accelerating rate of depreciation suffered by a leasehold interest. Similarly, the other common approach of the reducing balance, whereby a stated percentage is written off each year, is also of no help since this approach writes off the value more heavily in the early years which is the opposite of the depreciation pattern of a lease.

### (b) Years' Purchase Dual Rate

The investor in a leasehold investment, if he provides a sinking fund, can regard the net income after sinking fund provision, as perpetual.

The valuer may now value any leasehold profit rent with this in mind. Assume, for example, that a valuation is required of a leasehold interest where the lease has 5 years to run at a fixed rent of £800 p.a. The current rental value is £1,500 p.a. The valuation approach is as follows:

*Stage 1. Determine Profit Rent*

| | |
|---|---|
| FRV | £1,500 p.a. |
| *less* | |
| Rent paid | £800 |
| Profit Rent (Net Income) | £700 |

*Stage 2. Assess Annual Sinking Fund*

As was shown in Chapter 8, the Annual Sinking Fund for the Redemption of £1 Capital Invested Tables shows the annual payments needed to reach £1. In this case, the sum to be recouped will not be known until the valuation is completed, so assume value of interest is £V. Hence:

| | |
|---|---|
| Capital Sum required in 5 years | £V |
| × ASF to replace £1 in 5 yrs at 3% | 0.1884 |
| Annual Sinking Fund Payment | £0.1884 V |

*Stage 3. Assess Income Net of Sinking Fund*

| | |
|---|---|
| Profit Rent | £700 |
| *less* | |
| ASF to replace £V in 5 yrs | <u>£0.1884</u>  V |
| Net Income | £700 – 0.1884  V |

*Stage 4. Value Net Income*

Since the income net of ASF is effectively perpetual notwithstanding that the lease will end, as has been demonstrated, the net income can be regarded as a perpetual net income as for freehold in possession. However, as the income is produced by a leasehold interest rather than a freehold, the investor has to collect rent and pay rent on to the landlord; he will have the trouble and expense of replacing his investment at the end of the lease; and he is subject to the restrictions on his actions which the covenants in the lease place upon him. Apart from these factors the investment is the same as the freehold investment in an identical property on which the yield will be known. Allowing for these adverse qualities the appropriate yield for the leasehold investment should be higher than the freehold yield. The amount of additional return required depends on the facts of the case but, as a general rule, the addition will be around 0.5% on a freehold yield of 4% rising to 2% on a freehold of 12%. There is, however, no fixed adjustment to be made: the adjustment is a matter of judgment coupled with knowledge of adjustments found in analysis of comparables.

Hence, suppose in this case that the freehold yield is 7.5%, the leasehold yield would then be say 8.5%. Hence:

| | | |
|---|---|---|
| Net Income | £700  – | 0.1884V  p.a. |
| × YP in perp at 8.5% | | <u>11.7647</u> |
| Capital Value | £8,235  – | 2.216V |
| *BUT* | | |
| the Capital Value was taken to be | | |
| £V, therefore | £8,235  – | 2.216V  = V |
| | | V = £2,561 |

Thus, bringing the valuation together:

| FRV | £1,500 p.a. | | |
|---|---|---|---|
| *Less* | | | |
| Rent Paid | 800 | | |
| Profit Rent | 700 | | |
| *Less* | | | |
| ASF to replace £V in 5 yrs at 3% | 0.1884 V | | |
| | £700 – | 0.1884 V | |
| × YP in perp at 8.5% | 11.7647 | | |
| Capital Value | £8,235 – | 2.216 V | |
| Capital Value (V), therefore | 8,235 – | 2.216 V = V | |
| Capital Value = | £2,561 | | |

This is a straightforward approach but it was shown in Chapter 8 that there are Tables known as Years' Purchase Dual Rates. As was shown, these Tables allow for the need to provide a sinking fund by incorporating within their formula an allowance for the appropriate sinking fund. They therefore perform the function of Stages 2 and 4 described above and their use will exclude the need for Stage 3. These Tables can therefore be applied directly to the Profit Rent. Hence:

| FRV | £1,500 p.a. |
|---|---|
| *Less* | |
| Rent Paid | £800 |
| Profit rent | £700 |
| × Years' Purchase allowing for a yield of 8.5% and annual sinking fund at 3% for 5 years | |
| (YP 5 yrs at 8.5% & 3%) | 3.658 |
| | £2,561 |

Clearly this is a more convenient method and the one adopted in valuing a leasehold interest at full rental value. Nonetheless the valuer is aware that the method is the same as the valuation of a freehold interest at full rental value.

## (c) Capital Replacement and Inflation

Mention was made earlier of the problem of inflation in the replacement of the capital invested. The problem is that an investor who invests capital in a leasehold interest and who replaces his original capital at the end may find that, with the erosion of the value of money over the period, the replaced capital is worth less, and perhaps significantly less, in real terms than at the start. As a general rule, if he compares investments in freehold property, he will find that their value has risen to maintain more or less their real value. The implications of this situation which have led to some criticism of the approach described above are considered in Chapter 12.

*(d)  Annual Sinking Fund and Tax on Rents*

The valuation set out above can be analysed to show that it achieves its goals as follows:

| | | |
|---|---|---|
| Profit Rent | £700 | |
| *Less* | | |
| ASF to replace £2,561 | | |
| in 5 yrs at 3% = (2,561 × 0.1884) = | 482 | |
| Income net of ASF | £218 = | 0.085 |
| | £2,561 | |
| | | × 100 |
| Remunerative yield | | 8.5% |

the investor achieves his desired return.

| | | |
|---|---|---|
| *OR* | | |
| Profit Rent | | £700 |
| Capital Invested | 2,561 | |
| Interest on Capital at 8.5% | 0.085 | £218 |
| Sum available for investment in ASF | | £482 |
| × Amt. of £1 p.a. for 5 yrs at 3% | | 5.3091 |
| | | £ 2,559 |

the investor replaces his capital.

This analysis ignores income tax. The reasons for ignoring income tax in determining the return on an investment were set out above. However, the annual sinking fund is a cost, and the ignoring of the effects of income tax in respect of this leads to a misleading value, as the following analysis where income tax is considered shows.

Assume that the investor pays income tax at a marginal rate of 40p in the £. Hence:

| Profit Rent | | £700 |
| --- | --- | --- |
| Capital Invested | £2,561 | |
| Interest at 8.5% | 0.085 | £218 |
| Sum available for ASF | | £482 |

Thus the Profit Rent is divided as to £218 for return on capital and £482 for annual sinking fund investment. As to the £218, although income tax at, say, 40% is payable thereon (as part of the £700 p.a. on all of which say 40% tax is payable) this can be ignored for analysis purposes since the aim is to obtain 8.5% before tax. As to the sinking fund, this is an actual investment exercise and the investor needs to put £482 each year into it. However, he must pay, say, 40% income tax on this since no tax relief is given on normal sinking fund investments. So he can only invest:

| ASF before tax | £482 |
| --- | --- |
| *Less* | |
| Income Tax at 40% | £193 |
| ASF after tax | £289 |

It is obvious therefore that the investor will not be able to achieve his goal, since he will not be able to pay £482 each year into the annual sinking fund and also meet his tax commitments. The true cost of the annual sinking fund therefore is not what he must invest but the annual income which, after payment of the tax, leaves £482 available to spend on the sinking fund.

This gross cost can be determined by multiplying the sum required by $\frac{100}{100-40}$ as is shown below.

Assume A earns £100 and pays tax of £40 (i.e. 40p in £).
Then A's income after tax is (100 – 40)p.
Thus for each £1 that A earns he will be left with (100 – 40)p out of each 100p.
Thus his net income (NI) will be $\frac{40-100}{100}$ of his gross income (GI).

If $\qquad \frac{GI \times 100}{100} - 40 = N1$

Then $\qquad GI = NI \times \frac{100}{100-40}$

The same will be true at any rate of tax. Thus, if tax is t, gross income can be reduced to net income by applying $\frac{100-t}{100}$

Net income can be grossed up to gross income by applying $\dfrac{100}{100-t}$

### (e) Years' Purchase Dual Rate Adjusted for Income Tax

Applying the tax factor to the leasehold income, where an annual sinking of £482 was seen to be necessary to replace the capital, the amount of profit rent needed to be earned each year to leave £482 after tax is:

$£482 \times \dfrac{100}{100-40} =$                  £803.33  gross cost

*less*

Tax at 40p in £1 = £803.33 × 04 =      <u>£321.33</u>
Net Income for ASF =                          £482.00

If the investment is now analysed allowing for the true cost of providing the sinking fund, the result is:

Profit Rent                              £700.00
*less*
Gross ASF to produce £482 p.a.      <u>£803.33</u>
Hence Negative return on capital      £103.33

Clearly the income has been over-valued if the price of £2,561 is to be invested by a tax-payer paying 40p in the £1 tax on income. Indeed, it is obvious that the price of £2,561 will only leave £482 p.a. available for ASF after the return of 8.5% has been taken out, if the investor is a non tax-payer. Similarly, the cost of providing a sinking fund must vary with the individual tax-payer's marginal rate of income tax, the higher the rate of tax the more gross income needed before tax to produce the required net income for ASF.

This presents a problem for the valuer called upon to prepare a "Market valuation" when it is clear that the potential buyers in the market have differing rates of tax which have a direct effect on their sinking fund costs. The solution is for the valuer to adopt an "average rate" of tax, whilst recognising that investors whose personal rates differ sharply from this have an advantage or disadvantage which renders the valuation unreal for their own circumstances. If required, he can easily produce a valuation for a particular investor's actual marginal tax rate.

Given that the valuer adopts an average rate of tax, what is an average rate? In any year, tax rates may vary from nil for tax exempt persons to rates which have been as high as 98% with many different rates between, and these rates tend to vary from year to year. Faced with such a variety of rates the valuer needs to select a rate which is likely to be most commonly met for the remaining years of the lease. This rate is unknown, but a useful guide is the standard rate of income tax for individuals and the corporation tax rate for companies. The standard rate in recent years has come down steadily from around 30% plus 15% investment surcharge to a basic rate in 1997 of 23% coupled with a single higher rate of 40% on incomes above £26,100. In similar fashion, corporation tax has come down from 52% for most companies to 33%, with a rate of 23% on small profits. This suggests that an average rate of around 35% might be used. However, if rates of tax showed a tendency to rise or fall then the average rate should adjust to the trend.

Thus, the valuer adopts a predicted average level of tax for coming years, and a rate of 35% will be adopted for this purpose. In any event, a small variation in the rate will only be significant if the term of years is short.

The valuer can now incorporate in the valuation approach a grossed up cost of an annual sinking fund by applying 100/100 – 35 = 1.54 to the required ASF investment. This can readily be incorporated in the Years' Purchase Dual Rate formula. As was shown in Chapter 8, the formula is:

$$\frac{1}{i + ASF}$$

i represents the gross of tax rate of return and needs no adjustment. ASF represents the annual sinking fund required to replace £1 of capital (after payment of tax).

It follows that ASF × 100/100 – t will represent ASF required before payment of tax, i.e. the gross ASF.

Hence applying the facts of the example to the formula:

i = 8.5% = 0.085
ASF = ASF to replace £1 in 5 yrs at 3% = 0.18835
ASF × 100/(100 – 35) = 0.18835 × 1.53846 = 0.28977
YP 5 years at 8.5% and A.S.F. at 3% (adjusted for income tax) = 1/(0.085 + 0.28977) = 2.6683

(Tables of YP Dual Rate adjusted for varying rates of income tax can be found in *Parry's Valuation Tables* – pages 30.1–52.32).

*Valuation*

| | | |
|---|---|---|
| Profit Rent | | £700 p.a. |
| × YP 5 yrs at 8.5 & 3% adjusted | | |
| for income tax at 35% | | 2.6683 |
| Capital Value | | £1,868 |

*Analysis*

| | | |
|---|---|---|
| Profit Rent | | £700 |
| *less* | | |
| Interest on Capital at 8.5% | £1,868 | |
| | 0.085 | £158 |
| Gross ASF | | £542 |
| *less* | | |
| Income Tax at 35% | | 190 |
| Net ASF | | £352 |
| × Amt. of £1 p.a. for 5 yrs at 3% | | 5.309 |
| Capital Sum Replaced | | £1,868 |

This shows that the adoption of a YP Dual Rate adjusted for income tax allows the investor to achieve his required return and also permits him to replace his capital after meeting his tax liabilities.

With some exceptions, therefore, it is suggested that YP Dual Rate adjusted for tax tables should be used for valuing leaseholds and other terminable incomes. The main exception is where the market comprises non-tax payers. It should be evident from the discussion so far that a person or organisation not liable to pay tax would be in a very advantageous position in relation not only to the actual yield from any investment but, with leaseholds, in relation to the rate of interest at which a sinking fund accumulates and to the amount which must be set aside; the gross annual sinking fund would equate to the net annual sinking fund. The shorter the lease, the greater the advantage. This advantage has been exploited by non-tax paying investors, in some cases to the point where they comprise the market (see Chapter 12).

### (f) Leasehold Interest where Income Less than Full Rental Value

So far, the valuation of a leasehold interest has been restricted to those situations where the leaseholder receives full rental value, and the situation then compared with the valuation of a freehold interest in similar circumstances. However, as for freeholds, a leasehold owner may be receiving less than full rental value either because he sub-let in the past at the prevailing FRV and rental values have risen, or he sub-let and took a premium.

Such a situation arose with freehold interests and, as was shown in such cases, the valuation values the existing rent whilst it is receivable – the "term", and then reverts to FRV when this will be receivable – the "reversion". The same approach is adopted when valuing a leasehold varying income, with a YP dual rate adjusted for tax being employed in place of YP single rate or in perpetuity, as the following example shows.

*Example 9–6*

A took a lease of shop premises 23 years ago for 42 years at a fixed rent of £10,000 p.a. 17 years ago he sub-let the premises for 21 years at a fixed rent of £16,000 p.a. A valuation is now required of A's leasehold interest when similar shops are letting on 15-year leases with 5-yearly rent reviews at £40,000 p.a.

*Term*
A is receiving £16,000 p.a. for the under-lease. This will be payable for the next 4 years. Hence:

| | |
|---|---|
| Rent Received | £16,000  p.a. |
| *less* | |
| Rent Paid by A | £10,000 |
| Profit Rent | £6,000  p.a. |

A is receiving this profit rent for four years. Assume that the appropriate freehold rate for such a shop investment is 5.5%. The appropriate leasehold rate would be say 6%. However, the rent is less than FRV, but assume this is not significant.

| | | |
|---|---|---|
| Profit Rent | £6,000  p.a. | |
| × YP 4 yrs at 6% and | | |
| 3% adj. for tax at 35% | 2.338 | £14,028 |

*Reversion*
After four years A will be able to re-let the shop at the prevailing rental value. For the reasons set out above when considering the valuation of a freehold interest let at less than FRV the current rental value is adopted as the rent receivable at reversion. Hence:

| Reversion to FRV | £40,000 p.a. | |
|---|---|---|
| *less* | | |
| Rent Payable by A | £10,000 | |
| Profit Rent | £30,000 p.a. | |

After four years, A will still have 15 years to run of his own lease. Hence, in four years' time the valuation will be:

| Profit Rent | £30,000 p.a. | |
|---|---|---|
| × YP 15 yrs at 6% and | | |
| 3% adj. for tax at 35% | 7.007 | £210,210 |

However, the value is required now, so that the value brought forward four years is:

| Value in 4 years | £210,210 | |
|---|---|---|
| × PV £1 in 4 yrs at 6% | 0.792 | £166,486 |

Note that, as before, the discount rate is the same as that adopted for the valuation of the reversion. Thus, bringing the valuation together it becomes:

*Term*

| Rent Received | £16,000 p.a. | |
|---|---|---|
| *less* | | |
| Rent Payable | £10,000 | |
| Profit Rent | £6,000 p.a. | |
| | | |
| × YP 4 yrs at 6% and 3% adj. for tax at 35% | 2.338 | £14,028 |

| *Reversion* to FRV | £40,000 p.a. | |
|---|---|---|
| *less* | | |
| Rent Payable | £10,000 | |
| Profit Rent | £30,000 p.a. | |

| | | | |
|---|---|---|---|
| × YP 15 yrs at 6% and 3% adj. for tax at 35% | 7.007 | | |
| × PV £1 in 4 yrs at 6% | 0.792 | 5.550 | £166,500 |
| | | | 180,528 |

| Valuation of Interest say | | £180,000 |
|---|---|---|

This method of deferring the reversion is the only one which may be adopted. There are no Tables in *Parry's* which combine YP for a term of years × PV for the term (unlike YP reversion to perpetuity

Tables) and the YP for the period from the time of valuation minus YP for the term will give the wrong result (unlike YP in perp minus YP Single Rate for the Term). This latter aspect is considered in Chapter 12.

This valuation method has a built-in technical error because the first of the two sinking funds will continue to appreciate in value over the succeeding 15 years thus producing a capital return of £14,028 × (amount of £1 for 15 years @ 3%) + £166,500 = £188,355 in 19 years. This error can be eliminated by using an approach similar to that set out at the beginning of this section (known as the double sinking fund method) or by using Pannells method (see Chapter 12).

## 3. Marriage Value

The method of valuation of a freehold let at less than full rental value and of a leasehold interest have been considered. These interests will arise simultaneously in the one property since the lease held at less than rental value creates the fact of the freehold terms and reversion and the lease profit rent. If the owner of one of the interests buys the other interest, the interests merge by operation of law so that the lease merges into the freehold to create a freehold in possession or, in valuation terms, a freehold at full rental value. It is interesting to consider the valuation implications of this situation.

*Example 9–7*

A owns the freehold interest in office premises. The offices were let 26 years ago for 40 years at £10,000 p.a. without review. The current rental value is £45,000 p.a. The freehold yield at FRV is 6%. A proposes to buy the leasehold interest following which he will re-let the offices at £45,000 p.a. on a new lease.

*Value of A's Present Interest*
Term

| | | |
|---|---|---|
| Rent received | £10,000  p.a. | |
| × YP 14 years at 9% (rent is fixed for long term) | 7.786 | £77,860 |
| Reversion to FRV | £45,000  p.a. | |
| × YP in perp def'd 14 yrs at 6% | 7.372 | £331,740 |
| | | £409,600 |

*Value of Leasehold Interest*

| | | |
|---|---|---|
| FRV | £45,000 p.a. | |
| less | | |
| Rent Paid | £10,000 | |
| Profit Rent | £35,000 | |
| × YP 14 yrs at 6.5 and 3% | | |
| adj. for tax at 35% | 6.45 | £225,750 |
| | | £635,350 |

*Value of Freehold after Purchase of Lease*

| | | |
|---|---|---|
| FRV | £45,000 p.a. | |
| × YP in perp at 6% | 16.666 | £750,000 |

Hence, if A buys the leasehold interest for £225,750, he will "lose" his existing interest valued at £409,600, but he will then have an interest worth £750,000. Thus there is additional value of £750,000 – (£225,750 + £409,600) = £114,650 which represents extra value arising from the merger (or "marrying") of the interests. This additional value is termed marriage value.

The reason that the marriage value arises is that the profit rent element when owned by the lessee is valued at 6.5% dual rate whereas in the freeholder's hands it is valued on a single rate basis at 6% in one case and 9% in another. This marriage value phenomenon will arise in nearly all cases where interests are merged and can act as an incentive for owners of interests to sell to each other, or to join together in a sale to a third party so as to exploit the marriage value.

The question which arises is who should take the benefit of the marriage value. Since the parties are interdependent they are of equal importance to the release of the marriage value, so as a general guide the marriage value is shared equally. Hence in *Example 9–7* above, A is able to offer the value of the lease (£225,750) plus half the marriage value (£57,325). In this way, both A and the leaseholder make a "profit" of £57,325.

Marriage value may also arise where owners of adjoining properties marry their interests to form a larger single site. This will be particularly so where the united interests create a development site but the individual sites are incapable of or are unsuitable for development. In these circumstances a person buying the interests (known as "assembling the site") can pay over and above normal value by sharing the marriage value among the parties.

Indeed, in some cases, site assembly will involve buying interests in adjoining sites and also buying up freehold and leasehold interests in particular sites, thus creating both forms of marriage value.

### 4. Freehold Interest where Building has Terminable Life

In considering the question of rental value in Chapter 5 it was stated that, as a general rule, the present rental value of a property is assumed, for the purpose of valuation, to continue unchanged in the future. Reference was also made in Section 1 of this Chapter to the fact that the majority of buildings are demolished when they reach the end of their economic life and at that stage the site value for redevelopment should exceed the value of the existing buildings and site. However, in circumstances where a building is approaching the end of its physical life, whether for structural or limited planning permission reasons, it may be that some loss of rental value will be incurred.

*Example 9–8*

Freehold shop premises built about 60 years ago are let on lease for an unexpired term of 30 years to a good tenant at £8,000 p.a. following a recent rent review.

It is considered that the buildings will have reached the end of their physical life at the same time as the current lease expires and that rebuilding will be required. The present cost of rebuilding is estimated to be £40,000 and the value of the site at £3,500 p.a., capital value £50,000.

The analysis of sales of properties similar in nature in the same district, but more modern and whose structural life is estimated to be 75 years or more indicates that they sell at 12.5 Years' Purchase of the present rental value.

What is the present market value of the property?

*Valuation – Method 1*

Applying the evidence of other sales it would appear that the property in question could properly be valued at a higher rate of interest than 8% because the buildings have only a 30-year life.

| | | |
|---|---|---|
| Rent on lease | £8,000  p.a. | |
| YP in perp at say 10% | 10.0 | £80,000 |

## Method 2

Assuming a reversion to site value at the end of the lease.

| | | | |
|---|---|---|---|
| That portion of present rent which will continue | £3,500 p.a. | | |
| YP for 30 years at 8% | 11.3 | £39,550 | |
| Remainder of rent | £4,500 p.a. | | |
| YP for 30 years at 8%[1] and 3% adj. for tax at 35% | 8.9 | £40,050 | |
| Reversion to site value | £3,500 p.a. | | |
| YP in perp at 7% def'd 30 years | 1.88 | £6,580 | |
| | | £86,180 | |
| Value, say | | | £86,000 |

## Method 3

Assuming continuance of rental value and making provision for rebuilding.

| | | | |
|---|---|---|---|
| Rent on lease | £8,000 p.a. | | |
| YP in perp at 8% | 12.5 | £100,000 | |
| *less* | | | |
| cost of rebuilding | £40,000 | | |
| × PV £1 in 30 years at 8% | 0.1 | £4,000 | £96,000 |

Method 1 has the merit of simplicity but the increase in the rate of interest is very much a matter of opinion.

Methods 2 and 3 are open to objection on a number of grounds.

In Method 2 it may be argued that it is difficult to know what will be the value of the site upon reversion after the useful life of the building, 30 years hence. The only evidence of the value of the site is present-day selling prices, which suggest the figure of £50,000 (rental value £3,500 p.a.) used in the example, but this value may be considerably modified in the future. A similar objection applies to the figure of £40,000 for cost of rebuilding in Method 3, which, of necessity, has been based on present-day prices although these may be very different 30 years hence.

---

[1] The adoption of the same yield reflects the quality of the covenant. If the covenant was considered to be poor, then a higher yield would have been adopted in this part of the valuation.

Another serious objection is the difficulty of predicting the life of a building with any degree of accuracy, particularly over a long period of years. Such an estimate of the life of the building, as is made in the example, must necessarily be in the nature of a guess unless considerable data are accumulated over a long period showing the useful life in the past of similar buildings. Even such evidence, over perhaps 40 or 50 years in the past, is not conclusive of what the useful life of buildings will be in the future.

There is an inherent error in Method 2. The use of 8% to value the continuing rent for the next 30 years implies some growth in the rent. It follows that some part of the additional rent has been valued twice.

On balance it would appear that, in many cases, there is probably less inaccuracy involved in assuming the present income to be perpetual and using a higher rate to cover risks, than would be involved in the more elaborate method of allowing for possible variations in income.

On the other hand, the separation of land and building values and making provision for building replacement has the advantage of giving to an investor a more direct indication of the actual rate per cent he is likely to derive from an investment than does Method 1, provided it is used in conjunction with analysis of sales made on similar lines.

In cases where it can confidently be assumed that existing buildings will be worn out in a few years' time, Method 2 may be the soundest.

In the case of properties where the building clearly has a very limited life but the site is incapable of being re-developed in isolation, whether for practical or for planning reasons, it may be impracticable to have regard to any future site value. In such a case, the existing income might be valued for the estimated building life as a terminable income with no reversion.

## 5. Premiums

A premium usually takes the form of a sum of money paid by a lessee to a lessor for the grant or renewal of a lease on favourable terms or for some other benefit. The term premium is frequently used by agents to describe the price required for a lease which is being sold, e.g. "Lease of shop premises for sale. Premium £8,000". This is a misuse of the word since what is sought is the price and, since this is not a premium under the lease, it is not a premium. The following comments refer to a premium payable as a term of a lease.

Where a premium is paid at the commencement of a lease, it is usually in consideration of the rent reserved being fixed at a figure less than the true rental value of the premises.

From the lessor's point of view this arrangement has the advantage of giving additional security to the rent reserved under the lease, since the lessee, having paid a capital sum on entry, has a definite financial interest in the property, which ensures that he will do his utmost throughout the term to pay the rent reserved and otherwise observe the covenants of the lease. Premiums are only likely to be paid where there is a strong demand for premises on the part of prospective tenants. On the other hand, in the case of leases of less than 50 years, the tax treatment of premiums is penal for lessors so that they may be discouraged from taking a premium (see Chapter 22).

The parties may agree first on the proposed reduction in rent and then fix an appropriate sum as premium, or they can decide the capital sum to be paid as premium and then agree on the reduction in rent which should be allowed in consideration of it.

The calculation of the rent or premium can be made from both parties' points of view when it will be found that a different answer will emerge. The parties will then need to agree a compromise figure.

Where the premium is fixed first, in order to arrive at the reduction to be made in the rent it will be necessary to spread it over the term either by multiplying the agreed figure by the annuity which £1 will purchase, or by dividing by the Years' Purchase for the term of the lease.

*Example 9–9*

Estimate the premium that should be paid by a lessee who is to be granted a 30-year lease of shop premises at a rent of £20,000 p.a. with 5-yearly revisions to 20,000 ÷ 25,000 = 80% of rental value (a "geared" rent review). The full rental value of the premises is £25,000 p.a.

*Note*
The parties have agreed that the calculation shall be based on the dual rate table of Years' Purchase at 8% and 3% adjusted for tax at 35% for the leasehold interest and 7% for the freehold interest.

**Lessee's Point of View**

| | |
|---|---|
| Present rental value | £25,000  p.a. |
| Rent to be paid under lease | £20,000 |
| Future profit rent | £ 5,000  p.a. |
| YP for 30 years at | |
| 8 and 3% adj. for tax at 35% | 9.77 |
| | £48,850 |
| Premium, say | £50,000 |

**Landlord's Point of View**

Present interest:

| | | |
|---|---|---|
| FRV | £25,000  p.a. | |
| YP in perp at 7% | 14.28 | £357,000 |

Proposed interest:

| | | |
|---|---|---|
| Rent for term | £20,000  p.a. | |
| YP 30 yrs at 7% | 12.41 | £248,200 |
| Reversion to FRV | £25,000  p.a. | |
| YP in perp at 7% def'd 30 yrs | 1.88 | 47,000 |
| | | £295,200 |

*add*

| | | |
|---|---|---|
| Premium | | x |
| | | £295,200 + x |

| | | |
|---|---|---|
| Present interest | = | Proposed interest |
| 357,000 | = | 295,200 + x |
| x | = | £61,800 |
| Premium, say | = | £60,000 |

Hence the Lessee's View suggests £50,000 and the Landlord's View £60,000. A compromise would be needed to arrive at the premium. Since the premium would be a compromise it would be impossible to analyse the premium to show the assumptions made by the parties.

*Example 9–10*

A shop worth £5,000 p.a. is about to be let on full repairing and insuring lease for 35 years. It is agreed between the parties that the lessee shall pay a premium of £10,000 on the grant on the lease. What should the rent be throughout the term assuming quinquennial geared rent reviews?

**Lessee's Point of View**

| | |
|---|---|
| Present rental value | £5,000 p.a. |

*Deduct:*

Annual equivalent of proposed premium of £10,000

£10,000 ÷ YP 35 yrs at 9% and 3% adj. for tax

| | |
|---|---|
| at 35% = £10,000 ÷ 8.66 = | £1,154 p.a. |
| Rent to be reserved | £3,845 p.a. |

**Landlord's Point of View**

Present interest:

| | | |
|---|---|---|
| FRV | £5,000 p.a. | |
| YP perp at 8% | 125 | £62,500 |
| Proposed interest | | |
| Rent Reserved | x | |
| YP 35 yrs at 8% | 11.65 | £11.65 x |
| Reversion to FRV | £5,000 p.a. | |
| YP in perp def'd 35 yrs at 8% | 0.845 | £ 4,225 |
| *add* | | |
| Premium | | £10,000 |
| | | £11.65 x + £14,225 |

| Present interest | = | Proposed interest |
|---|---|---|
| £62,500 | = | £11.65 x + £14,225 |
| £x | = | £4,144 |
| Rent | = | £4,144 p.a. |

In this case the Lessee's View suggests £3,846 p.a. and the Landlord's View £4,144 p.a. No doubt a compromise rent around £4,000 p.a. would be agreed.

In some cases, the parties may agree to a premium being paid in stages, perhaps a sum at the start of the lease, and a further sum after a few years have passed.

The calculation of the premiums or rent follows the same approach as before, the future premium being discounted at the remunerative rate from the landlord's point of view (the landlord will receive it), and the accumulative rate from the tenant's point of view (it is a future liability for the tenant).

## 6. Surrenders and Renewals of Leases

Where the term of a lease is drawing to a close, lessees frequently approach their lessors with a view to an extension or renewal of the term so as to be able to sell the goodwill of the businesses which they have built up or so as to convert the lease into a more marketable form.

The usual arrangement is for the lessee to surrender the balance of his present term in exchange for the grant of a new lease, probably for the term on which he originally held, or for some other agreed period.

The Landlord and Tenant Act 1954, which gives tenants of business premises security of tenure, has not seriously affected this practice. It has undoubtedly strengthened the hand of the lessee in negotiation; but tenants traditionally prefer to hold their premises under a definite lease rather than to rely merely on their rights under the Act which in the end depend upon litigation, and an intending purchaser of a business held on a short lease will be reluctant to rely merely on rights under this Act. As for landlords, a surrender and renewal may give them the opportunity to obtain terms in a new lease not obtainable on a renewal under the 1954 Act, and this could make such a proposal attractive. They may, therefore, be prepared to give way on rental arguments for the new lease in order to achieve better overall lease terms.

If the true rental value of the premises exceeds the rent reserved under the present lease, it is obvious that the lessee will have to compensate the lessor for the proposed extension. The form which this compensation shall take is a matter of arrangement between the parties. It may be agreed that the fresh lease shall be granted at the same rent and on the same terms as at present, the lessee paying a premium to the lessor. More likely, the payment of a capital sum may be dispensed with in consideration of the lessee paying an increased rent throughout the proposed new term, or the parties may agree on the payment of a certain sum as premium and also on an increased rent throughout the term, with the possible additional obligation on the lessee of making some capital improvement to the premises when the new lease is granted.

The valuer usually acts for one side or the other but may be called upon to decide as between the parties the appropriate figure of increased rent or premium.

The calculation should be made from both the lessee's and the lessor's point of view, but the lessee must be credited with the improved value of the property for the unexpired term of the existing lease. For example, suppose the balance of the lessee's term is seven years and he is occupying the premises at a profit rent of £5,000 a year, it is clear that he has a valuable interest in the property which he will be giving up in exchange for the new lease. He is therefore entitled to have the value of the surrendered portion of his term set off against any benefits which he may derive from the proposed extension.

From both the lessor's and the lessee's point of view the principle involved is that of (i) estimating the value of the party's interest in the property, assuming no alteration in the present term was made; and (ii) estimating the value of the party's interest, assuming that the proposed renewal or extension were granted. The difference between these two figures should indicate the extent to which the lessee will gain or the lessor lose by the proposed extension.

If a premium is to be paid, the above method will suggest the appropriate figure. If, instead of a premium, the parties agree on the payment of an increased rent, the required figure may be found by adding to the present rent the annual equivalent of the capital sum arrived at above, although the method used in the examples below is to be preferred.

The terms for the extension or renewal of a lease are seldom a matter of precise mathematical calculation. An estimate of rent or premium made from the freeholder's standpoint usually differs from one made from the lessee's and the figure finally agreed is a matter for negotiation. A valuer acting for either lessor or lessee will carefully consider all the circumstances of the particular case, not only as they affect his own client but also as they affect the opposite party. He will make valuations from both points of view as a guide to the figure which the other side might be prepared to agree in the course of bargaining.

In the three examples which follow, the calculation of premium or rent has been made from both the freeholder's and lessee's points of view. A valuer acting for either party would follow this practice in order to establish not only his own client's position but also the likely requirements of the other party. The effect of income tax at 35% on the open market value of the lessee's present and future interests has been taken into account.

*Example 9–11*

A lessee holding a shop on a full repairing and insuring lease for 40 years, of which six years are unexpired, desires to surrender his lease and to obtain a fresh lease for 25 years at the same rent with geared 5-yearly reviews. The rent reserved under the present lease is £10,000. The true rental value is £25,000 p.a. The rent at reviews will be 40% of rental value.

What premium can reasonably be agreed between the parties?

**(1)  Lessee's point of view**
*Proposed Interest:*

| | | |
|---|---|---|
| Profit rent | £15,000  p.a. | |
| YP 25 years at 8% and 3% | | |
| adj. for tax at 35% | 8.18 | £122,700 |

*Present Interest:*

| | | | |
|---|---|---|---|
| Profit rent | £15,000  p.a. | | |
| YP 6 years at 8% and 3% | | | |
| adj. for tax at 35% | 3.15 | £47,250 | |
| On this basis Gain to Lessee | | | £ 75,450 |

**(2)  Freeholder's point of view**
*Present Interest:*
*Next 6 years*

| | | | |
|---|---|---|---|
| Rent reserved | £10,000  p.a. | | |
| YP 6 years at 7.5% | 4.694 | £46,940 | |

*Reversion*

| | | | |
|---|---|---|---|
| to Full rental value | £25,000 | | |
| YP in perp def'd 6 yrs | | | |
| at 7.5% | 8.639 | £215,980 | £262,920 |

*Proposed Interest:*

| | | |
|---|---|---|
| Next 25 years Proposed rent | £10,000 | |
| YP 25 years at 7.5% | 11.15 | £111,500 |

*Reversion*

| | | | |
|---|---|---|---|
| to Full rental value | £25,000 | | |
| YP in perp def'd 25 years | | | |
| at 7.5% | 2.19 | £ 54,750 | £166,250 |
| On this basis Loss to Freeholder | | | £96,670 |

Depending on the negotiating strength of the parties, the premium will be fixed somewhere between £96,670 and £75,450.


*Example 9–12*

Assume that in the previous example it was agreed that the lessee should pay an increased rent throughout the new term in lieu of a premium; what should that rent be? It is assumed that the rent payable on review will bear the same proportion to rental value as at the start of the lease (a "geared review").

## (1) Lessee's point of view:

*Present Interest* as before                                   £47,250

*Proposed Interest*
FRV                                                           £25,000  p.a.
*less*
Rent Reserved                                                       x
Profit Rent                                                   £25,000  – x
YP 25 yrs at 8.5 and
3% adj. for tax at 35%                                           8.18
                                                      £204,500 – 8.18 x

Present interest        =    Proposed interest
47,250                  =    204,500 – 8.18 x
x                       =    19,223

New Rent, say                £20,000 p.a.

## (2) Freeholder's point of view:

*Present Interest*
as before                                                    £262,920

*Proposed Interest*
Rent under Lease                          x
YP 25 yrs at 7.5%                       11.15                 11.15  x

*Reversion* to FRV                   £25,000  p.a.
YP perp def'd
25 yrs at 7.5%                          2.19                  54,750
                                                      11.15 x + 54,750

Present interest        =    Proposed interest
262,920                 =    11.15 x + 54,750
x                       =    18,669
New Rent, say                £18,700 p.a.

Depending on the negotiating strength of the parties, the rent will be fixed between £20,000 and £18,700.

The above approach from the freeholder's point of view rests upon the assumption that, so long as the capital value of his interest is not lowered, he will be satisfied with the arrangement.

Where, as part of a bargain for a renewal or extension of a lease, the lessee is to make an expenditure upon the property which will benefit the value of the lessor's reversion, the lessee must be given credit for the value due to his expenditure which will enure to the lessor at the end of the term, and the sum must be taken into account when considering the cost of the new lease to the lessee.

*Example 9–13*

A warehouse in a city centre is held on a lease having five years unexpired at £32,000 a year.

The present rental value is £40,000 p.a. The lessee is willing to spend £100,000 upon improvements and alterations affecting only the interior of the building, which will increase the rental value by £12,000 p.a., provided the lessor will accept a surrender of the present lease and grant a new lease for a term of 30 years. The lessee is willing to pay a reasonable rent under the new lease, or a premium. The lessor is agreeable to grant the new lease provided that the rent is fixed at £35,000 p.a., that a proper premium is paid and that the new lease contains a covenant that the lessee will carry out the improvements. What premium, if any, should you advise the lessee to offer if the new rent is £35,000 p.a.?

**(1) Lessee's point of view:**

*Proposed Interest*

| | | |
|---|---|---|
| Rental value | £40,000  p.a. | |
| *add* | | |
| value due to outlay | £12,000 | |
| | £52,000 | |
| *less* | | |
| Rent payable | £35,000 | |
| Profit rent | £17,000  p.a. | |
| YP 30 years at 9% and 3% | | |
| adj. tax at 35% | 8.17 | £138,890 |
| Deduct expenditure on | | |
| improvement | | £100,000 |
| Value of proposed interest | | £ 38,890 |

*Present Interest*

| | |
|---|---|
| Rental value | £40,000 |
| Rent paid | £32,000 |
| | £ 8,000 |

| | | |
|---|---|---|
| YP 5 years at 9% and 3% | | |
| adj. tax at 35% | 2.63 | £21,040 |
| *Gain to lessee* | | £17,850 |

## (2) Freeholder's point of view:
*Present Interest*

| | | |
|---|---|---|
| First 5 years | £32,000 | |
| YP 5 years at 8% | 3.99 | £127,680 |

## Reversion
*Note*: Since the estimated increase in rent due to improvements represents 12% on the sum expended, and since this is greater than the rate of interest which might reasonably be expected from such a property as this when let at its full rack rental, it would be worth the landlord's while, at the end of the present lease, to carry out the improvements at his own expense in order to obtain the increased income. The reversion can therefore be valued on the basis that the landlord would carry out the improvement work on the expiry of the lease, as follows:

| | | |
|---|---|---|
| b/fwd | | £127,680 |
| Reversion to rental value after improvements | 52,000 p.a. | |
| YP in perpetuity at 8% | 12.5 | |
| | £650,000 | |
| *Deduct* | | |
| cost of improvements | £100,000 | |
| | £550,000 | |
| PV £1 in 5 years at 8% | 0.68 | £374,000 |
| Value of Present Interest | | £501,680 |
| *Proposed Interest* | | |
| *First 30 years* | £35,000 | |
| YP 30 years at 8% | 1,126 | £394,100 |
| *Reversion* | | |
| to full rental value | £52,000 | |
| YP perp deferred 30 years at 8% | 1.24 | £ 64,480 |
| | | £458,580 |
| *Loss to freeholder* | | £ 43,100 |

Depending on the negotiating strength of the parties, the payment would be agreed at around £30,000.

Suppose that it had been decided between the parties that no premium should be paid and that the rent under the new lease should be adjusted accordingly. What would be a reasonable rent in these circumstances?

**(1)  Lessee's point of view:**

| | | |
|---|---:|---:|
| Rental value after improvements have been carried out | | £52,000  p.a. |
| *Deduct*  annual equivalent of | | |
| Cost of improvements | £100,000 | |
| and | | |
| Present Interest (as before) | £21,040 | |
| | £121,040 | |
| ÷ YP 30 years at 9% and 3% adj. tax at 35% | 8.17 | £14,800 |
| *Reasonable rent from lessee's viewpoint* | | £37,200  p.a. |

**(2)  Freeholder's point of view:**

| | | |
|---|---:|---:|
| Value of present interest, as above | | £501,680 |
| *Deduct* | | |
| value of proposed reversion to full rental value after improvements have been carried out | £52,000  p.a. | |
| YP in perp def'd 30 years at 8% | 1.24 | £64,480 |
| Value of proposed term | | £437,200 |
| ÷ YP 30 years at 8% | | 11.26 |
| *Reasonable rent from freeholder's viewpoint* | | £38,900  p.a. |

Again a negotiated settlement could be reached at an initial rent of around £38,000 p.a.

## 7.  Traditional Valuation Techniques and DCF

The method of valuation adopted in this chapter is what is termed by most as the "traditional method" applying the all risks yield. The Royal Institution of Chartered Surveyors published an Information Paper in 1997 entitled "Commercial Investment

Property: Valuation Methods". Although the Paper "is not advocating the revocation of traditional techniques" it is critical of them and advocates a wider use of discounted cash flow techniques (discussed in Chapter 10). The Paper is essential additional reading for valuers involved in the valuation of commercial property.

# Chapter 10

# Discounted Cash Flow

## 1. Generally

The principle underlying the valuation of a future income flow within the investment method of valuation is that the future income should be discounted at an appropriate rate of interest to determine its present value. The valuation therefore represents the discounted future cash flow. This same principle has been adopted to evaluate investment proposals in business generally, with variations developed to meet different situations, under the general term of discounted cash flow (DCF) techniques.

In recent years these DCF techniques have been seen to be of assistance in decisions relating to property investment and appraisal. Their principal application has been in making comparisons between choices of investment. However, arguments that a DCF approach may be more reliable in making a valuation were given impetus by support for this approach in the Mallinson Report published in March 1994.[1] A detailed consideration of the use of DCF is contained in the Information Paper published by the RICS in May 1997[2] which is essential reading for any member of the RICS, and recommended reading for those who are not.

DCF calculations involve the discounting of all future receipts and expenditures similar to the investment method of valuation, but they can readily be used to allow for inflation, taxation and frequent changes in the amount of income and receipts as may be required. In this way, explicit assumptions are made compared with the implicit assumptions contained in the traditional all-risks yield approach.

---

[1] "Report of the President's Working Party on Commercial Property Valuations" published by the Royal Institution of Chartered Surveyors, March 1994.
[2] "Commercial Investment Property: Valuation Methods", an Information Paper published by the Royal Institution of Chartered Surveyors.

Two principal methods of DCF calculations have emerged, the Net Present Value (or NPV) method and the Internal Rate of Return (or IRR) method.

## 2. Net Present Value Method

In this method the present value of all future receipts from a proposed investment is compared with all future outgoings. If the present value of receipts exceeds the present value of outgoings, then the investment is worthwhile since the return must exceed the yield adopted in the calculation (the discount rate). If the same calculations are applied to different investment possibilities, the investment which produces the greatest excess of present value of receipts over outgoings is the most profitable. This is clearly of great assistance in reaching a decision on the choices available although it does not follow that the most profitable will be pursued. For example, it may be that the "best scheme" requires a large amount of capital to be invested whilst other considerations make a scheme with a smaller capital outlay more attractive to the investor.

### The Discount Rate

As has been said the future receipts and outgoings are discounted at an appropriate rate of interest. The question then arises as to what is the appropriate discount rate. When calculating the worth of an investment to an individual investor (see Chapter 12) the discount rate will be the investor's target rate. When the method is used to determine a general market value a discount rate must be adopted which is likely to be at or around the majority of individual target rates.

The discount rate is commonly arrived at by adopting a risk-free rate and making allowances for the risks associated with the property in question.

The general consensus is that a risk-free rate can be taken as the gross redemption yield on medium-term government gilts. This has ranged from 7% to above 10% over recent years. To this must be added a yield to reflect the general risks of investing in property, such as liquidity, depreciation, changes in law, and then a further addition is made to reflect the risks specific to the property in question such as strength of covenant, location, risks of voids. So the discount rate may be taken to be risk-free yield (say) 7.0%, add for property market risks (say) 2.0% and for risks specific to the property in question (say) 1.0%, producing a discount rate of 10.0%.

## Rate of Growth

The DCF calculation makes specific assumptions as to the rate at which the various elements of the calculation will change in the future. Given that inflation is the norm, the changes will be the rate at which they will grow in the future. The rates of growth may be different. For example, it may be assumed that rents will grow faster than building costs.

There are methods of forecasting which are available to the Government and other economic monitors which assist in providing estimates of growth but such estimates are not commonly accessible to the general market. The valuer will therefore often rely on his judgment as to growth rates. In the case of rental values, where the valuer knows the all risks yield appropriate to a property and also the discount rate, an implied rate of growth of the rental values can be determined by formula to determine the predicted rental value at each rent review.

The all risks yield equals the discount rate minus the annual sinking fund to recoup the growth in rental value over the review period at the discount rate. So, assume that the all risks yield is 6%, the discount rate is 10.0%, and there are five-yearly rent reviews. Then:

$$6\% = 10.0\% - (\text{asf 5 yrs @ 10\% } (0.1638) \times \text{growth over 5 yrs})$$

| | |
|---|---|
| Growth over 5 years | = 24.62% |
| Annual growth rate | = 4.5% |
| (Amount of £1 for 5 years @ 4.5% | = 1.2462) |

## Cash Flow Period

When valuing a property by either the conventional method or by DCF it is apparent that early income has a higher discounted value than later income. A simple inspection of the YP single rate table shows, for example, that YP 20 yrs @ 6% is 68.82% of YP in perp at 6%; that YP 20 yrs @ 10% is 85.14% of YP in perp @ 10%; and so on. This demonstrates that a high proportion of value is likely to arise over the early years. As a result it is normal to prepare a detailed calculation of rents and costs for say 15 to 25 years and then to adopt the market value at the end of that period. This is the exit value.

*Example 10–1*

Value the freehold interest in shop premises which were recently let for 20 years at £20,000 p.a. with five-yearly reviews. All risks yield is 6%.

*1. Conventional Method*

| | |
|---|---:|
| Rent Reserved and Full Rental Value | £ 20,000  p.a. |
| YP in perp @ 6% | 16,667 |
| Value | £333,334 |

*2. DCF*

Discount Rate 10.0% Growth Rate 4.5% p.a.

| Years (£) | Cash Flow | YP @ 10% | PV @ 10% | Capital Value (£) |
|---|---|---|---|---|
| 1–5 | 20,000 | 3.7908 | | 75,816 |
| 6–10 | 24,884 | 3.7908 | 0.6209 | 58,570 |
| 11–15 | 30,961 | 3.7908 | 0.3855 | 45,249 |
| 16–20 | 38,522 | 3.7908 | 0.2394 | 34,959 |
| 20–perp | 47,929 | *16.6667 | 0.1486 | 118,704 |
| Net Present Value | | | | £333,294 |

*YP in perp at 6% to establish exit value.

In the valuation of a rack rented investment such as this, the two approaches give the same result (the difference in the examples arises from rounding off). However, the DCF approach might be favoured if, at the end of the lease, it is assumed that the tenant will vacate, the premises will need to be refurbished, and this, coupled with a rent free period on a new letting, will lead to a loss of one year's rent.

## Example 10–2

Same facts as Example 10–1. On expiry of the lease the freehold owner will need to spend £80,000 on the premises at current prices. The work will take six months. It is expected to take three months to re-let and a three-month rent free period is likely. Building costs are expected to rise at 3% p.a.. The rental value will be unchanged by the work.

| Years | Cash Flow (£) | YP @ 10% | PV @ 10% | Capital Value (£) |
|---|---|---|---|---|
| 1–20 | (as before) | | | 214,594 |
| 21 | (146,400) | — | 0.1452 | (21,257) |
| 22 | 49,547 | 16.667 | 0.1385 | 114,373 |
| Net Present Value | | | | £307,710 |

This example shows how expenditure can be allowed for. If there were several items it would be prudent to split the cash flow column into "cash in" and "cash out" columns for clarity, and similarly with the capital sums.

A further example where DCF can be used is in the valuation of a reversionary interest where explicit assumptions as to rental growth can be adopted.

## Example 10–3

Same facts as in Example 10–1 save that the property was let 12 years ago, the current passing rent fixed two years ago is £17,000 p.a., and the lease has eight years unexpired.

### 1. Conventional Method

| | | |
|---|---|---|
| Term | | |
| Rent Reserved | £17,000 p.a. | |
| YP 3 yrs @ 6% | 2.6730 | £45,441 |
| Reversion | | |
| to FRV | £20,000 p.a. | |
| YP in perp def'd 3 yrs @ 6% | 13.9937 | £279,874 |
| Value | | £325,315 |

2. *DCF*

| Years | Cash Flow (£) | YP @ 10% | PV @ 10% | Capital Value (£) |
|---|---|---|---|---|
| 1–3 | 17,000 | 2.4869 | | 42,277 |
| 4–8 | 22,824 | 3.7908 | 0.7513 | 65,003 |
| 8–perp | 28,397 | 16.6667 | 0.4665 | 220,787 |
| Net Present Value | | | | £328,067 |

The rent after four years is the current FRV of £20,000 p.a. increased at 4.5% p.a. It is assumed that there is no void on expiry of the lease.

As the values are close in this example it may be said that there is no merit in changing from the conventional method. Even so, the DCF approach does require the valuer to be more explicit, which may help to consider factors underlying values more carefully. In some cases where the conventional method is difficult to apply, such as where there are very few comparables or where the pattern of receipts and payments is irregular, the DCF approach can make the task of producing a value easier. The DCF approach is the recommended method to determine worth (see Chapter 12).

There are several DCF programmes available which take away the labour of preparing the calculation and which allow for greater sophistication with the calculation, such as rents payable quarterly in advance, allowances for tax and other items not easily accommodated by the conventional method.

### 3. Internal Rate of Return Method

As an alternative to the NPV method, an analysis can be carried out which discounts all future receipts and payments of a project at a discount rate whereby the discounted receipts equal the discounted payments. This discount rate will then show the actual rate of return on the capital invested in the scheme – the "internal rate of return". At this point the NPV will of course be nil.

The approach to this method is to adopt the same approach as illustrated in Example 10–1, but, by trial and error, to use different rates of interest until the correct rate is found.

A shop as in Example 10–1 has been offered for sale at £300,000. Calculate the Internal Rate of Return (IRR).

Adopting the same assumptions regarding estimated growth and exit value, it is clear that the IRR must be more than 10% as, at 10%, the Net Present Value is above £300,000. Try 12%.

*Example 10–4*

| Years | Cash Flow (£) | YP @ 12% | PV @ 12% | Capital Value (£) |
|-------|---------------|----------|----------|-------------------|
| 1–5 | 20,000 | 3.6048 | | 72,096 |
| 6–10 | 24,884 | 3.6048 | 0.5674 | 50,897 |
| 11–15 | 30,961 | 3.6048 | 0.3220 | 39,308 |
| 16–20 | 38,522 | 3.6048 | 0.1827 | 25,370 |
| 21–perp | 47,929 | 16.6667 | 0.1037 | 82,837 |
| Total | | | | £270,508 |

At 10% the NPV is £33,294 above £300,000, and at 12% it is £29,492 below. By interpolation, try 11%.

| Years | Cash Flow (£) | YP @ 12% | PV @ 12% | Capital Value (£) |
|-------|---------------|----------|----------|-------------------|
| 1–5 | 20,000 | 3.6959 | | 73,918 |
| 6–10 | 24,884 | 3.6959 | 0.5935 | 54,583 |
| 11-15 | 30,961 | 3.6959 | 0.3522 | 40,302 |
| 16-20 | 38,522 | 3.6959 | 0.2090 | 29,756 |
| 21-perp | 47,929 | 16.6667 | 0.1240 | 99,053 |
| Total | | | | £297,612 |

This is close to £300,000 so that the IRR can be taken to be 11%. If an even more accurate figure is required this can be achieved by further interpolation.

At 10% NPV = £33,294 above: at 11% NPV = £2,388 below: range £35,682. Assume a constant rate of change, then the exact rate is:

$$10\% + \frac{33,294}{35,682} \text{ of } 1\% \text{ i.e. } 0933\% = 10.933\%$$

This is not of course absolutely exact since the difference in discount rates is based on geometric rather than linear expansion. However, it is probably sufficiently accurate for most cases.

This can be expressed as a formula which provides for linear interpolation. The formula is:

$$R1 + \left\{ (R2 - R1) \times \frac{NPVR1}{NPVR1 + NPVR2} \right\}$$

Where R1 is the lower rate and R2 the higher rate

Hence $\qquad 10 + \left\{ \dfrac{1 \times 33{,}294}{35{,}682} \right\} = 10.933$

This example shows one of two things. If the valuer is correct in the choice of the all risks yield at 6% and rental growth of 4.5% per annum then the price is below its true value. A purchaser would therefore realise a yield above target yield.

On the other hand if the value is really £300,000 this would be because the all risks yield is above 6%, or the market's estimate of rental growth is below 4.5% – or a combination of the two.

The IRR is also known as the Equated Yield since it provides the yield which will be obtained at any chosen price after making explicit assumptions as to changes in rents and outgoings and other matters in the future.

This contrasts with the yield determined in a similar manner so that discounted income flow equals the NPV but without making any adjustment for changes in the future. This is known as the Equivalent Yield and is applied in the conventional valuation methods.

### 4. Comparison of NPV with IRR Approaches

As has been shown, the NPV method compares the future costs and receipts on a discounted present capital value basis, whereas the IRR method shows the return earned on an investment. In the above examples the application of each method to the same investment shows that the investment is worthwhile on either basis. This will not always be so as sometimes it will be found that one method shows an investment proposal to be profitable whilst the other method suggests that it will show a loss, or that in deciding between investments A and B, NPV method favours A whilst IRR favours B. The reasons for this will lie in the choice of discount rate or in the pattern of costs and revenue.

The resolution of this contradiction can be found in the standard tests on DCF which provide various ways of overcoming the problem. As a general rule, however, the IRR approach will be favoured by a valuer who is using DCF for analysis since it will show the yield of the investment which will enable him to compare this with yields on other investments which he will know. On the other hand he will prefer the NPV method if he wishes to compare the value of an investment proposition with others since the investment method of valuation and the NPV method are closely allied in their approach.

## 5. Application of DCF

The principal use of DCF lies in relation to investment decisions whereby a choice between alternatives can be made. In the case of development this may be a choice between development propositions on different sites or between different development schemes on the same site. For example, a landowner may be able to consider developing a site with 100 houses over three years or 140 flats over four years or various other permutations. The analysis of each proposal by DCF will show him which scheme is the more profitable.[3]

Again an investor may be offered the choice of different investments showing different rental patterns. Although he can value each investment by the investment value approach, he will be able to make a further comparison between them by DCF analysis to show the IRR of each investment. In so doing he may if he wishes allow for projected rental increases. This approach has been developed by the determination of the "equated yield" which is considered further in Chapter 12.[4]

---

[3] For an example of the application of DCF to a development property see the chapter on "Development Properties" in *Valuation: Principles into Practice*, 4th Edition, Editor W.H. Rees (Estates Gazette).

[4] For further reading on DCF the reader is referred to *The Valuation of Property Investments*, 5th Edition by N. Enever and D. Isaac (Estates Gazette) and to the RICS Information Papers on "Calculation of Worth" and "Commercial Investment Property: Valuation Methods".

# Residual Method

## 1. Concepts of the Residual Method

The various methods of valuation commonly employed are set out in Chapter 2. A more detailed consideration of the direct comparison and investment methods is made in Chapters 4 to 10 which now allows further consideration of the residual method.

The residual method is adopted in the valuation of development property. This may be of bare land which is to be developed or of land with existing buildings which are either to be altered and improved, an exercise commonly termed refurbishment, or to be demolished and redeveloped with entirely new buildings.

The method works on the premise that the price which a purchaser can pay for such property is the surplus after he has met out of the proceeds from the sale or value of the finished development his costs of construction, his costs of purchase and sale, the cost of finance, and an allowance for profits required to carry out the project. This can be expressed as follows:

|        | Proceeds of Sale                    |
|--------|-------------------------------------|
| *less* | Costs of Development, and Profits   |
| *equals* | Surplus for land                  |

These elements can be considered in turn in relation to an example.

*Example 11–1*

Value the freehold interest in 10 Main Road. The property comprises a house now vacant. Permission has been given to build a shop with offices above. The shop will have a frontage of 6 m and an internal gross depth of 20 m. Two floors of offices will be built over the shop with separate access from Main Road and providing 90 m² net internal floor space per floor. The house will be demolished.

The offices are expected to let at £100 per m² and the shop at £40,000 p.a. and the resultant investment should sell on the basis of a 7.5% return.

## Proceeds of Sale

These are known as the Gross Development Value (GDV) and arise from the disposal of the developed property. In the case of houses they will be the price anticipated for each unit determined by the direct comparison method. In commercial developments they will be the anticipated price which will be obtained on a sale, usually after they have been let to create an investment; they will thus be determined by the investment method as previously described.

It should be remembered that the valuation approach is to determine the surplus available after meeting costs. The proceeds of sale are the whole of the anticipated money to be realised from the development. It is true that they will not be receivable until the work is completed which may be some considerable time in the future. Nonetheless it would be incorrect to discount the proceeds to their present-day value. This is because the cost of holding the property is taken as a cost of the development and therefore to discount the proceeds of sale would be to double deduct.

What the valuer is seeking to establish is the ultimate size of the development "cake" which he can then cut up into the various slices needed for costs and profits and so establish how much of the cake remains as the land slice.

Where a purchaser intends to occupy the property, or to let the property and then retain the investment, he will not actually sell and so receive the moneys. Nonetheless it is necessary to determine the realisable value of the development to carry out the residual valuation.

Applying this approach to Example 11–1; the Proceeds of Sale will be:

| *Full Rental Value* | | |
|---|---|---|
| Shop | £40,000 | p.a. |
| Offices 180 m$^2$ at £100 per m$^2$ | £18,000 | |
| | £58,000 | p.a. |
| YP in perp at 7.5% | £13,333 | |
| | £773,333 | |
| GDV = say, | £775,000 | |

## 3. Costs of Sale

The main costs incurred in a sale of the interest will be the agents' fees, including advertising costs, and legal fees in the conveyance.

In practice, the general level of fees on a sale is in aggregate around 3% of sale price.

In the case of investment properties it may be considered desirable or even necessary in some cases to let the property before selling. If so, the agents' and legal fees incurred in the letting should be brought into account as well. Agents fees will normally be around 10% of the rents obtained or perhaps around 15% if two or more agents are instructed. The aggregate fees will probably be around 20% of the rents obtained if promotion costs are included.

Where a purchaser intends to retain the property, it has been shown that the realisable value needs to be determined nonetheless. Since it is the net realisable sale proceeds which are required, the costs of sale which would be incurred are incorporated in the valuation.

*Costs of Sale*

| | |
|---|---|
| *Sale Costs* | |
| Agents' and Legal Fees, say 3% | |
| of £775,000 = | £23,250 |
| | |
| *Letting Costs* | |
| Agents' and Legal Fees and | |
| promotion say 20% of £58,000 | £11,600 |
| Total Costs | £34,850 |

Where appropriate, VAT on these fees must also be allowed for (i.e. when the developer is unable to recover VAT – see Chapter 22).

## 4. Costs of Development

The major items normally met are the actual cost of building the development, and funding costs. Certain other miscellaneous items may be met on occasions.

### (a) Cost of Building

In the preliminary stages the costs of development will need to be estimated. Estimates are normally based on the prevailing costs of building per square metre of the gross internal floor area. As the scheme details become more advanced it may be appropriate to prepare a priced specification or even a priced bill of quantities. In addition to the actual costs of building, the professional services of the design team are payable. The membership of the team depends on the nature and scope of the development but commonly includes an architect and quantity surveyor and often one or more engineers. In the more complex schemes, structural, electrical, and heating and ventilation engineers will be needed. The fees payable depend on the circumstances but usually vary between 8% and 14% of the building costs, with 12% as the average.

### (b) Miscellaneous Items

All sites are different and have unique features which may require various special costs. Typically such items include costs of demolishing existing buildings, which will depend on the nature of construction and the salvage value; costs of obtaining possession, either by compensation to tenants on the site or even buying in minor interests; costs of agreeing compensation to neighbours such as buying rights over the land like easements or agreeing party wall rights and compensation; costs of providing above average quality boundary works; exceptional costs such as site clearance or filling of uneven land, and diversion of services; and off-site costs such as highway improvements required as a condition for the grant of planning permission. As can be seen, there are many possible problems which may need to be overcome, this list being far from exhaustive. Where possible, an estimated cost should be adopted. As these items are sometimes difficult to predict so that they are not known at the time of valuation, it is possible to allow some general sum for such contingencies if the need for an allowance can be forecast. On the other hand, it can be said that these are part of the general risks of development which are reflected in the allowance for profits, which also allows for the risk of the unknown.

### (c) Costs of Finance

Considerable sums of capital need to be spent on the carrying out of a development. Normally this money is raised from banks or other lending institutions, or it may be loaned as part of an overall deal with an investing institution such as an insurance company or pension fund, particularly in the case of medium to large scale commercial developments. The cost of borrowing the money which will be repaid on the completion and sale of the development is the interest charged at an agreed rate plus in many cases a commitment or fund-raising fee of around 1% of the money to be provided. The rate of interest depends on the prevailing rates being charged and will also vary with the status of the borrower and the risks attached to the development scheme. They commonly range between 1% to 4% above the base rate of one of the clearing banks. These rates have shown considerable fluctuation in recent years within a range of around 6% to 20%. The valuer will need to be aware of current rates at the time of the valuation.

In some instances the developer may have raised money on a long-term basis which means that the rate of interest is now low compared with prevailing rates, or he may be able to provide money from his own resources. Nonetheless the prevailing borrowing rate should be adopted in the valuation as this is the opportunity cost of the capital and it reflects the market for the site. The rate of interest chosen will vary according to the type of scheme and the size of the likely developer. A small scheme such as the one in this example will attract small developers who have, in general, a higher cost of interest than would be the case for a major developer with access to institutional funding. If the preferential rate is adopted, or none at all, this will produce a value to that person which is not necessarily the open market value.

Once the rate of interest is known, the interest costs can be determined. The money to be borrowed relates to two items, building costs and land costs. As to land costs, these will be incurred at the start, so that the money is borrowed at the start and interest runs for the whole period of development. On the other hand, money required for building works will only be needed in stages and, as a rule of thumb which experience shows is normally reasonably accurate, it may be assumed that the whole of the building money is borrowed for half the period of development. When the details of the development are known, it may be necessary to change this rule of thumb and adopt an S curve reflecting uneven building cost draw downs. In the case of housing development the developer may obtain revenue from sales of houses as the development proceeds with consequent savings in the money to be borrowed. The implications of this are considered in Chapter 16.

Turning again to Example 11–1, the Costs of Development are:

(a) *Building Costs*
Shop 1,200 m$^2$ (gross internal)
at £300 per m$^2$ =                    £360,000
Offices say 205 m$^2$ (gross
internal) at £600 per m$^2$ =          £123,000
                                       £483,000

*add*
Professional Services, say 12%         £57,960          £540,960

(b) *Demolition Costs*
Assume                                                  £ 1,200

(c) *Costs of Finance*
Assume development will
take 1 year and interest is 8%

Demolition Costs £1,200 for
1 year at 8%                           £96

Building Costs £540,960 for
say 6 months at 8%                     £21,638          £21,734
Total Costs of Development                              £563,894

As the value of the land is not yet determined, it is inappropriate to
calculate interest on this element at this stage. This will therefore be
considered at the end of the calculation.

## 5. Development Profits

As for any risk enterprise, a person undertaking a development
will seek to make a profit on the operation. Target levels of profit
will depend on the nature of the development and allied risks, the
competition for development schemes in the market, the period of
the development and the general optimism in relation to that form
of development. Consequently, it is not possible to lay down firm
limits of required profit levels. The profit is usually related to the
costs involved but sometimes to the development value. The profit
is the gross profit to the developer before meeting his general
overheads and tax. Hence developers may seek say 20% gross
profit on the capital invested, namely building costs and land costs,
or say 15% of the development value.

Although the profits are related to costs, at this stage in the
valuation the land costs are unknown so profits on these cannot be

calculated at this stage. The better approach is therefore to calculate profit as a percentage of the proceeds of sale and then to check this calculation later as a reference to the return on total costs.

Hence turning to Example 11–1

$$\text{Developer profit} = 15\% \times £775,000 = £116,250$$

## 6. Surplus for Land

At this stage the valuer has determined the net proceeds of sale, and the total cost of development and profits thereon. The difference between these figures represents the sum available to spend on land costs including interest. In some cases building costs will exceed net proceeds of sale. If so, this shows that there is negative value for that development and thus that the land is not suitable for development, unless some other form of development would be profitable.

Turning to Example 11–1 the surplus available for Land Costs is:

| | | |
|---|---|---|
| Proceeds of Sale GDV | | £775,000 |
| *less* | | |
| Costs of Sale @ say 3% | | £23,250 |
| Net Proceeds of Sale | | £751,750 |
| *less* | | |
| *Costs of Development* | | |
| (i)   Building costs | £540,960 | |
| (ii)  Demolition | 1,200 | |
| | £542,160 | |
| (iii) Interest on (i) & (ii) | £21,734 | |
| (iv)  Total letting costs | £11,600 | |
| (v)   Developer profit @ 15% GDV | £116,250 | £691,744 |
| Surplus for Land Costs | | £60,006 |

The land costs comprise three items. First there is the price to be paid for the land, the very purpose of the valuation. Secondly, there are the professional fees and perhaps stamp duty in relation to the purchase. These fees will generally be for an agent and for legal services in the conveyance. These, together with stamp duty, are likely to be approximately 3.75% of the price paid. The third item is the interest on the money borrowed to purchase the land to be repaid on sale, calculated for the period of development.

All of these items relate to the actual price of the land which in turn depends on the three items. The simple way of apportioning the surplus between them is to express the land price as a symbol, say x, and then solve the subsequent equation.

Turning to Example 11–1

|     |                                               |          |
|-----|-----------------------------------------------|----------|
| (a) | Land Price                                    | 1.0000x  |
| (b) | Fees on Land Purchase at                      |          |
|     | 3.75% of price =                              | 0.0375x  |
| (c) | Finance                                       |          |
|     | Interest at 8% for period of                  |          |
|     | development on (a) and (b)                     |          |
|     | = $1.0375x \times 1$ yr at 8% =               | 0.083x   |
|     | Land Surplus Total                            | 1.1205x  |
|     | But the Land surplus is                       | £60,006  |
|     | $\therefore 1.1205x = £60,006$                |          |
|     | $\therefore x =$                              | £53,553  |

Hence Value of Land is, say,                           <u>£53,500</u>

An alternative way of calculating the net value of the land is as follows:

|                                         |          |
|-----------------------------------------|----------|
| Surplus for land costs                  | £60,006  |
| *Less*                                  |          |
| Allowance for interest                  |          |
| $\times$ PV of £1 in 1 yr @ 8%          | 0.92593  |
|                                         | £55,561  |
| *Less*                                  |          |
| acquisition costs @ 3.75% $\div$        | 1.0375   |
|                                         | £53,553  |

Hence Value of Land is, say                            <u>£53,500</u>

Thus the Valuation is completed. The whole valuation is:

*Proceeds of Sale (GDV)*
FRV

|                                           |                |     |          |
|-------------------------------------------|----------------|-----|----------|
| Shop                                      | £40,000 p.a.   |     |          |
| Offices 180 m$^2$ at £100 per m$^2$       | £18,000        |     |          |
|                                           | £58,000        |     |          |
| YP in perp at 7.5%                        | 13.333         |     |          |
|                                           |                | say | £775,000 |

*less*
Costs of Sale @ 3%                                             <u>£23,250</u>
Net Proceeds of Sale                                          £751,750
*less*
Costs of Development
(i)   Building costs
      Shop 120 m² @ £300 per sq ft      £360,000
      Offices 205 m² @ £600 per m²      <u>£123,000</u>
                                        £483,000
(ii)  Professional fees @ 12%, say       <u>£57,960</u>
                                        £540,960
(iii) Demolition Costs                   <u>£1,200</u>
                                        £542,160
(iv)  Cost of Finance
      £1,200 for 1 yr @ 8%                   £96
      £540,960 for say 6 months
      @ 8%                               <u>21,638</u>
                                        £563,894
(v)   Agent's and legal fees on letting
      @ 20% of £58,000                   £11,600
(vi)  Developer profit @ say
      15% GDV                           <u>£116,250</u>        <u>£691,744</u>
Land Surplus                                                  £60,006
*less*
Allowance for interest 1 yr @ 8%
× PV of £1 in 1 yr @ 8%                                  ×  <u>0.92593</u>
                                                           £ 55,561
*less*
Acquisition costs @ 3.75%                               ÷  <u>1.0375</u>
                                                           £ 53,553

Hence Value of Land is, say                                <u>£53,500</u>

This example is comparatively simple and no allowance has been
made for VAT, delays in disposals, phased developments or
inflation. VAT would normally be recoverable if the developer has
waived the exemption available to property and has registered the
building for VAT purposes (see Chapter 22) or is zero rated (e.g.
new residential developments). Where VAT is not recoverable then
an appropriate allowance should be added to the development
costs.

Where delays are anticipated in the disposal of the property, additional interest should be allowed for on both the development costs and the land. Thus, if it is assumed that the letting and sale of the property in the above example would take six months to complete then the effect on the value would be as follows:

| | | |
|---|---|---|
| Net proceeds of sale | | £751,750 |
| *less* | | |
| Development costs as above | | |
| (including interest) | £563,894 | |
| Interest for 6 months @ 8% | £22,556 | |
| Agents fees on letting | £11,600 | |
| Developer profit @ 15% GDV | £116,250 | £714,300 |
| Land Surplus | | £37,450 |
| *less* | | |
| Allowance for interest 18 months | | |
| @ 8% | | |
| = × PV of £1 in 18 months @ 8% | | × 0.8903 |
| | | £33,342 |
| *less* | | |
| Acquisition costs @ 3.75% | | ÷ 1.0375 |
| | | £32,137 |
| Hence Value of Land is, say | | £32,000 |

Where there is a phased development then each phase can be valued separately, but a further allowance must be made for interest on the land value for the second and subsequent phases to reflect the additional holding cost of the land. If it is assumed that three shops with offices over are to be developed but with work starting on the second shop six months after the start of the first, and of the third shop six months later still, the calculations would be as follows:

| | | |
|---|---|---|
| Site value for Shop 1, say | | £32,000 |
| Site value for Shop 2 | £32,000 | |
| × PV of £1 in 6 months @ 8% | × 0.962 | £30,784 |
| Site value for Shop 3 | £32,000 | |
| × PV of £1 in 1 yr @ 8% | × 0.926 | £29,632 |
| | | £92,416 |
| Hence Value of Land is, say | | £92,500 |

No allowance has been made for inflation because of the normal valuation assumption that all values and costs are taken at present values. In times of significant inflation this approach may not be correct and this will be further considered in Chapter 16.

In the calculation above of the land value of £32,000, the profit was taken as £116,250, being 15% of the Gross Development Value (GDV). The return on total costs equates to:

$$116,250 \div [(714,300 - 116,250) + 37,450]$$
$$= 116,250 \div 635,500$$
$$= 18.29\%$$

If the valuer wishes to use return on cost rather than return on GDV to arrive at the profit allowance, then this can be done by substituting the relevant figure for the £116,250 used in the example [i.e. say 18.29% × (563,894 + 22,556 + 11,600) = 18.29% × 598,050 = £109,383] and the resultant land surplus will include the next tranche of developer profit [i.e. £751,750 − (£598,050 + £109,383) = £751,750 − £77,433 = £44,317 land surplus including developer profit]. This figure can then be reduced by dividing by 1.1829 to give the land surplus excluding developer profit, i.e. £37,465 [the difference from the figure in the example, which is £37,450, is the result of rounding off to 18.29%. If the actual figure of 18.29268% is adopted, the land surplus is £37,450.]

An alternative approach to the profit on cost basis is to apply the required percentage to the building costs and then to apportion the Land Surplus algebraically adopting x as the land value and applying the required percentage to x as well as the acquisition cost and interest. For example, adopting this approach to the above calculations:

| | | |
|---|---|---:|
| Land Surplus (including developer profit) = | | £44,317 |
| Let land value be x. | | |
| Then | | |
| (a)  Land Price = | 1.0000x | |
| (b)  Fees at 3.75% = | 0.0375x | |
| (c)  Interest for 18 months | | |
|         @ 8% of (a) + (b) = | 0.1266x | |
| | 1.1645x | |
| (d)  Profit @ 18.29% = | 0.2130x | |
| | 1.3775x  = | 44,317 |
| | x  = | £32,172 |

(the difference from the value of £32,450 again arises from rounding off the percentage figure)

It is clear that this valuation contains many figures all of which are based on estimates of cost or value or derived from such estimates. As with any estimates, one person's estimate may differ from another so in that way they are all variables. Given a calculation based on a large number of variables, the actual range of answers which can be produced is wide. This uncertainty is the method's weakness but it is one which is acceptable so long as the estimates are prepared with as much information as is available to narrow possible errors.

This method of valuation is disliked by the Lands Tribunal in compensation cases (see Chapter 26) because it is not tested by "haggling in the market". In the open market, however, the residual method will continue to be the main cornerstone of many opinions of value, particularly those involving land for development or redevelopment. Nevertheless, the method is weak because of the speculative nature of the main items, a small variation in any of which can make a large difference to the answer. Care must therefore be taken to see that the items in the valuation are as correct as possible and as verifiable as it is possible for them to be.

Because of the mathematical nature of the calculations, this method of valuation is suitable for use by computers. Many land development programs have been produced which not only produce a value for the land but also produce cash flow charts and schedules, as well as feasibility studies and sensitivity analysis which allow for rapid re-calculations and objectivity testing.

# Chapter 12

# Developments in Valuation Methods

## 1. Introductory

In recent years critical attention has been given to valuation methods both from within the valuation profession and also from other quarters, particularly the financial sector. This has come about for various reasons. There have been sharp variations in property values which have led to property transactions attracting considerable attention. There has been a growing involvement by institutional and overseas investors, who have invested significant funds in property. Public interest in and awareness of property has been fostered by press commentaries, with most of the national newspapers carrying regular property articles. The importance of property to the economy is now clearly recognised: a fact which was highlighted by the collapse of the property market after 1973 and 1990 when many companies were at risk, even though they were not in business as property companies. Indeed, many companies have substantial property interests as a natural consequence of their activities, and some companies such as retailers and banks have a property portfolio rivalling the major property companies and even institutions in value and quality.

Given the importance of property, it is natural that the ways in which valuers work will be closely scrutinised. A valuation may be a vital factor in deciding whether shares or bonds will be purchased, companies remain solvent or expensive projects be carried out. Most importantly, a valuation may affect members of the public at large who could suffer seriously if the valuation is misleading.

Consequently, the methods by which valuers arrive at a valuation, both as to assumptions and the detailed workings, have been increasingly examined. This is especially so in the case of the investment method of valuation. This is evidenced by criticism of methods expressed in books, journals and other publications, and also by considerable academic research.

This body of critique has led to many suggestions as to alternative approaches to certain valuations or considerable adaptation of the general principles. However, until these methods receive wide acceptance such as to oust the methods set out in previous chapters they cannot be set forth as the appropriate method to adopt. Indeed, given the variety of views expressed, it would be difficult to select any one against the others. This chapter therefore will limit itself to a consideration of the aspects which give rise to criticism of investment valuations and some brief comments on the solutions offered so that the reader may be aware of the ideas being expressed.

## 2. Relationship of YP Single Rate in Arrear and in Advance

The original valuation methods were predicated on the assumption that rents were paid and received annually in arrear and the Valuation Tables were based on this premise. However, current practice is for rents to be paid quarterly in advance, but no change was made to the basic valuation approach.

If it is assumed that rent is payable either annually in arrear or annually in advance, the difference between YP (rent in arrear) and YP (rent in advance) is obviously $1 -$ (PV for nth year). The YP (rent in advance) can be derived from $1 +$ YP (rent in arrear) for $(n - 1)$ yrs, e.g. YP 3 yrs at 10% (rent in advance) = YP 2 yrs at 10% + 1 = 1.7355 + 1 = 2.7355. This compares with YP 3 yrs at 10% (rent in arrear) of 2.4869, a difference of 0.2486 or 10%.

It follows that as the years are extended, the difference between the YPs grows as the PV falls. For example YP 50 yrs at 10% (rent in arrear) is 9.9148 whereas YP 50 yrs at 10% (rent in advance) is 1 + YP 49 yrs at 10% (rent in arrear) = 1 + 9.9063 = 10.9063, an increase of 0.9915 on 9.9148, again a difference of 10%. In fact, the difference will always be exactly 10% at a yield of 10% and it can be shown that the percentage increase at any rate will always be the rate adopted. Hence, for example, YP 30 yrs at 6% (rent in advance) will be 6% greater than YP 30 yrs (rent in arrear), YP 40 years at 7% (rent in advance) will be 7% greater than YP 40 years at 7% (rent in arrear) and so on.

However, rent paid annually in advance is probably as untypical as rent paid annually in arrear. In practice, where the rent is generally quarterly in advance, the method of approach is as follows:

*Example 12–1*

Value a rent of £1,000 p.a. receivable for 10 years, rents payable quarterly in advance. The rent is regarded as arising from a 12% investment.

(a) *Conventional approach (YP SR, rent annually in arrears)*

| | |
|---|---|
| Rent | £1,000 p.a. |
| YP 10 yrs at 12% | 5.65 |
| | £5,650 |

(b) *Valuation approach reflecting rent in advance*

One method is to adapt the Tables by treating the income as being 4 payments per annum with an effective rate of 0.25 of the annual rate, subject to adjustment as described above of $(1 - V)$ or adopting $1 + YP (n - 1$ year). Hence:

| | | |
|---|---|---|
| Rent per quarter | | £ 50 |
| YP $(4 \times 10) - 1$ quarters at 3% per | | |
| quarter (= YP 39 yrs at 3%) | 22.808 | |
| add 1 = | 1.000 | 23.808 |
| Capital Value | | £5,952 |

This alternative approach may be criticised in that it is not a true comparison. A yield of 12% means that in one year the investor earns 12% on his capital. The interest may be added at the end of each year or may be added half yearly or quarterly or at other periods. If the interest is added at periods of less than a year it will normally be on a compound basis. Thus, where the annual yield is 12%, if interest is added quarterly and is compounded, the quarterly rate will not be 3% since this will produce an annual yield of 12.55%.

| | |
|---|---|
| Amt. of £1 for 4 periods at 3% = | 1.1255 |
| *less* | |
| £1 invested | 1.0000 |
| Interest after 1 year = | 0.1255 |

The problem derives from what is meant by the yield. If a yield of 12% is meant to represent not only a nominal annual yield but the effective annual return, or annual percentage rate (APR) or annual equivalent rate (AER), then the quarterly equivalent is not 3% but 2.874%. If this appropriate quarterly effective rate were adopted, then there would be no difference between interest being paid quarterly in arrears or annually in arrear. This fact is understood by the market and all valuations are carried out on the same conventional basis of the assumption of rent annually in arrears, which is the basis for other new property investments. It means that such a property investment will automatically provide a better rate of return than a non-property investment even though this is not directly apparent but this will be known to all investors. When one property investment is compared with another the same "error" occurs in both and therefore it is effectively self-cancelling and since this yield is calculated on the same "error" basis the result is in any event correct.

The comparison of the treatment of rent in arrear and in advance is explored fully in various works referred to at the end of this chapter under Further Reading.

### 3. Taxation of Incomes

In Chapter 6 when identifying the outgoings commonly found in property, reference was made to the taxation of incomes, and in Chapter 9 it was established that, in the investment method of valuation, income is valued gross of tax payable on rents. The principal reason for this is that the tax payable depends on the status of the recipient of the income. Hence, the net income reflects this status as indeed will the net yield to any investor. If net yields are adopted as a measure of comparison they will no longer act as a means of comparison between respective investments since they will reflect, in addition to the qualities of the investment, a unique feature of an individual taxpayer/investor.

However, valuations may be made on a net of tax basis and these will be of use in determining the value of an investment to an individual. It is interesting to consider such approaches, often referred to as a "true net" approach.

Consider first a freehold interest let at full rental value. The conventional gross of tax approach is as follows:

*Example 12–2*

Value the freehold interest in commercial premises recently let at their full rental value of £10,000 p.a. The yield is 10%.

| | |
|---|---|
| FRV (and rent received) | £10,000 p.a. |
| YP in perp at 10% | 10.0 |
| Capital Value | £100,000 |

Now value the interest allowing for tax. Assume that the investor pays tax at 40% on rents.

| | |
|---|---|
| FRV (and rent received) | £10,000 p.a. |
| *less* | |
| Income Tax at 40% 04 × 10,000 = | 4,000 |
| True net income | £6,000 |

Before continuing the valuation, consider the appropriate yield. It is agreed that this is a "10% investment". But this is the gross of tax yield. The net of tax yield depends on the tax payable, and for an investor paying 40% tax, he will lose 40% of any return in tax leaving him with 60% of 10% = 6% as the net yield. This is the appropriate yield to adopt in a net of tax valuation. Hence:

| | |
|---|---|
| Income net of tax | £6,000 |
| YP in perp at 6% | 16.67 |
| Capital Value, say | £100,000 |

It is seen that each method produces the same value. This will be true of any valuation of a freehold at full rental value and at any rate of tax.

The same will be true of a leasehold interest let or enjoying full rental value as the following example shows:

*Example 12–3*

Value the leasehold interest in commercial premises sub-let at full rental value of £12,000 p.a. The leasehold interest has 10 years to run at £2,000 p.a. This is a 10% investment. The investor pays tax at 40% on rents.

| | |
|---|---|
| FRV (and rent receivable) | £12,000 p.a. |
| *less* | |
| Rent Payable | 2,000 |
| Profit Rent | £10,000 p.a. |
| YP 10 yrs at 10% and 3% | |
| (Tax at 40%) | 4.075 |
| Capital Value | £40,750 |

Now value the investment on a net of tax basis.

| | | |
|---|---|---|
| FRV (and rent receivable) | £12,000 | p.a. |
| *less* | | |
| Rent Payable | 2,000 | |
| Profit Rent | £10,000 | p.a. |
| *less* | | |
| Income Tax at 40% = 04 × 10,000 = | 4,000 | |
| Net Profit Rent | 6,000 | p.a |

Before continuing the calculation, it is clear that the net yield should be adopted of 6%. As to the sinking fund, as the net of tax profit rent is being valued, there is clearly no need to gross up the cost of the income tax element. Hence:

| | | |
|---|---|---|
| Net Profit Rent | £6,000 | p.a. |
| YP 10 yrs at 6% and 3% | 6.792 | |
| | £40,750 | |

Thus far it is clear that a valuation of a freehold or leasehold interest producing full rental value gives the same result whether valued gross or net of tax. From this it may be concluded that such investments present no advantage to one taxpayer as against another.

This should not be confused with the advantage to a non-taxpayer, as against a taxpayer, which a short leasehold interest offers for quite different reasons, as explained below.

The valuation of a freehold or leasehold interest where the present rent receivable is not at full rental value – a reversionary investment – does, however, produce different results as between gross and net approaches.

*Example 12–4*

Value the freehold interest in commercial premises with a rental value of £10,000 p.a. They are currently let for the next five years at £4,000 p.a. The appropriate yield is 10%.

| Term | | | |
|---|---|---|---|
| Rent receivable | | £4,000 p.a. | |
| YP 5 yrs at 10% | | 3.791 | £15,164 |
| Reversion to FRV | | £10,000 p.a. | |
| YP in perp at 10% | 10.0 | | |
| × PV of £1 in 5 yrs at 10% | 0.6209 | 6.209 | 62,090 |
| | | | £77,254 |

Now value the investment on a net of tax basis.

Assume the investor pays tax on rent at 40%

| Term Rent receivable | | £ 4,000 | |
|---|---|---|---|
| *less* | | | |
| Income Tax at 40% of £4,000 | | 1,600 | |
| Net Rent receivable | | £2,400 p.a. | |
| YP 5 yrs at 10% gross, 6% net | | 4.212 | £10,109 |
| Reversion to FRV | | £10,000 | |
| *less* | | | |
| Income Tax at 40% of £10,000 | | 4,000 | |
| Net FRV | | £6,000 | |
| YP in perp at 6% | 16.667 | | |
| × PV £1 in 5 yrs at 6% | 0.74726 | 12.45 | 74,700 |
| | | | £84,809 |

It is apparent that the net of tax valuation gives a higher value. Indeed, as the rate of tax is increased it will be found that the value rises on a net of tax basis. It should be noted that in five years' time, when the interest becomes a freehold interest let at full rental value and adopting current rental value and yields, it will be valued at £10,000 p.a. × 10 YP or £6,000 p.a. × 16.67 YP = £100,000. Clearly, therefore, the difference between gross and net of tax valuations arises because there is a term at a lower rent.

The reason for this can be seen in the mathematics in that, in valuing the term, although the rent is reduced by 40%, the YP is not increased correspondingly. Similarly, in determining the present value of the reversion, the value at reversion is the same but the present values differ at different rates. Indeed, the difference between the terms is more than offset by the difference in present values.

At the same time, it should be stressed that the net of tax valuation shown above is incomplete if a full allowance for taxation is required in that no allowance has been made for capital gains tax. However, if an allowance for capital gains tax of 40% were adopted, in the net of tax approach the value still lies above the value on a gross of tax basis, as the following shows:

*Example 12–5*

| | | |
|---|---|---|
| Term (as before) | | £10,109 |
| Reversion to Net FRV | £6,000 | |
| YP in Perp. at 6% | 16.67 | |
| | £100,000 | |

*less*

Capital Gains Tax (CGT)
Net value of interest = x
∴ Gain = (100,000 – x)

| | | |
|---|---|---|
| ∴ Tax = 0.4 (100,000 – x) = | 40,000 – 0.4x | |
| ∴ Reversion net of CGT | 60,000 + 0.4x | |
| × PV £1 in 5 yrs at 6% | 0.747 | 44,820 + 0.299x |
| Capital Value | | £54,929 + 0.299x |

But Capital Value = x
∴ x = 54,929 + 0.299x
∴ x = £78,360

*Note*

The CGT element has ignored indexation relief and other elements of the CGT computation for simplicity. A more detailed estimate of the liability could be made.

The full allowance for tax does therefore indicate that there will be a different value for investors with differing rates of tax. If a full allowance is to be adopted to analyse an investment to an individual, it is probably more appropriate to use discounted cash flow techniques, discussed in Chapter 10. An allowance for tax in determining the open market value of an interest, on the other hand, is inappropriate since, as has been shown, the value would then depend on the choice of the rate of tax. Of course, if it can be shown that a group at a certain rate of tax is dominating the market for a specific type of investment then it may become appropriate to adopt a net of tax approach applying that group's tax rate. Such a situation has arisen in the case of short leasehold investments, as described below.

It has been stated frequently that rates of tax vary. At one extreme an individual or company may pay tax at high rates whilst, at the other extreme, charities, and in particular pension funds, pay no income tax whatsoever – hence, their description as "gross funds".

In Chapter 9 the valuation of leasehold interests was described, including the need to allow for the gross cost of an annual sinking fund. Such a cost represents a high proportion of profit rent when the term is short. However, if a gross fund invests in such interests, it is not faced with a tax liability so that the annual sinking fund costs do not need to be grossed up to allow for tax. The effect of this is that it can afford to pay significantly more for a short leasehold interest (say up to 15 years) than a taxpayer or, alternatively, if it can buy such an interest on the basis allowing for tax then it will obtain a higher than anticipated yield since the grossed up element remains as a return to capital.

This has led to gross funds tending to be a major factor in this part of the investment market. However, it cannot be said that all short leasehold investments should be valued ignoring the tax element; rather that it may be appropriate to value grossing up the cost for tax and also ignoring this factor, with the price payable lying in between the two resultant figures as the vendor and gross fund purchaser share the benefits of the latter's tax-free status on the value. In practice, many gross funds adopt a single rate YP approach or a DCF approach to the valuation of short leasehold investments, and it would be appropriate when valuing investments likely to appeal to gross funds to consider these different approaches.

## 4. Freehold Interests

Two aspects of the valuation of freehold interests cause the greatest controversy, the yield to be adopted and the effects of inflation.

### (a) Nature and Function of Yields

The choice of, and the function of, the yield appropriate to a particular investment are both areas for comment. The fundamental issue turns on the function of the yield itself.

As a matter of basic principle, the conventional approach treats the yield as being a measure of comparison between various investments available and the yield chosen reflects all the different qualities between the investment in question and others that are available. Thus it reflects the potential for future growth, the strength of covenant, the likely performance in an inflationary economy, and any other factors which are thought to be relevant. It is referred to as an "all-risks" yield.

Another school of thought argues that the yield itself should be analysed to reflect these different and distinct qualities. That view has been expressed in various articles, books and theses where a "real value approach" is advocated.

It is not the role of this book to attempt to evaluate the different methods advocated but it is important for the student in particular to be aware of them.

## (b) Equated Yields

The valuation of a freehold interest calls for the determination of the appropriate yield for the class of investment under consideration. Thus it may be said that the yield on first-class shops is 4% or modern factories 8% and so on. What is meant by this is that where such a property is let at its full rental value on a normal market basis then that is the yield to be adopted in arriving at the Years' Purchase.

This is sufficient for valuation purposes but the converse of the method may be inadequate if an analysis is to be made of investments allowing for differing rent review patterns or projections of inflation in rents.

Indeed there may be difficulties in valuing. For example, if by analysis of similar properties let on 21-year leases with seven-yearly reviews the yield is known to be 5.3%, what is the appropriate yield to adopt if one further similar property is found to have been let on a 21-year lease with three-yearly reviews? In the absence of any comparables on this basis the valuer must use his judgment. The yield is probably lower, but by how much? 1%, 0.5%, 0.25%? It is fair to say that in many cases the valuer's judgment, in conjunction with the investor's judgment, will determine the price offered in the market. But in some instances, and particularly where the investor is a fund or institution, a more reasoned approach would probably be demanded.

A solution to the valuation problem, and also to that of how to allow for projections of inflation, can be found in a technique that has been developed and is referred to as the equated yield.

The authors have found that the expression equated yield has been given different definitions so it should be made clear that in this instance the definition is taken to be the discount rate which needs to be applied to the projected income (allowing for growth) so that the summation of all the incomes discounted at this equated yield rate, equates with the capital outlay.

This definition follows the internal rate of return concept of DCF as set out in Chapter 10. Thus an investment can readily be analysed allowing for actual and projected rent at whatever may be the review dates against the market price to show the IRR or equated yield. Alternatively, given the actual and projected rents and review dates, and given the equated yield appropriate to the investment, the price which should be paid to maintain these factors can be determined.

Hence, in the valuation problem posed above, if the analysis of the investments where the rent reviews were seven-yearly is established, the appropriate price/value for three-yearly reviews can be determined and thus the initial yield for valuation purposes.

*Example 12–6*

Similar properties have been letting for £1,000 p.a. on leases for 21 years with seven-yearly reviews. Rental growth is anticipated @ 12% p.a. The properties have been selling @ £18,750 (= 5.33% return). Value a similar property recently let @ £1,000 p.a. for 21 years with three-yearly reviews.

**Step 1: Find equated yield of properties sold @ £18,750.**

Try 16%

| Years | A £1 @ 12% | Expected Rent (1,000 × A £1) | PV £1 @ 16% | PV × YP 7 yrs @ 16% | PV of Rents |
|---|---|---|---|---|---|
| 0–7 | — | 1,000 | 1.0 | 4.0386 | 4,039 |
| 8–14 | 2.2107 | 2,211 | 0.3538 | 1.4289 | 3,159 |
| 15–21 | 4.8871 | 4,887 | 0.1252 | 0.5056 | 2,471 |
| 22–perp | 10.8038 | 10,804 | 0.0443 | 0.8306 [1] | 8,974 |

|  |  |
|---|---|
|  | 18,643 |
| *Less* Price Paid | 18,750 |
| NPV | −107 |

*Note*[1]: 0.8306 = YP in perp @ 5.33% (18.75) × PV 21 yrs @ 16% (0.0443).

Try 15%

| Years | A £1 @ 12% | Expected Rent | PV £1 @ 15% | PV × YP 7 yrs @ 15% | PV of Rents |
|-------|------------|---------------|-------------|---------------------|-------------|
| 0–7 | — | 1,000 | 1.000 | 4.1604 | 4,160 |
| 8–14 | 2.2107 | 2,211 | 0.3579 | 1.5639 | 3,458 |
| 15–21 | 4.8871 | 4,887 | 0.1413 | 0.5879 | 2,873 |
| 22–perp | 10.8038 | 10,804 | 0.0531 | 0.9956 [1] | 10,756 |

| | |
|---|---|
| | 21,247 |
| *Less* Price Paid | 18,750 |
| NPV | +2,497 |

*Note* [1]: 0.9956 = YP in perp @ 5.33% (18.75) × PV 21 yrs @ 15% (0.0531)

$$\text{IRR} = 15\% + \frac{2,497}{2,497 + 107} = \text{say } 15.95\%$$

## Step 2: Apply IRR to the property to be valued.

| Years | Expected Rent (1,000 × A £1) @ 12% | PV £1 @ 15.95% | PV × YP 3 yrs @ 15.95% (2.2477) | PV of Rents |
|-------|-----------------------------------|----------------|--------------------------------|-------------|
| 0–3 | 1,000 | 1.0 | 2.2477 | 2,248 |
| 4–6 | 1,405 | 0.6415 | 1.4419 | 2,026 |
| 7–9 | 1,974 | 0.4115 | 0.9250 | 1,826 |
| 10–12 | 2,773 | 0.2640 | 0.5934 | 1,645 |
| 13–15 | 3,896 | 0.1693 | 0.3805 | 1,483 |
| 16–18 | 5,474 | 0.1086 | 0.2441 | 1,336 |
| 19–21 | 7,690 | 0.0697 | 0.1567 | 1,205 |
| 22–perp | 10,804 | 0.0447 | 0.0447x | 482.9388x |

| | |
|---|---|
| | 11,769 + 482.9388x |
| *Less* Price Paid | 1,000x |
| NPV | 0 |

$$1,000x = 482.9388 x + 11,769$$
$$\therefore 517.0612x = 11,769$$
$$x = 22.761 \text{ YP } (= \text{YP in perp @ } 4.393\%)$$

*Note*: x = YP in perp @ initial yield.

Hence initial yield on property to be valued is 4.393%.
The price that should be paid is £22,760.

It can be seen that the valuer making a subjective judgment as to the appropriate reduction to the prevailing yield would need to have reduced the yield by close to 1%, but this would be a matter of chance. On the other hand, the calculations have produced a mathematical answer which is not necessarily acceptable as a true valuation. For this to be so it must be assumed that rents are unaffected by the review pattern. However, in practice, if identical properties are offered on leases with either three-year or seven-year review patterns, it is most unlikely that the initial rent would be £1,000 p.a.

The more likely situation is that, if properties let on 21-year leases with seven-year reviews command an initial rent of £1,000 p.a. and sell for £18,750, the valuer would be required to determine the approximate initial rent for a lease with three-yearly reviews so as to maintain the value @ £18,750.

The calculation is the same as for Step 2 above, save that the initial rent is x, the value after 21 years is £18,750 × Amount of £1 for 21 years @ 12% (10.804) × PV £1 in 21 yrs @ 15.95% (0.0447) = 9055.103, and the Price Paid is £18,750. This produces a result of:

$11.769x + 9055.103 = £18,750$
$x =$ say £824

Thus, the initial rent should be £824 p.a.

There are several Tables available which provide a simple means of adjusting rents to match changes in rent review patterns. However, these are rarely applied in practice. The most common practice is to apply a value of thumb (usually add 1% to the rent for every year that the review pattern exceeds the normal).

## (c) Effects of Inflation

In recent times the impact of inflation has been experienced widely. It has led to widespread increases in the prices of many goods, and property has not been immune from its effects. The term inflation is often used imprecisely, so that any increase in price is treated as an effect of inflation. Thus, in property, as rents increase they are said to rise with inflation. If inflation is measured as the increase in the retail price index then the growth in rents of many types of property has sometimes been in excess of inflation and sometimes behind it.

Whatever may be the causes of changes in rents, such changes have led people to examine the investment method of valuation and to question whether adequate, or even any, allowance is made in the valuation.

Taking the simplest example of a property recently let at rental value, the valuation method is to multiply the rent passing by a YP in perpetuity. As was explained in Chapter 9, the yield from which the YP is derived reflects the likely future increases in rental value, be they real or nominal. In this way, the valuation can be said to reflect fully the likely effects of inflation, used in its widest sense. Nonetheless it has been argued that a more explicit allowance should be made for growth. No generally accepted approach for making such an allowance has emerged, although the DCF method or methods derived from DCF techniques enable explicit allowance to be made quite readily. As the recent past has demonstrated, there is no guarantee that rents will rise in the future and care must be taken not to confuse short-term changes with long-term trends. In so far as commercial property is concerned, technical changes make property prone to obsolescence which will have an adverse effect on future values.

## (d) Valuation of Varying Incomes

It was shown in Chapter 8 that the valuer may be required to value a freehold interest where the rent receivable is currently less than full rental value. The method adopted was to capitalise the present rent until the full rental value could be obtained, the Term, and then to capitalise the rental value receivable thereafter, the Reversion. An alternative approach is sometimes adopted whereby the rent currently receivable is capitalised into perpetuity, and the incremental rental receivable at the expiry of the lease is capitalised separately. The present rent represents a hard core of rent which will continue into perpetuity whilst future rents are incremental layers. Hence, it is commonly referred to as the hardcore or layer method. This can be illustrated diagrammatically as below.

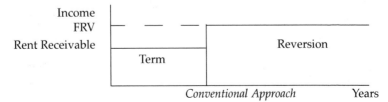

*Conventional Approach*                    Years

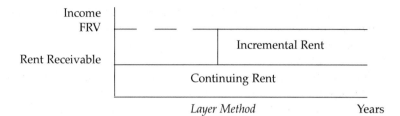

*Layer Method*

The valuation approach by the hardcore method involves applying a Years' Purchase in perpetuity to both the continuing rent and the incremental rent, the capital value of the incremental slice then being deferred by applying the appropriate Present Value. The problem which arises, and the major criticism of the method, is the choice of the appropriate yield for each slice. Where the Term and Reversion would be valued at the same yield, then that same yield can be applied throughout the hardcore method, and an identical result will emerge since each pound of future rent is being discounted at the same rate.

*Example 12–7*

Value the freehold interest in commercial premises let at £5,000 p.a. for the next 3 years. The full rental value is £8,000 p.a. The yield at FRV is 8%

*Conventional Approach*
Term

| | | | |
|---|---|---|---|
| Rent Received | | £5,000 p.a. | |
| YP 3 yrs at 8% | | 2.577 | £12,885 |
| Reversion to FRV | | £8,000 p.a. | |
| YP in perp at 8% | 12.5 | | |
| × PV £1 in 3 yrs at 8% | 0.7938 | 9.923 | £79,384 |
| Capital Value | | | £92,269 |

*Layer Method*

| | | | |
|---|---|---|---|
| Continuing Rent | | £5,000 p.a. | |
| YP in perp at 8% | | 12.5 | £62,500 |
| Incremental Rent | | £3,000 p.a. | |
| YP in perp at 8% | 12.5 | | |
| × PV £1 in 3 yrs at 8% | 0.7938 | 9.923 | £29,769 |
| Capital Value | | | £92,269 |

However, where it is felt that the rate for the term should be different from the reversion, perhaps lower because the covenant of the tenant makes the rent receivable more certain of receipt, or higher because the rent receivable is fixed for several years, then the layer method is likely to produce a different result. For example, where the term rent is valued at a lower rate, then the results will only coincide if the incremental rent is valued at an appropriate marginal rate.

Take the rents in *Example 12–7*, but assume that the term rent is valued at 6%.

*Conventional Approach*

| Term | | | |
|---|---|---|---|
| Rent Received | £5,000 p.a. | | |
| YP 3 yrs @ 6% | | 2.673 | £13,365 |
| Reversion to FRV | £8,000 p.a. | | |
| YP in perp @ 8% | | | |
| × PV £1 in 3 yrs @ 8% | | 0.7938 | £79,384 |
| | | | £92,749 |

*Layer Method*

| Continuing Rent | | | £5,000 p.a. | |
|---|---|---|---|---|
| YP in perp at 6% | | | 16.667 | £83,335 |
| Incremental Rent | | | £3,000 | |
| YP in perp at 18.9% | 5.29 | | | |
| × PV of £1 in 3 yrs @ 18.9% | 0.593 | 3.137 | | £9,411 |
| | | | | £92,746 |

Thus, if the incremental rent is valued at approximately 18.9% and the continuing rent at 6% the same results will emerge as the conventional approach. In practice, the likelihood is that the valuer, whilst raising the yield in valuing the increment to reflect that it is the top slice, is unlikely to adopt what appears to be a very high rate of return. This leads to the criticism that, in the hardcore method, there is no yield adopted which can be checked by direct comparison with the yields of similar investments unless identical investments can be produced.

Similarly, where the term rent would be valued at a higher yield, the layer method tends to be difficult to apply. Assume again the facts in Example 12–7, but assume that the rent of £5,000 p.a. is fixed for 15 years, following which the property can be let at £8,000 subject to regular reviews. Clearly, the appropriate yield to apply to the £5,000 p.a. could be significantly higher than 8%, based on medium-term fixed interest investments. On the other hand, it would be unrealistic to value £5,000 into perpetuity at such a high rate.

The argument in favour of the hardcore method is that it is simpler to apply, particularly where the rent varies at renewal stages.

The choice between the methods must be made by the individual valuer. This presents no problem in practice unless different results emerge, when it is left to the respective valuer to justify whatever approach he has chosen.

A further argument is advanced that the yield for the term and reversion should be the same in all circumstances since this is the yield for the investment as a whole. If this is accepted, then the case for the hardcore method becomes one of convenience only. This latter view is now the more generally acceptable one.

## 5. Leasehold Interests

A considerable amount of criticism has been expressed of the method of valuing leasehold interests. The criticism turns on two major areas, one being the concept of capital recoupment and the other of the inherent errors to be found in the valuation of varying profit rentals.

As was explained in Chapter 9, the valuation of a leasehold interest recognises that a lease is a wasting asset and makes appropriate adjustments by the provision of a sinking fund for recoupment of capital. This leads to the adoption of a dual rate Years' Purchase.

The questions that are raised include how valid is it to allow for replacement of capital? If valid, why not replace the capital in real terms allowing for the erosion of value due to inflation? Why allow for replacement of capital by an annual sinking fund at a leasehold redemption policy rate of around 3% interest when it can be demonstrated that few people actually take out such policies, and if they did, would they do so at that yield? If not valid, why use a dual rate Years' Purchase?

If these criticisms are valid then considerable doubt is thrown upon the method of valuation described in Chapter 9. Before considering the overall effects of the criticisms it is useful to consider the criticisms in turn.

## (i) Replacement of Capital

The purpose within the valuation method of replacing the capital which must ultimately disappear in the case of a leasehold investment is to isolate this factor, which is unique to leasehold interests as against freehold interests, and so to allow a ready comparison between freehold and leasehold interests. The alternative is to reflect this unique feature in the yield. This would lead to adjustments in the yield which would be partly subjective in the absence of direct comparables. The adjustments in the case of short leasehold terms would need to be considerable if it is accepted that investors who buy at prices allowing for a sinking fund accept such prices as being reasonable.

*Example 12–8*

Value the leasehold interest in commercial premises held for 5 years at £1,000 p.a. They are underlet for the term remaining at £11,000 p.a. The appropriate yield is 7%.

*Conventional Approach*

| | | |
|---|---:|---|
| FRV (and rent receivable) | £11,000 | p.a. |
| *less* Rent payable | 1,000 | |
| Profit Rent | £10,000 | p.a. |
| YP 5 yrs at 7% and 3% (Tax at 35%) | 2.78 | |
| | 27,800 | |

| | | |
|---|---:|---|
| Value without ASF at 3% | | |
| Profit Rent (as before) | £10,000 | p.a. |
| YP 5 yrs at 23% single rate | 2.80 | |
| | £28,000 | |

If it is felt that 23% is unrealistic and a lower rate is adopted, the value will rise which suggests that short leasehold interests are under-valued by the conventional approach. Such a view is supported in practice where gross funds are in the market, as described above, but only because of the taxation effect on both the income and the sinking fund yield. Also, in short-term leases the inflation effect is not so marked and these investments can be directly compared with dated gilt-edged securities.

If a longer term lease is considered, then the increase in yield is lessened. For example, if the term were 50 years at 7% and 3% adj. IT at 35% = 11.956 which is approximately equal to YP 50 years at 8.2% single rate.

A further justification for incorporating a sinking fund is that it follows the accounting convention of writing off wasting assets. Indeed, it would be possible to write off the value of the lease by one of the other depreciation methods such as straight line instead of sinking fund.

The valuation would then become:

| | | |
|---|---|---|
| FRV | £11,000 | p.a. |
| *Less* | | |
| Rent payable | 1,000 | |
| Profit rent | £10,000 | p.a. |
| *Less* | | |
| Depreciation over term | | |
| of lease (5 years) = | 0.2x | |
| | £10,000 − 0.2x | |
| YP 5 years @ 7% | 4.10 | |
| | x = 41,000 − 0.82x | |
| | x = £22,527 | |

*(ii) Replacement of Capital in Real Terms*

It is true that the conventional method allows for an annual sinking fund to replace the original capital invested. In times of inflation there will be an erosion of capital value in real terms. Clearly, if £10,000 had been invested in a lease 20 years ago which now expires, then the £10,000 now available under the sinking fund is not putting the investor in the same position as he was when he made his original investment in real terms.

The YP dual rate can be adjusted quite simply if it is decided to replace the capital in real terms. It would be necessary initially to determine either the predicted rate of inflation or the predicted rate of growth in an equivalent freehold interest. The latter approach is more logical since the aim is to put freehold and leasehold interests on an equal footing. Once the rate of inflation or growth has been determined, then the annual sinking fund should be calculated to replace not £1 invested but the amount of £1 at the chosen rate over the period of the lease.

If this is done the leasehold interest is more secure than it would otherwise have been so that the remunerative rate should be reduced from what it would otherwise have been, possibly to the level of the yield on an equivalent freehold interest. The YP difference is shown below.

*Example 12–9*

Assume a lease with 40 years to run and a freehold yield of 9% and
leasehold yield of 10%
*Conventional YP*
YP 40 yrs at 10% and 3% (Tax at 35%) = 8.305

*Inflation Proof YP*
Assume inflation at 7% p.a. leasehold rate now 9%

Formula =

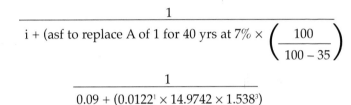

$$\frac{1}{i + (\text{asf to replace A of 1 for 40 yrs at 7\%} \times \left( \dfrac{100}{100 - 35} \right)}$$

$$= \frac{1}{0.09 + (0.0122^1 \times 14.9742 \times 1.538^3)}$$

$$= \frac{1}{0.09 + 0.3054}$$

$$= \qquad 2.529 \text{ YP}$$

1 = asf for 40 years at 3%
2 = value of £1 in 40 years @ 7%
3 = tax adjustment at 35%

This difference in Years' Purchase suggests that the conventional
method greatly over-values leasehold interests. There is, however,
a fallacy in the argument that the annual sinking fund under-
provides without an allowance for inflation.

The fact is that a leasehold interest is a wasting asset by which is
meant that it ultimately wastes away to nil. However, in the period
between the start and finish of a lease the value may rise
considerably before falling away. The reason for this is that, if rental
values increase faster than the Years' Purchase for succeeding years
falls, then the capital value of the leasehold interest will rise in
nominal terms. The effect of inflation is therefore reduced. If the
investor sells his interest before the capital value starts to fall he
will obtain more than he paid at the start.

Indeed, an inspection of the YP dual rate Tables will show that it
is only in the last few years that there is a significant falling away
of the value of the YP figures such as to outweigh the inflation on

rents. Thus it is only in the case of short leases that the ignoring of the effects of inflation in the sinking fund can be criticised, but the impact of any allowance is far less over such short periods.

*Example 12–10*

As for *Example 12–9*, but the lease has four years to run.

*Conventional YP*
YP 4 yrs at 10% and 3% (Tax at 35%) = 2.138

*Inflation Proof YP*

$$\frac{1}{0.09 + (0.239 \times 1.311 \times 1.538)} = 1.749$$

### (iii) ASF at Low Rate of Return

It is suggested that few investors actually take out leasehold redemption policies. Instead it is said that they either invest their redemption funds in alternative investments showing a higher return than the low rates of around 2% to 3% obtainable or make no such provision at all. This may perhaps be so but such criticism misses the essential point of the annual sinking fund allowance, that it is a notional allowance to allow comparison between freehold and leasehold investments, as explained before. Given that the notional allowance makes such a comparison possible, it remains that only by means of insurance policies can an investor, if he so chooses, obtain a guarantee of a future specific capital payment, and the rates offered under such policies are naturally low. It is true that, in the case of short terms, there are investments which guarantee more than 2% to 3%, and it is in these cases where the dual rate YP is less likely to be adopted as has been discussed.

### (iv) Why Use Dual Rate Years' Purchase?

The answer to this question is provided by the comments on the above points. Nonetheless it should be recognised that there is a body of opinion which argues for the abandonment of the dual rate approach and its replacement by a single rate Years' Purchase or DCF or other methods.

## (v) Valuation of Variable Incomes

The usual method employed to calculate the capital value of a leasehold interest, where the income varies during the term of the lease, is inherently incorrect because of the use of the dual rate tables. The method envisages more than one sinking fund being taken out to provide for the redemption of capital, whereas in practice a purchaser would in all probability take out a single policy to cover the whole term. The concept of a variable sinking fund is not in itself incorrect and a valuation could be made on this basis if a method were employed which would enable the varying instalments to compound to the required amount over the full unexpired term.

The normal method of approach used in the valuation of varying incomes for a limited term does not allow for the correct compounding of the sinking fund accumulations. This type of valuation is made in two or more stages, so that each separate sinking fund instalment will be based on the number of years in the appropriate stage of the valuation and will not be based over the full length of the term. Consequently, the sum provided by these sinking funds over the full length of the term will not equal the capital sum required for the redemption of capital.

The following examples illustrate the error which occurs when the normal method of valuation is used:

*Example 12–11*

The valuation of a leasehold interest on an 8% and 2.5% basis. The lease has an unexpired term of 10 years and produces a profit rent of £1,000 p.a.

| | |
|---|---:|
| Profit Rent | £1,000  p.a. |
| YP 10 years at 8% and 2.5% | 5.908 |
| Capital Value | £5,908 |

*Example 12–12*

If the unexpired term of 10 years in *Example 12–11* is considered in two stages of say, four years and six years the valuation will take the following form:

| | | | |
|---|---|---|---|
| Profit Rent | | £1,000  p.a. | |
| YP 4 years at 8% and 2.5% | | 3.117 | £3,117 |
| Profit Rent | | £1,000  p.a. | |
| YP 6 years at 8% and 2.5% | 4.227 | | |
| PV in £1 in 4 years at 8% | 0.735 | 3.107 | £3,107 |
| Capital Value | | | £6,224 |

In these two examples, the profit rental of £1,000 p.a. has been capitalised in each case for a total period of 10 years at the same remunerative rate of 8%; each example should therefore produce the same result. The difference of £316 is caused by the error which arises in Example 12–12. The method of valuation used in this example is based on the dual rate basis and is made in two stages; an incorrect sinking fund instalment will therefore be implicit in the valuation and will introduce an error in calculation.

The degree of error which occurs in Example 12–12 is approximately plus 5.25%. If the 10-year term had been divided into three stages (with the income varying in the third and sixth year) the error would increase to over 8%.

### The Effect of Tax

If the incidence of income tax on the sinking fund element is taken into account, this must have some effect on the degree of the error. In the following valuations the figures of Years' Purchase used in *Examples 12–11* and *12–12* are adjusted to allow for the effect of income tax, assuming tax at 50% for the purpose of the Examples:

### Example 12–11(t)

| | | |
|---|---|---|
| Profit Rent | | £1,000  p.a. |
| YP 10 years at 8% and 2.5% (Tax at 50%) | | 3.87 |
| Capital Value | | £3,870 |

### Example 12–12(t)

In this example the 10-year term is again considered in two stages:

| | | |
|---|---|---|
| Profit Rent | | £1,000  p.a. |
| YP 4 years at 8% and 2.5% | | |
| (Tax at 50%) | 1.78 | £1,780 |
| Profit Rent | | £1,000  p.a. |

| | | | |
|---|---|---|---|
| YP 6 years at 8% and 2.5% | | | |
| (Tax at 50%) | 2.54 | | |
| PV £1 in 4 years at 8% | 0.735 | 1.87 | £1,870 |
| Capital Value | | | £3,650 |

These examples illustrate that when an allowance is made to reflect the incidence of tax, it will not always increase the error which exists, although this is often thought to be the case. They show that, in making the allowance for income tax, the adjustment has had a compensatory "pull" on the degree of the error which is reduced in the last example to approximately –5.5% from +5.25%.

In many instances the effect of the tax adjustment will be more marked and will usually over-compensate for the "normal" error to such an extent that a much greater inaccuracy will be introduced. In some cases an error which is well in excess of 10% may be involved.

### The Double Sinking Fund Method

The following method, which is one developed by A.W. Davidson, enables this form of valuation to be made without involving any significant error. An attempt has been made to use the conventional methods of valuation and to avoid any involved mathematical approach.

### Example 12–13

In this example the same leasehold interest is again considered and the term is divided into two stages as in Example 12–12(t). It should be noticed that the dual rate principle is not used.

| | | | |
|---|---|---|---|
| Let the Capital Value = P | | | |
| Profit Rent | | £1,000 p.a. | |
| less asf to replace P in | | | |
| 10 years at 2.5% | 0.0893P | | |
| adj. for tax at 50% | × 2.0 | 0.1786P | |
| Spendable Income | | £1,000 – 0.1786P | |
| YP 4 years at 8% | | 3.312 | |
| | | £3,312 – 0.5915P | |

| | | | |
|---|---|---|---|
| Spendable Income | | £1,000 | – 0.1786P |
| YP 6 years at 8% | 4.623 | | |
| × PV £1 in 4 yrs at 8% | 0.735 | 3.398 | |
| | | | £3,398 – 0.6068P |
| | | | £6,710 – 1.1983P |

*Plus Repayment of the capital replaced
by the Single Rate SF
P × PV of £1 in 10 years at 8%

| | |
|---|---|
| | 0.4632P |
| | £6,710 – 0.7351P |

$\therefore$ P $\quad= 6{,}710 - 0.7351P$

1.7351 P $\quad= £6{,}710$

P $\qquad = £3{,}867$

The spendable income has been valued on the single rate basis. This means that the "capital" has been replaced by allowing for a sinking fund at 8% in the capitalisation. Thus two "capital recoupments" have taken place, one at 2.5% and one at 8%.

*As the valuation has been reduced by allowing for the capital to be replaced twice, a figure equal to the deferred capital value has been added back to the valuation.

It will be seen that the capital value of £3,867 is higher than that produced in Example 12–12(t) of £3,650. Thus, removing the inherent error has increased the value by 5.9% to produce the same value as in Example 12–11(t) which is correct.

A simplified approach has been produced by P. W. Pannell which is probably sufficiently accurate for most cases. This again applies a single rate YP, the result being adjusted by applying the ratio of the YP dual rate to the YP single rate.

*Example 12–14*

The same facts are adopted as in Example 12–13.

| | | | |
|---|---|---|---|
| 1st 4 years | | | |
| Profit Rent | | £1,000 p.a. | |
| YP 4 yrs at 8% | | 3.312 | £3,312 |
| 2nd 6 years | | | |
| Profit Rent | | £1,000 p.a. | |
| YP 6 yrs at 8% | 4.623 | | |
| PV £1 in 4 yrs at 8% | 0.735 | 3.398 | 3,398 |
| | | | 6,710 |

$$\frac{\times \text{YP 10 yrs at 8 \& 2.5\% (IT 50\%)}}{\text{YP 10 yrs at 8\%}} \quad = \quad \frac{3.868}{6.710} \quad = \quad \frac{\underline{0.576}}{\underline{£3,865}}$$

As can be seen, the answer is almost exactly the same as that produced by the double sinking fund method (the difference being attributable to the rounding off of decimal places) and is much easier to produce. This chapter has considered some of the questions raised over valuation methods currently adopted. Its sole purpose has been to air some of the issues raised. The comments in the chapter do not seek to accept or reject the views being expressed, nor is it suggested that all points of controversy have been introduced.

The hope is that the reader will appreciate that methods of valuation are under scrutiny and that this may result in changes in the approaches adopted to valuation. However, until such changes achieve general acceptance the authors believe that they have set out the principles and current practice of valuation in the preceding chapters.

## Further Reading

*Research Report on Valuation Methods* (RICS – Andrew Trott).

*Valuation of Commercial Investment Property: Valuation Methods* (RICS Information Paper, 1997).

*Calculation of Worth* (RICS Information Paper, 1997).

*The Valuation of Property Investments* by Nigel Enever and David Isaac (Estates Gazette, 1995).

*Property Investment Appraisal* by Andrew Baum and Neil Crosby (Routledge, 2nd Edition, 1995).

# Some Practical Points

## 1. Generally

It has already been emphasised that the primary purpose of valuation is the ascertainment of present open market value. This term has now been formally defined by the RICS and ISVA in the *Appraisal and Valuation Manual* (known as the Red Book)[1] and the full definition should always be referred to when performing any valuation task. There is, however, nothing different or special about this definition because it conforms to the commonly held understanding of an open market value, being the price at which an interest in a property can be expected to sell as between a willing vendor and a willing purchaser, both of whom are fully informed regarding the interest in question, who are not forced to sell and are free to deal elsewhere if they choose.

But although most of the valuer's problems involve consideration of present market value, there are often other factors in the problem to which his attention must also be directed.

It is often necessary to have regard to future trends of value, to consider if prices at which interests in properties have been sold are reasonable or likely to be maintained, and to examine the possibilities of changes in rental value. This has now been formalised by the Red Book as ERP, i.e. Estimated Realisation Price. This definition does not require a valuer to use a crystal ball but requires him to take into account those matters which appear now to create the possibility of a change in value in the immediate future; by way of a simple example, the valuation of a short lease will change over a comparatively short time because of the wasting nature of the asset being valued. However, the valuer can also have regard to his fears that the market is unreasonably high due to a shortage of supply which might be amended in the short term, or unreasonably low due to the absence of buyers caused by a

---

[1] See Chapter 20.

particular event, for example the massive fall in values which occurred on the Stock Exchanges throughout the world on Black Monday in 1987.

When valuations are made for certain purposes, e.g. for taxation or in connection with compulsory purchase for public or private undertakings, the valuation, although based on market value, may be regulated by statutory provisions as to the date at which the valuation is to be assumed to be made and as to the factors which may or may not be taken into account in making it. For these purposes, it is normally permissible to use post-dated evidence because the object of the exercise is to establish the correct value of the property as at the relevant date and not what price a valuer might have placed upon the property in advance of that date. This concept of acceptable post-dated evidence has been accepted by the courts and also in the Privy Council case of *Melwood Units Property Ltd* v *Commissioner of Main Roads*.[2] The same principle of utilising post-dated evidence has also been accepted in connection with rent reviews for the same reason. However, the further that the valuer moves away from the valuation date the more likely it is that the post-dated evidence will not be relevant because of changes in the market between the relevant dates.

Again, where properties are purchased for investment, the valuer may be asked to advise on policy, to suggest what reserves should be created for future repairs, or in respect of leasehold redemption, and to advise generally on the many problems involved in good estate management.

The knowledge required to deal with all these matters can only be gained by experience but the valuer will find it both necessary and of great advantage to keep abreast of current affairs (international, national and local) and of yields in other investment markets. The valuer then values in the light of this information and after making a careful study of sale prices, yields and rental values, particularly in areas with which he is acquainted and where there is an opportunity of seeing the properties which are the subject of recorded transactions.

As stated in Chapter 7, it is usually possible to obtain evidence of yields for different kinds of properties from the property market itself but this is not always possible, and sometimes the valuer has

---

[2] [1979] 1 All ER 161.

to look at similar investments which might be quoted on the Stock Exchange. A good example of this relates to the valuation of ground rents. These are often compared with undated Gilts issued by the British Government which at the time of writing has an average yield of 5% to give a YP of 20. It is then usual to add to this yield to reflect the relatively high cost of collecting relatively low ground rents and the cost of selling in the absence of a guaranteed market (the risk factor in a ground rent is virtually non-existent because of the security offered by the overriding property). There is in addition another Stock Exchange Investment known as PIBS (Permanent Income Bearing Shares). These are fixed interest securities which have been marketed by the leading building societies on the basis that they have no redemption date and that they rate last in terms of their repayment in the case of a failure by the building society concerned. In addition, they suffer from the risk that the interest will not be paid, if by doing so the building society (or bank) will be forced into losses. They have been issued by a total of 10 building societies and banks ranging from First Active to the Halifax Bank. They are quoted on the Stock Exchange and prices are published in the financial press and they are sold in specified amounts, usually in units of 1,000 but sometimes in units of as much as 50,000. As at the time of writing, the highest yield was provided by First Active which was 7.99% and the lowest yield was provided by the Newcastle at 6.32%. The prices of these shares has fluctuated with changes in interest rates. For example, the Halifax Bank has issued three blocks of shares; the first was at 13.625%, the second was at 12% and the third was at 8.75% and their current yields are 6.95%, 7.32% and 6.89%, thus showing that these long term investments fluctuate in value.

The comparison of PIBS with ground rents shows that PIBS are easier to buy and sell with little cost and the income has no cost in terms of collection; however, they carry more risk than ground rents because of the possibility of non-payment. It therefore follows that a ground rent yield should have a great similarity to a PIBS yield and it is suggested that ground rents should yield something in the order of 1% more than PIBS.

The valuer needs to obtain as much information and evidence as possible so as to render the valuation as accurate as possible. Even so, it should be emphasised that a valuation is only an expression of one valuer's opinion and it is not surprising that other valuers may hold different views. For this reason a valuation is at best an estimate of the price which will be achieved, but the best estimate must be the one based on the best evidence.

Some interesting observations on valuation in general including limits of accuracy were made in *Singer & Friedlander* v *John D Wood & Co*[3] and these have led to a general proposition that a valuation may be accurate if it falls within a reasonable range of the true value. This acceptable bracket or margin of error has been crucial in determining negligence by valuers in a stream of cases. The bracket is often taken to be 10% of the true value (above or below) but the Court may regard this as too small or too large.[4]

## 2. The Analysis of Rents and Prices

In estimating the market value of an interest in property the valuer must consider all the evidence available.

The property may be let at a rent, in which event its reasonableness or otherwise must be considered in relation to the rents of other similar properties.

The rate per cent at which the net income is to be capitalised will be determined by reference to the rate shown by analysis of the sale price of other similar properties and investments.

The keeping of accurate records of rents and sale prices, and of analyses of the latter to show investment yields is therefore of very great assistance to the work of the valuer.

An example is given in Chapter 5 of an analysis of rents made to determine the rental value of offices.

Similar methods, usually related to the superficial area of premises, can be used for the making and keeping of records in connection with a variety of properties.

Although in many offices manual records are still maintained, computers have added considerably both to the ease of maintaining records and to the ways in which information can be analysed, manipulated and retrieved. National databases have been organised by the Estate Gazette (EGi) and other internet service agencies.

---

[3] [1977] 2 EGLR 84.
[4] See for example, *Arab Bank plc* v *John D Wood Commercial Ltd* [1998] EGCS 34 and also the article 'Close Brackets' in the *Estates Gazette* of 4 April 1998 p. 133 which considers the result of research at Reading University into margin of error cases where the bracket varies widely.

In analysing sale prices the usual steps in making a valuation are reversed. The purchase price is divided by the estimated net income to find the figure of Years' Purchase, from which can be determined the rate per cent yield which the purchaser price represents.

In the case of perpetual incomes the "yield" or rate per cent at which the purchaser will receive interest on his money can be found by dividing 100 by the figure of Years' Purchase. In the case of incomes receivable for a limited term it must be found by reference to the valuation tables or computer programmes.

*Example 13–1*

Freehold property producing a net income of £2,000 p.a. has recently been sold for £20,000. Analyse the result of this sale for future reference.

Analysis:

$$\frac{\text{Purchase Price £20,000}}{\text{Net Income £2,000}} \quad \text{represents 10 YP}$$

Rate at which purchaser will receive interest on money

$$= \frac{100}{10 \text{ YP}} = 10\%$$

This is not the only analysis if the £2,000 does not represent the Full Rental Value. An alternative analysis would involve a term and reversion calculation.

*Example 13–2*

Shop premises held on lease for an unexpired term of 52 years at a rent of £15,000 p.a. have been sold for £135,000: the rental value is estimated to be £27,500 p.a.

Analysis:

The profit rental is £12,500 p.a. The purchase price is £135,000 which, divided by £12,500, represents 10.8 Years' Purchase. On reference to the dual rate tables for a term of 52 years this is seen to equate to 8% and 3% with tax at 35%.

*Example 13–3*

Similar, but somewhat smaller, premises nearby are held on lease for an unexpired term of 45 years at £7,500 p.a.; the lease was recently sold for £100,000. On the evidence of the previous example, at what do you estimate the rental value?

*Answer*

It would seem proper to apply the rate per cent derived from the previous analysis.

$$\frac{\text{Sale Price}}{\text{YP for 45 years @ 8\% and 3\% (tax at 35\%)}}$$

|   |   |   |   |   |
|---|---|---|---|---|
| = | $\dfrac{10,000}{10.353}$ | = | £9,659 | profit rent |

add                                                    7,500  lease rent

Rental Value, say                      £15,000  p.a.

Although this means of ascertaining rental value must be resorted to where other evidence is not available, rental values should be estimated, if possible, by analysing the results of recent lettings of comparable properties and an estimate based on the analysis of a capital transaction should be avoided if at all possible. The object of analysing a capital transaction is usually to find the remunerative rate of interest.

In many instances comparisons of rental value will only be of value if restricted to a particular locality; this is obviously so in the case of shops, where wide variations in rent can occur within a distance of a few hundred yards.

On the other hand, with experience, the valuer may be able to identify a level of values which will be similar in locations of similar qualities within a region or even nationally. For example, in towns close to a common factor such as a motorway, the levels of office rent will tend to display regional rather than narrowly local characteristics. Similarly, shops in major shopping locations will tend to display a similar Zone A rental value.

In the case of industrial or warehouse properties the area for comparison purposes may be similarly extended; although evidence of value of a particular type of property in one town may have limited significance in relation to another town, say 30 miles away, even in respect of identical property.

Whatever the general pattern, a comparison derived from nearby property will always be more significant.

In estimating the yield likely to be required from a particular type of investment, evidence over a much larger area can usefully be employed. Indeed, there are two broadly defined markets found in practice – a national and a local market. Prime properties, such as first-class shops or offices, or well designed modern warehouses or factories, tend to produce a similar yield nationally. On the other hand, secondary properties are more prone to local factors influencing the yield so that they form a local market.

It is in the analysis of past sales and the application of the evidence derived therefrom to a particular case that the skill and experience of the valuer is called into play. He must consider carefully the extent to which the property to be valued is similar to those that have been sold. The correctness or otherwise of the rent must be determined; trends of value since the evidence was accumulated must be considered.

In some cases the prices at which properties have changed hands may be above or below *"market value"*. The vendor may have been anxious to dispose of a property quickly and have taken a rather lower price than might have been expected; or a high price may have been given by a buyer in urgent need of a certain type of accommodation.

When analysing sales for the purposes of Leasehold Reform Valuations, care must be exercised in excluding any part of the value attributable to the right to acquire the freehold or an extended lease. Indeed, it may sometimes prove impossible to obtain *"untainted evidence"* from the market because of the existence of these rights. In some cases the valuer should use his own knowledge and experience to distinguish between that part of the property which represents the true value of the leasehold interest concerned and that which effectively represents the lessees share of the marriage value which is being paid to the outgoing lessee. This is a special situation where open market evidence is not to be used for a statutory valuation.

It follows that, whilst the analysis of transactions may show a level of value (or rents) for a particular type of property in a particular area, this level may not accord with the valuer's own expectations based on his knowledge and experience. In such a case, the valuer should advise the client that the level of value may not be sustained if there was in existence some facts or factors which may not have been taken into account. An example is where a property has been subject to a sale and leaseback, the agreed rent may have been above open market value, and therefore the analysis of the transaction could lead the unwary valuer to over value a similar property.

The duty of the valuer is to question evidence before applying it, and not slavishly to follow other people's assessments without understanding the basis of their assessment.

### 3. Special Purchasers

A valuation is prepared adopting certain assumptions as to the market conditions. In some instances the valuer will be aware that some people are in the market for whom the property in question will be of special interest. Typical examples are where marriage value exists – as described in Chapter 9.

The valuer must always consider if there are special purchasers since they are likely to pay the highest sum and one which is in excess of the value to the general market (*"an overbid"*). Even when the existence of the special purchaser is known, it is often difficult to determine the price to which such a purchaser will go.

In Chapter 9 the price which might be agreed in marriage value situations was discussed. But in other cases the only way of discovering the over-bid is by marketing the property in a manner designed to draw out the final bid. The valuer must therefore be able to comment on the appropriate method of offering the property to the market. For example, it may be appropriate to sell at public auction, or by public tender, or by a closed tender (when the property is offered only to a selected group) or of course by a general offer for sale by private treaty.

The situation of special purchasers creating valuation difficulties is commonly found where leasehold interests are offered for sale in respect of shops in prime locations. In general, the major retailers will be established in such a location and they will be determined to remain whilst the centre continues to thrive. Hence the chance of obtaining a shop rarely arises since the leases will be renewed by the existing tenants. Thus, when a trader does decide to move out and offers the balance of his leasehold interest he will find that there is considerable competition to purchase the lease. The valuer frequently discovers that if he applies the conventional methods of calculation, particularly the investment method of capitalising the profit rent with YP Dual Rate adjusted for Income Tax, the calculation figure will be far less than the price likely to be obtained. The most appropriate method might be a profits approach, perhaps valuing the profits for several years ahead beyond the term of the lease offered, since traders may be prepared to buy the right to be able to earn such profits. Whatever method is adopted it will often be found that the price paid defies analysis on any normal valuation method since it is the cost of getting into the centre. These exceptional prices offered by special purchasers are sometimes termed colloquially *"key money"* or *"foot in value"*, the expressions' meanings being self-evident. This value phenomenon suggests that the prevailing rents which are accepted as rental values are falsely low, a situation which can arise in any artificial market, which a centre can become in the absence of actual market transactions.

The valuer may also experience the situation where a price is achieved considerably in excess of any valuation and in contradiction of all previous market evidence. The reasons may be many, such as the inexperience of the purchaser, a mistake being made, the property having a special sentimental value and so on. The valuer needs to recognise such fluke transactions and to disregard or distinguish them when considering them as market evidence.

### 4. RICS/ISVA Appraisal and Valuation Manual

This has already been referred to at the beginning of this chapter. It is known as the Red Book. It was published in 1995 and it has been the subject of a number of updates since publication. Its use is mandatory on members of the RICS and the ISVA and the IRRV. This manual is not directly concerned with valuation theory or the method of valuation. It is, however, directly concerned with the mechanics of practice, including advice on referencing, the assembly, interpretation and reporting of information and matters which valuers should include within their reports. The purpose of the manual is to provide a uniform level of service to be provided by valuers to their clients and particular notice should be taken of the need to clarify the instructions on which the valuer is acting. In effect it sets out 'best practice' for valuers to follow.

The practice statements within the Red Book also include a large number of definitions, together with guidance notes thereon. These apply to the majority of valuations prepared by valuers and no valuation governed by the Red Book should be prepared without a reference to these definitions and guidance notes. They are considered in more detail in Chapter 20.

It is probable that, in time, the Red Book will be supplanted by international valuation standards such as the European standards of valuation contained in a Blue Book.

## 5. The Cost of Building Works

It is often necessary, in connection with the valuation of land or buildings for various purposes, to estimate the cost of development works, or of erecting new buildings, or of making alterations to buildings.

The following are examples of such cases:

(i)   In valuation of land likely to be developed in the future it is usually necessary to form some estimate of the cost of the necessary development works, including roads, sewers, public services and the provision of amenities such as tree planting, open spaces and the like.

(ii)  In connection with the development or redevelopment of urban sites for various purposes, estimates of the cost of building works are required.

(iii) Similar estimates are necessary where a valuer is called on to advise as to the sum for which existing premises should be insured against fire.

The method to be adopted in arriving at appropriate costs of buildings and development works will depend upon the accuracy desired and the information available.

In estimating the development value of large areas of land it is necessary to make assumptions as to the value which the land, or various parts of it, will command in the future. These estimates will be based on the evidence available at the time when the valuation is made, but will necessarily depend on factors which may vary considerably in the future. Any greater degree of accuracy in estimating the cost of necessary development works will therefore be out of place.

In many cases, estimates have to be made of the cost of works to be carried out in the future. In view of the fluctuations in costs that may occur before the works are carried out, undue refinement or accuracy in the methods used might again be out of place. However, in the case of, for example, major town centre developments where the planning stages are often prolonged, it may be necessary to project both income and costs into the future. Indices of past changes in rents and of building costs will provide some assistance but some additional technique which enables various assumptions on future changes to be tested will also be applied.

In many instances, both in connection with building estates and with the development of individual sites, the particulars available of the type of building to be erected or which, in fact, can be erected on a particular site may be very scanty. In the absence of detailed plans and specifications any estimate of cost can only be arrived at by means of an approximation.

It is not possible to formulate any general rule in this connection. The circumstances of each case must dictate the degree of accuracy to be attempted, and the usefulness of the result achieved will depend largely upon the experience of the valuer in applying his knowledge, derived from other similar cases, to the one under review.

### Bills of quantities

Where detailed drawings and a specification are available, it is possible for a bill of quantities to be prepared and for this bill to be priced by a quantity surveyor or contractor. This is the most accurate method of arriving at cost of works, but is often not practicable.

An approximate bill of quantities may be prepared whereby only the principal quantities in various trades are taken off, the prices to be applied being increased beyond what is customary, to include for various labours which normally would be measured and valued separately.

In the case of works of alteration, approximate quantities may be the only way of arriving at a reasonably accurate estimate where the works are extensive or present unusual difficulties.

### Unit comparisons

The most common approximate method of comparing building costs is per square metre of gross internal floor area.

The gross internal floor area of a proposed building can be arrived at with an acceptable degree of accuracy from sketch plans or from a study of the site on which the building is to be erected, taking into account the likely requirements of the local planning authority, the restrictions imposed by building regulations and any rights of light or other easements.

The gross internal floor area having been determined, the total cost can be estimated by applying a price per square metre derived from experience of the cost of similar buildings or from a publication such as *Spon's Architects and Builders Price Book* which is revised annually. Costs are based on typical buildings and do not include external works. Adjustments will, therefore, be required if the building is not typical, if site conditions are abnormal and for external works. It must be remembered that costs taken from publications are necessarily based on tenders prior to publication and are therefore prone to inaccuracy, particularly if there has been a significant change in building costs since those tenders.

Gross internal floor area is usually determined by taking the total floor area of all storeys of the building measured inside external walls and without deduction for internal walls. The method is set out in detail in the *Code of Measuring Practice* published by the RICS and ISVA (Fourth Edition published in 1993).

It should be noted that gross internal floor area will commonly differ from net lettable area.

*Other methods of comparison*
In some cases the gross external area of the building is utilised but the valuer should be careful to exclude the reference to "*gross external floor area*". The unit of comparison being used is the gross building area and not the gross floor area. The most usual case where this method is used is in relation to the valuation for insurance purposes of houses and flats and for most mortgage purposes it is usual to use the latest edition of the BCIS Guide to House Rebuilding Costs published by the Royal Institution of Chartered Surveyors.

In the case of building estates, the cost of road and sewer works is often estimated on a linear basis. This price should provide for the roads to be completed to the standard required by the local authority for adoption.

In the absence of a detailed development scheme, it is usually desirable to prepare a sketch plan of the proposed development in order to estimate the cost of the development works required. It may sometimes be possible, however, to make a comparison between one estate and another, whereby the total development cost per unit may be estimated by reference to the development costs of other land or similar character in the neighbourhood.

The usefulness of these methods necessarily depends upon the extent of the information available and upon the experience of the valuer, who must make full allowance for any differences between the properties from which his experience is derived and the particular property to which such figures are to be applied.

*Alterations and repairs*

The use of an approximate bill of quantities in the case of alterations has already been referred to. A similar degree of accuracy may often be arrived at by preparing a brief specification of the works required to be done and placing a *"spot"* amount against each item. Similar methods may be used in regard to repairs.

## 6. Claims for Damages

### (i) Dilapidations

There is a limit provided by Section 18 of the Landlord and Tenant Act 1927 to the amount which a landlord can recover from his tenant by way of damages for breaches of the covenant to repair. This limit is known as the *"diminution in the value of the landlord's reversion"*.

It should be noted that this section only applies to the breach of the covenant to repair and not for breaches of other covenants, such as a breach of the covenant to decorate in the last year of the term or breach of the covenant to remove internal partitions or to reinstate the premises. In the latter cases regard should be had to the case of *Joyner* v *Weeks*[5] which held that damages were payable based on the cost of the work, irrespective of whether the landlord actually suffered any loss.

For the purposes of calculating the diminution in value, it is necessary for the valuer to produce two calculations. The first calculation is the value of the property on the basis that there had been no breach of the covenant to repair and the second the value of the building as it actually existed on the term date of the lease.

---

[5] [1891] 2 QB 31.

Some items of disrepair will not have an effect on value because the item concerned will be superseded by works which the landlord proposes in any event to carry out in order to modernise or upgrade the building. This concept of supersession was considered in the case of *Mathe*r v *Barclays Bank plc.*[6] In many cases the two valuations will have the appearance of two residual calculations. The calculation for the building in disrepair will show higher building costs, a possibly longer period of development, higher fees and higher interest than the valuation of the building as if in repair, but some of the items put into the calculations will be neutral to both valuations. Therefore they require a lesser degree of accuracy in their ascertainment. This would apply to the rental value and yield in particular.

It sometimes happens by virtue of the nature of the landlord's interest, the value in repair may be negative but likewise the value in disrepair may be a greater negative. The difference between these two figures will still represent the diminution in the value of the landlord's reversion. This was the situation in the case of *Shortlands Investments Ltd* v *Cargill plc.*[7]

In some cases the value of the property in repair and the value of the property in disrepair may be the same and therefore no damages will be payable to the landlord. An example of this is where a large multiple retailer leaves first class premises with the interior being in disrepair. It is usual for each of the large multiple retailers to have its own internal style of fitting and therefore it will usually strip out the entire interior of its new shop unit so as to fit it to its own style. Consequently, it follows that the state of the repair of the interior would not in that case lead to any reduction in value.

There will, of course, be no claim for dilapidations if the building is to be demolished. (For a full consideration and understanding of dilapidations, regard should be had to the specialist text books in this area).[8]

---

[6] [1987] 2 EGLR 254

[7] [1995] 1 EGLR 51.

[8] *Handbook of Dilapidations* (Sweet & Maxwell); *West's Law of Dilapidations* (Estates Gazette, 10th Edition, 1995); *Dowding's Dilapidations: The Modern Law and Practice* (2000) by N. Dowding and K. Reynolds (Sweet & Maxwell).

## (ii) Other Claims

The level of damages payable in many actions often relates to the difference in value between a property in a known state as compared to in an assumed hypothetical state. One example is in connection with the calculation of damages for negligence resulting from a defective structural survey. The valuer's duty in these cases is to advise on the loss in value to the plaintiff and not to consider other heads of damage which may as a matter of law also apply. Frequently the difference in value will relate to the cost of necessary repair, for example where structural movement has been missed or dry rot has been missed. It is usually impossible to find comparable evidence for a property in a given state of disrepair. Therefore the valuer will start with the value in repair and then will adjust that figure to represent the cost of dealing with the defective items, financing same, and profit to reflect the risk inherent in the purchase of an affected property as well as the reward to the person carrying out the work in recompense for his inconvenience and, often, entrepreneurial input. It is very unusual for a difference in value to be represented solely by the cost of work without the other incidental cost being added to it, but regard must also be had to the possibility of betterment resulting from the work having been carried out.

An example of the betterment point is where the surveyor has failed to point out that a 40-year old house needed a new roof. The comparable for the house in good repair would often be one where the roof is in an acceptable, but not new, condition. Once the house has been re-roofed then it would go up in value to reflect the absence of any cost relating to maintaining the roof over the foreseeable future. Similarly, in some areas, purchasers welcome the opportunity to reconstruct the interior of their property and are not particularly put off by the need to redecorate completely or to renew fixtures or fittings. Indeed, in some areas perfectly new and modern fixtures and fittings are removed from the premises to reflect the purchaser's different taste.

The approach to all of these valuations is the same. The valuer should produce two valuations in the same way as when considering diminution in the value of the landlord's reversion. The first valuation reflects the true actual position and the second valuation reflects the position before the event which has triggered the litigation was known. Loss in value is clearly valuation 2 minus valuation 1 and in both valuations all relevant factors must be taken into account, many of which will be common to both sets of calculations.

### 7. Management of Investments

The valuer may be required to advise on the policy to be adopted in the management of investment properties.

No attempt is made here to deal with general estate management issues or with the type of problem that can arise where, for example, a major estate forms part of the wider investment portfolio of a large financial institution. However, an attempt is made to comment on issues with a valuation connotation which are common to most types of estate, large or small.

Speaking very generally it may be said that good estate management should aim at securing the maintenance and maximisation of income and capital value.

To advise on a group of investment properties of varying types involves consideration of a number of factors, including:

(a)  The possibilities of increase or decrease in future capital value.
(b)  An examination of present rents with a view to possible increase or reductions.
(c)  The security of the income. This relates to both security in real terms and security in actual terms. In connection with the former, consideration of (a) above will apply. However, there is a second consideration which is the actual security of payment of the money and this will depend on the quality of the covenant and the ability to seek payment from a predecessor in title. This latter point is bound up with a consideration of Privity of Contract.

    Where the actual tenant is the first lessee then the security of income will depend on the quality of that tenant's covenant and any guarantor who may have been a party to the lease. However in many cases the lease will have been assigned and then it may be possible to go back to a previous tenant, or guarantor, to obtain from him any unpaid rent or to require him to pay damages for any breach of covenant. This is now governed by the Landlord & Tenant (Covenants) Act 1995 which came into effect on 1 January 1996.

    There are two sets of rules depending upon whether the lease was entered into prior to 1st January 1996 or after that date. In the former, Privity of Contract still applies, but for arrears of rent this only relates to a maximum of six months' rent which is due prior to the issue of a notice under the Act; any additional rent above six months' will not be recoverable. Where a lease is granted after 1 January 1996 then there will be no Privity of Contract unless an authorised guarantee agreement was entered into. The value of an investment would depend on the security of the income and therefore different yields might be applicable to identical properties where the rules concerning Privity of Contract are different.

(d) Consideration of outgoings.

(e) The provision of reserves for future repairs or other capital expenditure, to enable the cost of repairs to be spread evenly over a period, and to provide sinking funds for wasting assets such as leasehold properties.

(i) **Maintenance of capital values:**

From time to time it will be desirable to review the properties comprised in a landed estate to decide whether some should be sold or otherwise dealt with and to consider whether or not they are likely to increase or decrease in value.

As an obvious and desirable objective is to maintain or enhance the capital value of the estate, properties likely to depreciate should be sold and others sought to replace them of a character which may increase in value or at least not depreciate. On the other hand, such properties may produce a high rate of return and if a high level of income is desired they may nonetheless be suitable for retention in the portfolio.

Opportunities may arise where other interests in the properties can be bought, or adjoining properties, which will create marriage value. For example, an estate may own the freehold interest in properties let on leases granted some years ago when rents were considerably lower than to-day's and where no provision was made for rent review. It will be advantageous to purchase any such leases which come up for sale so that the lease can be extinguished and a new one granted on improved terms. This will not only increase income but probably increase the capital value of the freehold interest by more than the cost of the lease as marriage value applies. Re-structuring (or re-arrangement of the terms, including rent) of leases is a further possibility. Alternatively, the freehold interest could be offered to the lessees on the basis of a shared marriage value with obvious benefit to both parties.

The possibilities of enhancement of capital value by re-development or refurbishment need to be kept in mind subject to the availability of finance for such works. In the case of large-scale developments they may be carried out by developers under ground lease so that external capital is imported into the estate.

There is also the possibility of investment in properties of a type where increases in value may be expected and which are not dependent upon the grant of planning permission.

For example, a block of shops offered for sale may be the subject of leases granted sometime ago at rents which at the end of the lease may be expected to be substantially increased. An investment in property of this type offers possibilities of an inherent increase in capital value in the future.

Property in blocks may also contain the possibility of realisation of "break-up" value, the converse of marriage value. This happens when the sum of the parts exceeds the value of the whole and where individual occupiers are likely to be interested in purchasing the properties they occupy.

(ii) **Examination of rents:**
A periodic examination of rentals should be made to ensure that when a property comes into hand the rent is increased to the market value. In other cases it may be found that, owing to a fall in values or for other reasons, a property is over-rented.

It will probably be wise to treat the excess over market value with caution; in making provision for the future it must be remembered that the only security for the enhanced rent is the tenant's covenant to pay, and in the event of his failure the rent will need to be reduced. The valuation implications are considered in Chapters 12 and 18.

The general tendency over the long term has been for rental values to increase, but this pattern was broken by a significant fall in rental value from 1989 to 1995. Before 1989, in order that landlords could secure the benefit of such increases, it was usual in the case of occupation leases to provide for upward only rent review clauses to operate at regular intervals, say every three, four or five years. The terms were generally a multiple of the review periods, for example, 15 or 25 years' lease with five-year reviews. Since 1989, leases have tended to be shorter, with tenant's break clauses within the term. In the case of building leases, similar periodic rent review clauses are incorporated but the rents commonly remain as a percentage of the full rental value so that they rise in parallel – they are "geared" to the full rental value.

(iii) **Outgoings:**
The present outgoings, particularly those liable to variation such as repairs and rates, need to be scrutinised. Where the landlord is responsible for outgoings, rating assessments should be checked to ensure they are correct, that void allowances are claimed where properties are empty and that adequate provision is made for repairs and insurance.

(iv) **Reserves:**
Certain outgoings, in particular repairs, will vary in amount from year to year. For instance, external painting may be required perhaps every five years. It is usually desirable that the available income from an estate should be reasonably constant, and in the years when such expenditure is not actually incurred an amount should be set aside for future use. So that if the cost of external painting every fifth year is estimated at say £5,000, a sum of £1,000 should be set aside each year to meet it, rather than have an excessively large outgoing in one year which may swallow up the rental income.

Where property is held on lease with full repairing covenants, or where there is a covenant to reinstate after alteration, provision should be made for future expenditure either by setting aside a certain amount of income each year or on the lines indicated below.

Where income is derived from leasehold properties, proper provision should be made for the replacement of capital at the end of the term. This may be done by means of a sinking fund policy or other amortisation plan or it may be possible to find a suitable investment – for example, ground rents with a reversion to rack rents which will come into hand at about the time when the leasehold interests run out. This portfolio mix can ensure that the estate overall maintains its rental and capital value.

The latter arrangement has certain advantages over the sinking fund method. The value of the reversion accumulates at a high rate of interest in most cases and also a sinking fund, if established, has to be paid out of taxed income, and reduces the net available revenue. It may be, however, that the capital resources of an estate do not permit the purchase of suitable investments and provision for the future has to be made out of income.

A valuer may have to advise a particular investor in landed property on the type of investment best suited to him. In such a case, the valuer would have to take account of such a factor as the investor's liability to income tax. The approach to value to a particular investor, is known as "worth". This is considered more fully in Chapter 12.

# Chapter 14

# Principles of the Law of Town and Country Planning

## 1. The Town and Country Planning Act 1947

The modern system of planning control, that is to say control of land use and development,[1] originated in the Town and Country Planning Act 1947. As is well known, there were several earlier planning statutes, but the system of control which they introduced was of limited scope and effect. The Act of 1947 is in a very real sense the starting point of planning as we know it in Britain.

It was intended in 1947 that there should be two sides to modern planning law – physical (development control) and financial (compensation and taxation). They were meant to be logically complementary; but that principle has long been abandoned, and for purposes of analysis they must be considered separately. The original rigorous balance between the two sides disappeared in 1952 with the abolition of the "development charge" system introduced by the 1947 Act; and the "compensation-betterment problem", as it has sometimes been called, of deciding upon which financial principles, if any, public planning in the long term should be based, was shelved.

The evolution of British planning law after 1952, had it been foreseen in 1947, would have caused astonishment. The earlier law had left the question of land valuation to the market. The prospect of demand for development was one factor, but not the only one, which might increase value. The law recognised the concept of market value, but did not define or regulate it because it was a valuation and not a legal concept; except that from 1919 onwards

---

[1] See *Planning Law and Procedure* (9th Edition, 1993) by A.E. Telling and R.M.C Duxbury (Butterworths); *An Outline of Planning Law* (10th Edition, 1991) by Sir Desmond Heap (Sweet & Maxwell); *A Practical Approach to Planning Law* (4th Edition, 1994) by V.WE. Moore (Blackstone Press); *Encyclopedia of Planning Law and Practice* (Sweet & Maxwell); *Statutes on Planning Law* (1995, Blackstone Press).

statutory rules were imposed to ensure that market value in compulsory purchase did not diverge from market value in free market transactions (Acquisition of Land (Assessment of Compensation) Act 1919). In regard to planning, all that was required before 1947 was that compensation should be paid to any landowner whose land was depreciated by a planning scheme. After 1947 all was changed. Totality of planning control thereafter meant that development value must be clearly differentiated from existing use value: the former would depend on planning permission coupled with demand (Lord Denning's words), the latter on demand alone.

The aim of the Act of 1947 was to transfer development value (i.e. the value of development potential as distinct from the actual cost of development) from landowners to the State, but not existing use value. A system of planning control without financial control would have led to the capricious result that, given the existence of market demand for development, a landowner would enjoy development value in the event of the grant of planning permission but be deprived of it in the event of refusal. A "development charge" therefore transferred to the State the whole of any development value which resulted from the grant of planning permission. Refusal of permission merely resulted in there being no development value anyway.

Abolition of the "development charge" in 1952 produced the capricious result referred to above, but over the years this has come to be taken for granted, so much so that a judge (Sir David Cairns, in the Court of Appeal) could say as a self-evident proposition: "There is no general right to compensation for the refusal of planning permission" *(Peaktop Properties (Hampstead) Ltd* v *Camden London Borough Council.*[2] Yet from 1952 to 1991 compensation for such refusals could be obtained in a restricted range of cases. The Planning and Compensation Act 1991, Section 31, has finally abolished such compensation, except for revocation and discontinuance orders (see below).

The 1947 Act recognised an owner's right to some compensation in respect of development value existing at 1 July 1948, but not in respect of development value accruing after that date. In theory, at any rate, the recognised basis for all dealings in land after 1 July 1948, was "existing use value".

---

[2] (1983) 46 P & CR 177.

"Existing use value" is a term which was first introduced into valuation practice by the provisions of the 1947 Act. It indicates the value which can be assigned to a property if it be assumed that no "development", within the meaning of the Town Planning Acts, will take place in the future except such forms of development as were formerly specified in the Third Schedule to the Town and Country Planning Act 1947, which has now been replaced by Schedule 3 to the Town and Country Planning Act 1990 as amended by the Planning and Compensation Act 1991.

Contrasting with "existing use value", therefore, is "development value", as described above. Before planning control existed, "development value" depended simply on market demand. If buyers would not offer more for a given property for purposes of development than for its existing use, then its "development value" over and above "existing use value" would be nil. But now "development value" depends on both planning permission and demand. The then Master of the Rolls, Lord Denning, said in *Viscount Camrose* v *Basingstoke Corporation* [3]: "It is not planning permission itself which increases value. It is planning permission coupled with demand."

## 2. Planning Authorities and Development Plans

The central planning authority is the Department of the Environment, Transport and the Regions. The Secretary of State does not usually administer planning control directly; but appeals are made to him from decisions of local authorities and he has the power to "call in" applications from them for decision at first instance. He makes orders and regulations, has default powers and co-ordinates policy by issuing circulars to guide and advise the local planning authorities, which handle detailed administration. Of these, county councils are "county planning authorities", concerned chiefly with "strategic" matters and district councils are "district planning authorities", concerned normally with routine control. Other English and Welsh "unitary" authorities do both, e.g. in London (boroughs and the City Corporation but *not* the Greater London Authority), Metropolitan districts, and National Parks.

---

[3] [1966] 1 WLR 110; [1966] 3 All ER 161.

In order that they do not make their decisions at random, they are required to make, and constantly revise, "development plans" for their area; and planning decisions should always be made with the appropriate development plan in mind even if for sound reasons they deviate from it. The system of development plans under the Town and Country Planning Act 1947 was gradually superseded by a new system, introduced by Part I of the Town and Country Planning Act 1968, because of the delays caused by the requirement that the Minister had to approve all plans, in every detail, before they could become effective.

### Structure Plans and Local Plans

The system which began in 1968 consists of two kinds of plan, to match the "two-tier system" of county and district planning authorities in the non-metropolitan counties (but not, of course, the six metropolitan counties and Greater London). First comes the structure plan, to formulate "general policies in respect of the development and use of land" which the country planning authority will already have made and will under the current legislation have a duty to update. This is done by alteration or by replacement. Formal proposals have to be adopted. Then come the local plans, which may provide in detail for the area covered by the structure plan.

The essence of a structure or local or old-style plan is a written statement. In addition, a local plan "shall contain a map illustrating each of the detailed policies", but a structure plan may only be illustrated by diagrams. The prerequisite for all these plans is a survey, which it is "the duty of the local planning authority to institute . . . in so far as they have not already done so".

### Unitary Plans

More recently the replacement in some local government areas of the "two-tier" by a "one-tier" system (Greater London and the six so-called metropolitan counties) has brought about a reversion to single or "unitary" development plans. These are now governed by the Town and Country Planning Act 1990, Part II, Chapter I (Sections 10–28).

Unitary development plans are to be produced, from dates appointed by the Secretary of State, by local planning authorities in

Greater London and the metropolitan counties, where there is only one "tier" of authorities. (Other "unitary" councils are now being set up.) Such plans "shall comprise two parts", namely a written statement of principles and a written statement of detailed proposals plus a map, a "reasoned justification", and "diagrams, illustrations or other descriptive or explanatory matter". The two written statements "shall be in general conformity"with one another.

### Purpose of Development Plans

The legal effect of a development plan is largely indirect. Its main function is to guide the authorities; in making policy decisions they must "have regard" to it (TCPA 1990, Section 54A). There are a few positive consequences: there is a requirement to consider development plans when compulsory purchase compensation is claimed on the basis that it includes "development value", and when it is proposed to serve certain kinds of "blight notice".

Mention may be made here of registers which local planning authorities are required to keep for public inspection. In addition to registers of local land charges, which include various orders, agreements and notices relevant to planning and compulsory purchase, there are registers kept specifically for planning. Thus there are registers of planning applications, of applications for consent to display advertisements and of caravan site licences, together with the decision of the planning authority on the applications; and there are also lists of buildings of special architectural or historic interest. Prospective purchasers and their solicitors should always consult these registers and lists in appropriate circumstances, just as they normally apply for an official search of the local land charges register.

### 3. Judicial Control of Planning Decisions

No one holding public office has unrestricted freedom in decision-making, i.e. arbitrary power. Subject to alteration by Act of Parliament, the "discretion" of all public authorities has express or implied limits, enforced by the courts. Activities outside those limits are *ultra vires* and the courts can invalidate them. But there are stringent restrictions on recourse to the courts in planning as in other matters within the scope of the *ultra vires* principle.

Many, but not all, decisions of the Secretary of State (as distinct from those of local planning authorities) can only be challenged by application within six weeks to the High Court; and even so the court can only quash such decisions on the ground that they are "not within the powers of (the relevant Act), or that the interests of the applicant have been substantially prejudiced" by some procedural default. This represents a tightened definition of the *ultra vires* principle itself. Local planning authorities and other public bodies are, however, liable to have their *ultra vires* decisions challenged by a more general procedure in the High Court, "judicial review", in accordance with the Supreme Court Act 1981, Section 31, in which the time limit is three months instead of six weeks (Rules of the Supreme Court, Order 53).

The House of Lords has stated that "the courts' supervisory duty is to see that (the authority) makes the authorised inquiry according to natural justice and arrives at a decision, whether right or wrong . . . they will not intervene merely because it has or may have come to the wrong answer, provided that this is an answer that lies within its jurisdiction".[4]

The reference to "natural justice" is a reminder that the Secretary of State's decisions are usually reached after first granting a hearing to objectors. Many of the statutory provisions require him to "afford . . . an opportunity of appearing before, and being heard by, a person appointed by the Secretary of State for the purpose" (i.e. an inspector) to objectors, appellants, claimants or "persons aggrieved". Such proceedings must be conducted in accordance with "natural justice", which comprises two basic rules, namely that the person presiding must not be biased and that both sides are given a proper hearing on the points at issue. Though part of an administrative process, this private hearing or public inquiry is said to be "quasi-judicial"; so that the final decision can be quashed by the courts if "natural justice" is not observed.

Some inquiries are subject to safeguards additional to "natural justice". These include planning appeal inquiries, planning enforcement appeals inquiries, and compulsory purchase order

---

[4] *Anisminic Ltd* v *Foreign Compensation Commission* [1969] 1 All ER 208 at p. 237. See also *Page* v *Hull University Visitor* [1993] 1 All ER 97, at p.100 *per* Lord Griffiths: "The purpose is to ensure that those bodies that are susceptible to judicial review have carried out their public duties in the way it was intended they should".

inquiries. The various sets of Inquiries Procedure Rules[5] governing these inquiries prescribe time limits and notice to be given to the parties and require written submissions to be made in advance by the authority stating the contentions on which they intend to rely. The Secretary of State has full discretion to make his eventual decision, so that he may reject any or all of the recommendations made by the inspector in his report; but he must hear any further representations if he disagrees on any finding of fact or considers any new issues or evidence of fact; and in the latter two cases he must re-open the inquiry if requested. In some cases the inspector who conducts the inquiry is himself empowered to take the decision instead of reporting back to the Secretary of State.

The "supervisory duty" of the High Court in regard to the *ultra vires* principle (natural justice included) is not an appeal jurisdiction; it does not relate to the merits of official decisions, as distinct from their general lawfulness, because those merits are questions of policy, not law.

## 4. Legal Meaning of "Development"

Section 55 (1) of the Town and Country Planning Act 1990 defines "development" as "the carrying out of building, engineering, mining or other operations in, on, over or under land, or the making of any material change in the use of any buildings or other land". Thus there will be development either if an "operation" is carried out, or if a "material change of use" is brought about. Often a project involves development because there will be one or more operations and a material change of use as well.

Section 55 lists specific matters which either are or are not "development". The latter include "in the case of buildings or other land which are used for a purpose of any class specified in an order made by the Secretary of State under this section", the use thereof "for any other purpose of the same class". Thus we have the Town and Country Planning (Use Classes) Order 1987, which lists 16 "use classes" in four groups A, B, C and D; and any change within a "use class" is not development at all, or in other words not "material".[6]

---

[5] In particular, see the Town and Country Planning (Inquiries Procedure) Rules, SI 1992 No 2038.
[6] But a planning permission may lawfully be granted on condition that a use of premises must not be subsequently changed even within a "use

Whether any work amounts to an "operation" or whether any change of use is "material" is a "question of fact and degree" in each case. Examples include such difficult questions as whether demolition is an "operation", and whether the abandonment or the intensification of a use is a "material change". Building a model village as a permanent structure has been held to involve an "operation" but not placing a mobile hopper and conveyor in a coal-merchant's yard. Placing an egg-vending machine on the roadside of a farm has been held to involve a material change of use; but not altering part of a railway station yard from a coal depot to a transit depot for crated motor vehicles. The burden of proof rests heavily on that party who alleges that a finding in relation to development is *ultra vires*.

Ownership of land, or of things placed on land, is irrelevant to planning except in special circumstances: what matters is the nature of what is done on or to the land. Questions which may be relevant include: (i) what is the actual area involved; (ii) whether there are multiple uses on a given area of land and whether these are of equal importance, or are major and minor uses, or are confined to separate parts of the premises, or are intermittent, alternating or recurring; (iii) whether a project may amount to development for two or more quite separate reasons, for example the making of a reservoir in such a way as to involve an engineering operation (the work of construction) and a mining operation (the work of excavation and removal of mineral substances).[7]

## 5. Planning Permissions

Section 57 of the 1990 Act states that planning permission is "required" for carrying out development (subject to certain special exceptions). Section 59 empowers the Secretary of State to make "development orders", for the purpose, *inter alia*, of actually granting permission, on a general and automatic basis, for certain forms of development. Practitioners have long been familiar with the Town and Country Planning General Development Order, known for short as the "GDO", which gave such permission for

---

[6] *(continued)*
class": *City of London Corporation* v *Secretary of State for the Environment* (1971) 23 P&CR 169.
[7] See *West Bowers Farm Products* v *Essex County Council* (1985) 50 P&CR 368.

specified classes of development. But with effect from 3 June 1995, this Order has been replaced by two separate Orders.

Of these, the General Permitted Development Order ("GPDO") largely re-enacts the classes of permitted development previously specified with some specialised additions, the whole being set out in Schedule 2 of the Order in 33 "Parts". The kinds of permitted development which apply most widely are to be found in Parts 1 (within the curtilage of a dwelling house), 2 (minor operations), 3 (changes of use) and 4 (temporary buildings and uses), also 6–8 (certain kinds of agricultural, forestry and industrial development as specified).

The other order is the General Development Procedure Order ("GDPO"). This prescribes the procedure for making applications to the local planning authority for planning permission. If a building is to be erected, an application may be made for "outline" permission, which means for approval in principle. If this is refused no time and expense need be wasted on detailed plans. If it is granted, separate application will need to be made for details to be approved – "reserved matters".

Any person may apply for planning permission; but an applicant who owns neither a freehold nor leasehold in all the land affected (normally a prospective purchaser) must notify all freeholders, leaseholders with seven years or more to run, and farm tenants, either directly or, if that is not possible, by local press publicity. There are also certain classes of controversial development which must be publicised. The persons notified by these methods may "make representations" which the authority must take into account.

A prospective developer who is not certain whether his project amounts to "development" at all may apply to the authority for a "certificate of lawfulness of proposed use or development" (1990 Act, Section 192) or a "certificate of lawfulness of existing use or development" (1990 Act Section 191), as the case may be. It is also possible for a "person interested in land" to make an agreement with the authority (enforceable against him or persons deriving title under him) regulating development of that land on a more general basis than for a normal planning permission (1990 Act (as amended by the Planning and Compensation Act 1991) Sections 106–106B).

On receiving an application the local planning authority must consult other authorities and government departments, as appropriate, and "have regard" to the provisions of the development plan.[8] Within two months they must notify their decision to the applicant. They may grant permission "unconditionally or subject to such conditions as they think fit", or refuse it (1990 Act, Section 70). Obviously if a developer acts on a planning permission he cannot continue any previous use of the land which is inconsistent with it. The possibility that a project could be regulated under some other statutory procedure does not preclude a refusal of planning permission, even if that other procedure might carry with it a right to compensation.

Planning conditions are subject to a test of validity both in principle and in detail. That is to say they must "fairly and reasonably relate to the permitted development" and they must be reasonable in their detailed terms. A condition that cottages to be built must only be occupied by "persons whose employment is or was employment in agriculture" seems to have satisfied both tests. A condition that a project of industrial development on a site next to a dangerously congested main road must include the provision of a special access road, and that this access road should be made available to members of the public visiting adjoining premises, seems to have satisfied the first test but not the second.[9]

Conditions which may be valid include those which cut down the use of other land of the applicant and also those which require a new use to cease after a stated time and thus take effect as temporary planning permissions. But permissions are normally permanent and "enure for the benefit of the land" (1990 Act, Section 75).[10]

Other time conditions, which are so frequent as to be virtually standard-form conditions, specify the time within which development must take place, or at least begin. There is a statutory three-year deadline in "outline" permissions for seeking approval

---

[8] This means that the decision "shall be made in accordance with the plan, unless material considerations indicate otherwise" (TCPA 1990, Section 54A, inserted by the Planning and Compensation Act 1991). Environmental impact assessments may be required (Section 71A).

[9] See: *Pyx Granite Co Ltd* v *Ministry of Housing and Local Government* [1959] 3 All ER 1; *Fawcett Properties* v *Buckingham County Council* [1960] 3 All ER 503; *Hall & Co Ltd* v *Shoreham-by-Sea Urban District Council* [1964] 1 All ER 1.

[10] They cannot therefore be "abandoned": *Pioneer Aggregates (UK) Ltd* v *Secretary of State for the Environment* [1984] 2 EGLR 183.

for all detail or "reserved matters", followed by a two-year deadline for starting development after final approval; alternatively there is an overall five-year deadline for starting development (if longer) as well as a five-year deadline for starting development under ordinary as distinct from "outline" permissions. The authority, however, can vary any of these periods. There is, moreover, an additional control by "completion notice". Where any of the above deadlines applies and development has duly begun in the time specified but has not been completed in that time, the local planning authority may serve a "completion notice", subject to confirmation by the Secretary of State (with or without amendments) specifying a time, not less than a year, by which development must be complete or else the permission "will cease to have effect" (1990 Act, Sections 91–94).

If he so wishes, the Secretary of State may direct that a planning application be "called in" (as it is usually termed), that is referred to him instead of being decided by the local planning authority. Appeal to the Secretary of State against a refusal of permission, or a grant made subject to conditions, or a failure to give any decision within the appropriate time-limit, must be made in writing within six months of the adverse decision or of the expiry of the time-limit. He may allow or dismiss the appeal or reverse or vary any part of the permission, and his decision is as free as if he were deciding at first instance. The procedure is now governed by Inquiries Procedure Rules (which have already been mentioned) and a hearing must be given if it is asked for (1990 Act, Sections 77–79).

The Secretary of State's decision on an application "called in" or on an appeal, or that made by an inspector on his behalf, is "final" and cannot be challenged in a court except in the circumstances described earlier in relation to the judicial control of planning decisions.

Planning permission can be revoked or modified (1990 Act, Sections 97–100). The authorities which do this must pay compensation for any abortive expenditure and for any depreciation in relation to development value which, having come into existence by virtue of the permission, disappears because of the revocation or modification.[11] Revocation or modification orders must be confirmed by the Secretary of State except in uncontested cases. If

---

[11] TCPA 1990 Sections 107–113.

permission is given automatically by the Town and Country Planning (General Permitted Development) Order 1995, referred to earlier, it may in effect be revoked or modified. If permission given by an "article 4 direction" under that Order is partly or wholly withdrawn, and a specific application is then made which is refused or only granted subject to conditions, compensation is paid in these cases also.

In so far as authorised development has actually taken place, even if only in part, revocation or modification orders and "article 4 directions" are ineffective. To put an end to any actual development or "established use" of land (except of course where it is the necessary consequence of acting on a planning permission that this should happen) requires a discontinuance order, which must also be confirmed by the Secretary of State (1990 Act, Section 102). Compensation must be paid for loss of development value and abortive expenditure and also the cost of removal or demolition; but as compliance involves physical action there is also an enforcement procedure in cases of recalcitrance, similar in essentials to ordinary enforcement of planning control. Discontinuance of mineral working, or its temporary suspension, is governed by a specially modified code of regulation and compensation (1990 Act, Schedule 9).

Public authorities are subject to planning control with certain reservations. The most far-reaching concerns the Crown, to which planning control does not apply at all,[12] though as a matter of practice the relevant government departments do normally consult local planning authorities when proposing to develop land. But Crown lessees and other persons using Crown land are subject to planning control. The Crown may obtain planning permissions or consents for the benefit of prospective purchasers, and local planning authorities may make tree preservation orders and issue enforcement notices in respect of Crown land so as to affect (where appropriate) persons other than the Crown (1990 Act, Part XIII, Sections 299–302).

Ordinary local authorities, however, have no such immunity, except that when any project which involves expenditure requires the approval of a government department such approval may also be expressed to confer "deemed" planning permission, with or without conditions, if needed (1990 Act, Sections 58 and 90). This

---

[12]*Ministry of Agriculture, Fisheries and Food* v *Jenkins* [1963] 2 All ER 147.

rule also applies to "statutory undertakers", that is the public utility authorities and similar public bodies; but with them there is also another factor, the difference between their "operational" and non-operational land (the latter being offices, houses, investment property and any other land which is not the site of their operating functions). "Operational" land has the benefit of special rules in planning law, for example in regard to compensation for restrictions on development. Local planning authorities are subject to special rules where they propose to carry out development and require the Secretary of State's approval if the development is not in accordance with the development plans (1990 Act, Part XV, Section 316; Town and Country Planning General Regulations 1992).

### Simplified Planning Zones

Sections 82–87 and Schedule 7 of the Town and Country Planning Act 1990 require local planning authorities to consider making simplified planning zone schemes, and the Secretary of State may direct them to do so. A scheme resembles a development order, in that it automatically grants planning permission for specified classes of development, with or without conditions, in the whole or part of a zone.

### Enterprise Zones

Finally, reference should be made to "enterprise zones". The Secretary of State is empowered to approve schemes designating these zones which are prepared by local authorities or new town or urban development corporations at his invitation. Planning permission is automatically granted for various kinds of development specified in each scheme; it may in some cases be "outline" permission.[13]

### 6. Enforcement of Planning Control: Enforcement Notices

Part VII of the 1990 Act deals with enforcement of planning control. It is not a criminal offence to develop land without planning permission. If this happens, the local planning authority should first consider whether it would be "expedient" to impose sanctions,

---

[13] TCPA 1990, Sections 88—89. There are also fiscal advantages.

"having regard" to the development plan and to any other material considerations. If it would, they may issue an "enforcement notice". This must specify the "breach of planning control" complained of, the steps required to remedy it, the date when it is to take effect, and the time allowed for compliance. "Breach of planning control" occurs when development takes place either without the necessary permission or in disregard of conditions or limitations contained in a permission.[14]

## Planning Contravention Notices

A local planning authority to whom "it appears ... that there may have been a breach of planning control" may serve a "planning contravention notice" on the owner or occupier of the land in question, or on the person carrying out operations thereon, requiring the recipient to furnish specified information which will enable the authority to decide what action (if any) to take by way of enforcement. Failure to comply within 21 days is a criminal offence, punishable by a fine up to level 3 on the standard scale; the giving of false information either knowingly or recklessly is punishable by a fine up to level 5 on the standard scale.[15]

## Stop Notices

The period specified in the enforcement notice before it takes effect is intended to allow for making an appeal, and the notice is "of no effect" while any appeal is going forward. This may encourage a recalcitrant developer to press on with his activities in the meantime, in the hope of creating a *fait accompli*. Local planning authorities therefore have the additional power, during this period, to serve a "stop notice" prohibiting any activity which is "specified in the enforcement notice as an activity which the local planning authority requires to cease, and any activity carried out as part of that activity or associated with that activity".[16]

---

[14] *Ibid*, Sections 171A; 172–3. The enforcement notice must not specify a date to take effect earlier than 28 days after service of copies of the notice on all owners and occupiers of the relevant land (and anyone else with an interest "materially affected").

[15] *Ibid*, Sections 171C–D.

[16] *Ibid*, Sections 173A–175; 183–4.

Appeal may be made against an enforcement notice by the recipient "or any other person having an interest in the land" within the time specified before it is to take effect (1990 Act, Sections 174 and 191-2). It must be made in writing to the Secretary of State, and may be on one or more of seven grounds:

(a) permission ought to be granted or a condition or limitation ought to be discharged;

(b) the alleged breach of planning control did not take place;

(c) the matters alleged do not amount to a breach of planning control;

(d) the alleged breach occurred more than four years ago (operations, or material changes of use *to* a single dwelling) or 10 years (other changes);

(e) copies of the notice were not served on the proper parties;

(f) the specified steps for compliance are excessive;

(g) the specified time for compliance is too short.

The Secretary of State must arrange a hearing or inquiry before an inspector, if either side requires it, and he may uphold, vary or quash the enforcement notice and also grant planning permission if appropriate. He may "correct any informality, defect or error" in the notice if "satisfied" that it is "not material", and may disregard a failure to serve it on a proper party if neither that party nor the appellant has been "substantially prejudiced". The Court of Appeal has stated that "an enforcement notice is no longer to be defeated on technical grounds. The Minister . . . can correct errors so long as, having regard to the merits of the case, the correction can be made without injustice". That was said in a case in which it was held to be at most an immaterial misrecital for an enforcement notice to allege development "without permission" when in fact a brief temporary permission had existed under the General Development Order. "The notice was plain enough and nobody was deceived by it."[17]

Further appeal from the Secretary of State's decision on an enforcement notice lies to the High Court on a point of law (1990 Act, Section 289). Apart from these appeals as a general rule anyone

---

[17] *Miller-Mead* v *Minister of Housing and Local Government* [1963] 1 All ER 459, *per* Lord Denning. On the question of who may be served with an enforcement notice, see *Stevens* v *London Borough of Bromley* [1972] 1 All ER 712.

may challenge the validity of an enforcement notice in appropriate legal proceedings, but normally no such challenge shall be made on any of the seven grounds in Section 174 otherwise than by appeal to the Secretary of State (1990 Act, Section 285). One obvious possibility of challenge is in defence to a prosecution, since although a breach of planning control is not a criminal offence, failure to comply with an enforcement notice is (1990 Act, Section 179). An owner who has transferred his interest to a subsequent owner can, if prosecuted, bring the latter before the court.

In addition to prosecution after failure to comply with an effective enforcement notice within the time specified in it, the authority also have the power, after that time, to enter on the land and carry out the steps prescribed by the notice, other than discontinuance of any use, and recover from the owner the net cost reasonably so incurred. He may in turn recover from the true culprit, if different, his reasonable expenditure on compliance (1990 Act, Section 178).

Anyone wishing to take precautions against being subjected to enforcement procedures can, in suitable cases, cover themselves by applying for a certificate of lawfulness of existing or prospective use or development (1990 Act, Sections 191-2).

## 7. Amenity and Safety

The other major concerns of planning law are amenity and safety. Amenity "appears to mean pleasant circumstances, features, advantages";[18] but it is not statutorily defined, nor is safety. The provisions of planning law governing amenity cover trees, buildings of special interest, advertisements, caravan sites and unsightly land; those governing safety cover advertisements and hazardous substances.

To grow or cut trees is not of itself development.[19] But local planning authorities are specifically empowered, "in the interests of amenity", to make "tree preservation orders" ("TPO") for specified "trees, groups of trees or woodlands", restricting

---

[18] *Re Ellis and Ruislip-Northwood Urban District Council* [1920] KB at p.370 *per* Scrutton, LJ.

[19] Conditions for replanting or preservation of trees should be imposed in planning permissions where appropriate (TCPA 1990, Section 197). For control of timber felling see the Forestry Act 1967, as amended.

interference with the trees except with the consent of the local planning authority. Dangerous trees, however, may be cut if necessary. There are also provisions governing replanting. Unauthorised interference with any protected tree calculated to harm it is a criminal offence (1990 Act, Part VIII, Section 210).

A TPO is made and confirmed by the local planning authority or the Secretary of State after considering any objections from owners and occupiers of the relevant land, though, if necessary, a provisional TPO taking immediate effect can be made for up to six months.[20] Regulations are prescribed governing the procedure for making TPOs, and their content. Standard provisions in TPOs lay down essentially the same procedure for applying for consents to interfere with protected trees as exists for making planning applications.[21]

"Amenity" is not expressly mentioned in relation to buildings of special interest, which are now governed by the Planning (Listed Buildings and Conservation Areas) Act 1990. Part II of the Act[22] refers to "areas of special architectural or historic interest, the character or appearance of which it is desirable to preserve or enhance", and requires local planning authorities to determine where such areas exist and designate them as "Conservation Areas". When one of these areas has been designated, "special attention shall be paid to the desirability of preserving or enhancing the character or appearance of that area" by exercising appropriate powers to preserve amenities under planning legislation, and also by publicising planning applications for development which in the authority's opinion would affect that character or appearance. All trees in conservation areas are protected in the same way as if subject to a TPO.

The phrase "special architectural or historic interest" applies chiefly to buildings, although trees and other objects may affect their character and appearance. The Secretary of State has the duty,

---

[20] TCPA 1990, Sections 198–202.
[21] TCP (Tree Preservation Order) Regulations SI 1969 No 17, amended by SI 1981 No 14 which includes a model TPO in standard form.
[22] Sections 69–80. Demolition of buildings generally in a conservation area requires a listed building consent (Section 74) which is discussed below. As to planning permissions in a conservation area, see *South Lakeland District Council* v *Secretary of State for the Environment* [1992] 1 PLR 143.

under Part I[23] of the Planning (Listed Buildings and Conservation Areas) Act 1990, of compiling or approving lists of such buildings, after suitable consultations, and supplying local authorities with copies of the lists relating to their areas. Such authorities must notify owners and occupiers of buildings included in (or removed from) these lists. The Secretary of State may, when considering any building for inclusion in a list, take into account the relationship of its exterior with any group of buildings to which it belongs and also "the desirability of preserving . . . a man-made object or structure fixed to the building or forming part of the land and comprised within the curtilage of the building". If a building is not "listed", the local planning authority may give it temporary protection by a "building preservation notice" while they try to persuade the Secretary of State to list it.

Except when for the time being a "listed building" is an ecclesiastical building used for ecclesiastical purposes[24] or an ancient monument,[25] it is a criminal offence to cause such a building to be demolished, or altered "in any manner which would affect its character as a building of special architectural or historic interest", without first obtaining and complying with a "listed building consent" from the local planning authority or the Secretary of State, unless works have to be done as a matter of urgency. A consent may be granted subject to conditions, contravention of which is also a criminal offence; and it is normally effective for five years.

The procedure for applying for listed building consents, and for appeals and revocations, is laid down on lines very similar to the procedure in ordinary cases of planning permission for development; and so is the procedure for listed building enforcement notices and purchase notices. Compensation is payable for depreciation or loss caused by revocation or modification of listed building consents or by the service of building preservation notices. If an owner fails to keep a listed building in proper repair, a local authority or the Secretary of State may first serve a "repairs notice" and, if this is not complied with

---

[23] Sections 1–68. See also the House of Lords decision in *Shimizu (UK) Ltd* v *Westminster City Council* [1997] 1 All ER 481.

[24] *Ibid*, Section 60. See *Attorney-General* v *Howard United Reform Church Trustees,Bedford* [1975] QB 41; [1975] 3 All ER 273.

[25] See the Ancient Monuments and Archaeological Areas Act 1979, and the National Heritage Act 1983.

after two months, may then compulsorily purchase the property. Local authorities can, on seven days' notice to the owner, carry out urgent works at the owner's expense to preserve any unoccupied building, or part of a building, which is listed.

Control of the display of advertisements is provided for, under the 1990 Act, Section 220, in the interests of amenity and safety, but not censorship. The details are laid down in regulations.[26] The use of any land for the display of advertisements requires in general an application to the local planning authority for consent, which in normal cases is for periods of five years. Appeal lies to the Secretary of State. There are several categories of display in which consent is "deemed" to be given, including the majority of advertisements of a routine nature and purpose; but "areas of special control" may be declared where restrictions are greater. If the authority "consider it expedient to do so in the interests of amenity or public safety" they may serve a "discontinuance notice" to terminate the "deemed" consent of most kinds of advertisement enjoying such consent; but there is a right to appeal to the Secretary of State. Contravention of the regulations is a criminal offence. Consent under the regulations is "deemed" to convey planning permission also, should any development be involved.

The control of caravan sites may also be regarded as a question of amenity. Such control involves questions of public health, and there is authority for the view that control for purposes of public health must not be exercised for purposes of amenity. But there can be little doubt in practice that although control is concerned with health and safety on the caravan site itself, it preserves amenity for the neighbourhood of the site.

Planning permission must be sought for caravan sites, but the detailed control of the use of each site is governed by a system of "site licences", obtainable from the local authority (Caravan Sites and Control of Development Act 1960). "There are two authorities which have power to control caravan sites. On the one hand, there is the planning authority . . . On the other hand, there is the site authority . . . The planning authority ought to direct their attention to matters in outline, leaving the site authority to deal with all matters of detail. Thus the planning authority should ask themselves this broad question: Ought this field to be used as a

---

[26] TCP (Control of Advertisements) Regulations, SI 1992 No 666.

caravan site at all? If "Yes", they should grant planning permission for it, without going into details as to number of caravans and the like, or imposing any conditions in that regard". Nevertheless, "Many considerations relate both to planning and to site . . . In all matters there is a large overlap, where a condition can properly be based both on planning considerations and also on site considerations" (*per* Lord Denning).[27]

It is the "occupier" of land who must apply for a site licence, which must be granted if the applicant has the benefit of a specific planning permission, and withheld if he has not; and it must last as long as that permission lasts, perpetually in a normal case. The practical question, therefore, is what conditions a site licence shall contain. They are "such conditions as the authority may think it necessary or desirable to impose", with particular reference to six main kinds of purpose. Appeal may be made to a magistrates' court against the imposition of any conditions, or a decision or refusal to vary them at any time after imposition, on the ground that as imposed or varied they are "unduly burdensome".[28]

There are several categories of use of land for caravans which are exempted from control, and also additional powers conferred on local authorities in special cases.

Next comes the question of unsightly land: neglected sites, rubbish dumps and the like. Local planning authorities are empowered, when "the amenity of any part of their area, or of any adjoining area, is adversely affected by the condition of land in their area", to serve a notice on the owner and occupier, specifying steps to be taken to remedy the condition of the land (Section 215). As with enforcement notices, two time limits must also be specified: a period (of 28 days or more) before the notice takes effect, and the time of compliance. Failure to comply is a summary offence (1990 Act, Section 216 as amended).

Appeal lies, at any time before the notice takes effect, to a magistrates' court on any of the following grounds:

---

[27] *Esdell Caravan Parks Ltd* v *Hemel Hempstead Rural District Council* [1966] 1 QB 895, at p.922.

[28] Caravan Sites and Control of Development Act 1960, Sections 1–12. A condition not related to health, safety or amenity will almost certainly be *ultra vires* rather than merely "burdensome": See *Mixnam's Properties Ltd* v *Chertsey Urban District Council* [1965] AC 735; [1964] 2 All ER 627 (rent control).

(a)  the condition of the land is not injurious to amenity;
(b)  the condition of the land reasonably results from a use or operation not contravening planning control;
(c)  the specified steps for compliance are excessive;
(d)  the specified time for compliance is too short. The magistrates may uphold, quash or vary the notice and "correct any informality, defect or error" if it is not material (1990 Act, Section 217). A further appeal lies to the Crown Court (1990 Act, Section 218), during which the notice is in suspense.

An additional set of controls over the use of land has been introduced into planning law by the Planning (Hazardous Substances) Act 1990. This deals with the placing on any premises of substances such as dangerous chemicals. The purpose of this control is to protect safety, and (subject to that) amenity. It is additional to existing controls upon the handling of such substances, in that its emphasis relates to the *land* as distinct from the *substances* themselves; but nevertheless it has been derived from those controls, specifically the Health and Safety at Work etc. Act 1974. "Hazardous substances" are defined in the Notification of Installations Handling Hazardous Substances Regulations 1982.[29] On 1 May 1984 the Use Classes Order and the General Development Order were amended so as to withdraw generally from the scope of those Orders any use of premises involving a "notifiable quantity" of any "hazardous substance", as defined in the above Regulations of 1982 (apart from certain limited types of permission preserved in what is now the GPDO).

The Planning (Hazardous Substances) Act 1990 introduced for England and Wales a new code, whereby the presence of a hazardous substance on, over or under land requires the consent of the hazardous substances authority unless "the aggregate quantity of the substance . . . is less than the controlled quantity".[30] The Secretary of State is empowered to define "hazardous substances" by specifying them in regulations, together with "the controlled quantity of any such substance" (as distinct from the "notifiable quantity" referred to above).

Control of land, the use of which involves hazardous substances is to be exercised whenever they are present in an appreciable

---

[29] SI 1982 No 1357.
[30] Section 4 of the Act.

amount ("controlled quantity"). The Act requires applications for "hazardous substances consents" to be made to "hazardous substances authorities" which are, by and large, the local planning authorities, including county councils where sites used for mineral workings or waste disposal are involved and in most National Parks, as well as certain urban development corporations and housing trusts. Central Government is also involved because the "appropriate ministers" are the authorities for the "operational land" of "statutory undertakers". The system of consents (with or without conditions), plus revocations, appeals, enforcement, etc., is broadly similar to planning control, and in fact was previously integrated with it; the purpose of the separation is to free this system of control of land use from being tied to the concept of "development" as against *safety* which is the true consideration.

There is a requirement for "appropriate consultations" to take place with the Health and Safety Executive of the Health and Safety Commission.

The Planning (Hazardous Substances) Act came into force on 1 June 1992, subject to some amendments by the Environment Protection Act 1990.[31]

There is also the Radioactive Substances Act 1960 which enacts the "duty of public and local authorities not to take account of any radioactivity in performing their functions". That Act was amended by Part V of the Environmental Protection Act 1990, which empowers the Secretary of State for the Environment to appoint inspectors to enforce safety requirements.

## 8. Planning Compensation

During and after the Second World War much attention was focused on the so-called "compensation–betterment" problem, which was concerned with the question of whether, and how much, compensation should be paid (and by whom) to owners of land deprived of development value by planning control, and with the converse question of whether, and how much, development value should be taken (and by whom) from owners of land enabled under planning control to realise such value. The attempts by

---

[31] Planning (Hazardous Substances) Regulations 1992 (SI 1992 No 656) and DOE Circular 11/92.

Parliament to grapple with this problem subsided gradually into incoherence over the years.

The position now is that "betterment", or development value, is taxed when it accrues to land as part of the general system of taxation of capital gains, instead of being regarded as something separate from other forms of capital gains. Planning compensation for depreciation of "betterment" has been abolished, in regard to the limited situations in which it was previously obtainable, by the Planning and Compensation Act 1991, Section 31 (including compensation for refusal of listed building consent), except for revocation and discontinuance orders under Part IV of the Town and Country Planning Act 1990,[32] and cases under Sections 28 and 29 of the Planning (Listed Buildings and Conservation Areas) Act 1990. These cases are not common in practice.

---

[32] Sections 107–112 and 115–118. The local planning authority for the area must pay this compensation, which covers loss of development value, abortive expenditure, etc., but it is assumed that planning permission (and hence development value) is available for rebuilding works not substantially increasing the volume of the original building, and also for converting houses into flats. A leading case illustrating these provisions is *Canterbury City Council* v *Colley* [1993] 1 EGLR 182, in which the owner obtained compensation in regard to the vacant site of a demolished house on the assumption that there was planning permission to rebuild it when such permission had in fact been revoked.

# Chapter 15

# Principles of the Law of Compulsory Purchase and Compensation

## 1. Legal Basis of Compulsory Purchase

There are numerous Acts of Parliament under which Government departments, local or public authorities, or statutory undertakings, may carry out schemes for the general benefit of the community involving the acquisition of land or interference with owners' proprietary rights.

Where an owner's property is taken under statutory powers he is entitled to compensation as of right, unless the Act which authorises the acquisition expressly provides otherwise. Where no interest in land is taken but a property is depreciated in value by the exercise of statutory powers, the owner's right to compensation depends on the terms of the Act under which these powers are exercised.

For a detailed account of the law on the subject the reader should consult one of the standard textbooks on the subject.[1] The present chapter is confined to a brief outline of the law such as will provide the necessary background to the principles involved in compulsory purchase valuations.

Compulsory purchase of land normally brings into play four main sets of statutory provisions, as follows. First, there is the authorising Act, normally a public general Act authorising a public body or class of public bodies (e.g. county councils) and former public bodies which are now private following privatisation to carry out some specified function, and going on to state:

---

[1] See *Compulsory Purchase and Compensation* (5th Edition, 1998) by Barry Denyer-Green (Estates Gazette); *Law of Compulsory Purchase and Compensation* (5th Edition, 1994) by Keith Davies (Tolley Publishing Co); *Guide to Compulsory Purchase and Compensation* (5th Edition, 1984) by J.K. Boynton (Oyez Publications); *Encyclopedia of Compulsory Purchase and Compensation* (Sweet & Maxwell).

(a)  whether such a body may acquire land for the purpose;
(b)  whether they may buy it compulsorily;
(c)  whether they may obtain power to do this by compulsory purchase order specifying the land required; and
(d)  if so, what procedure is to be followed when making the CPO.

There is now a standardised procedure laid down by the Acquisition of Land Act 1981, though alternative procedures are occasionally specified instead. Second, therefore, is the Act of 1981 (or such alternative code as the authorising Act may prescribe), which governs the making of the CPO; and it may be said that the great majority of acquisitions are made under that Act. Third is the Compulsory Purchase Act 1965, which has to all intents and purposes replaced the Lands Clauses Consolidation Act of 1845 and governs the actual procedure for acquisition after the CPO has sanctioned it, supplemented by provisions in the Compulsory Purchase (Vesting Declarations) Act 1981 and in the Land Compensation Act 1973. Fourth is the Land Compensation Act 1961, which contains the current rules for assessing compensation in so far as it relates directly to land values; these, too, are supplemented by provisions in the Land Compensation Act 1973.

Disputes over compulsory purchase fall broadly into two main classes, depending on whether or not they relate to the assessment of compensation. If they do (and also in one or two special cases to be mentioned below) they must be brought before the Lands Tribunal, a specialised body set up under the Lands Tribunal Act 1949 and staffed by valuers and lawyers. Otherwise they should normally be brought before the High Court. Appeal lies to the Court of Appeal not only from the High Court but also from the Lands Tribunal (though on a point of law only, by way of case stated, and within six weeks of the Tribunal's decision).

## 2.  Compulsory Purchase Procedure

Any acquiring authority who are empowered by an appropriate authorising Act to select and acquire compulsorily the particular land they need by CPO procedure must normally make the CPO in accordance with the procedure laid down in the Acquisition of Land Act 1981. This involves making the order in draft, and submitting it to a "confirming authority", which will be the appropriate Minister or Secretary of State unless of course he himself is acquiring the land. There must be prior press publicity and notification to the owners and occupiers of the land, and the hearing of objections by an inspector from the Ministry or Department concerned. Statutory inquiries procedure rules for hearings and inquiries are in force, closely parallel with those discussed above in relation to planning appeals. The order may be confirmed, with or without modifications, or rejected. If confirmed it takes effect when the acquiring authority publish a notice in similar manner to the notice of the draft order and serve it on the owners and occupiers concerned. The order cannot be challenged (except perhaps on the ground of nullity) apart from the standard procedure for appeal to the High Court within six weeks on the ground of *ultra vires* or a procedural defect substantially prejudicing the appellant.

The CPO will lapse, in relation to any of the land comprised in it, unless it is acted on within three years. When the authority wish to act on the order, they must serve a notice to treat on the persons with interests in the land to be acquired, requiring them to submit details of their interests and their claims for compensation. When the compensation is agreed in each case, it and the notice to treat together amount to an enforceable contract for the sale of the land. This is then subject to completion by the execution of a registered or unregistered conveyance in the same way as a private land transaction.[2]

There is, however, an alternative procedure at the authority's option whereby the notice to treat and the conveyance are combined in a "general vesting declaration". The authority must notify the owners and occupiers concerned, in the same notice as that which states that the CPO is in force (or a separate and later notice), that they intend to proceed in this manner by making a vesting declaration not less than two months ahead. This, when made, will by unilateral action vest the title to the land in the authority at a date not less than 28 days after notification to the owners concerned. It will by and large have the same consequences as if a notice to treat had been served.[3]

Freeholds and leaseholds, both legal and equitable, are capable of compulsory acquisition. Leasehold tenancies with a year or less to run, including periodic tenancies, are not subject to acquisition and compensation but allowed to run out, after due service of

---

[2] Compulsory Purchase Act 1965, Sections 4, 5 and 23.
[3] Compulsory Purchase (Vesting Declarations) Act 1981.

notice to quit if necessary; although if the authority desire possession quickly they can take it subject to payment of compensation for the loss caused.

An authority cannot normally, without clear statutory authorisation, take rights over land in the limited form of an easement or other right less than full possession (even a stratum of land beneath the surface). For example, in *Sovmots Ltd* v *Secretary of State for the Environment*[4] the House of Lords quashed a compulsory purchase order for acquisition of a lease and a sub-lease (not the freehold) of certain property, together with new easements of access and support which would be required because only the upper part of a building was to be taken, on the ground that appropriate statutory authority was lacking. For local authorities the necessary statutory authority was provided by the local Government (Miscellaneous Provisions) Act 1976, Section 13 and Schedule 1. But if the authority acquire a dominant tenement they acquire the easements appertaining to it, as in private conveyancing; and if they acquire a servient tenement they either allow the easements and other servitudes over it to subsist without interference or else pay compensation for "injurious affection" to the dominant land if they do so interfere.

If part only of an owner's land is to be acquired, this is "severance". The owner of "any house, building or factory" or of "a park or garden belonging to a house" can require the authority to take all or none; but the authority can counter this by saying that to take part only will not cause any "material detriment", and any such dispute has to be settled by the Lands Tribunal. Similar rules now apply to farms.[5]

Unjustifiable delay by the authority after service of a notice to treat may, in an extreme case, amount to abandonment of the acquisition.[6] But the Planning and Compensation Act 1991, Section 67, amplifying the Compulsory Purchase Act 1965, Section 5, provides that a notice to treat expires at the end of three years unless it has been acted on (e.g. by entry, or settlement of compensation, or reference to the Lands Tribunal, or substitution of a general vesting declaration). As for making actual entry on the

---

[4] [1979] AC 144; [1977] 2 All ER 385.

[5] Compulsory Purchase Act 1965, Section 8; Land Compensation Act 1973, Sections 53–8.

[6] *Grice* v *Dudley Corporation* [1957] 2 All ER 673.

land, the authority are not normally entitled to do this until completion and the payment of compensation, unless they first serve a "notice of entry";[7] entry before payment of compensation entitles an owner to receive interest on the compensation to be paid.

Many acquisitions by authorities are made by agreement, but are mostly under the shadow of compulsory powers and consequently involve the same rules of compensation. Obligations owed to third parties, as in restrictive covenants, do not normally involve the expropriated owner in liability, and the third party should seek compensation from the authority if there is any breach in such a case. On the other hand an owner must not increase the authority's liability to compensation by creating new tenancies and other rights in the land or carrying out works on it after service of the notice to treat, if any such action "was not reasonably necessary and was undertaken with a view to obtaining compensation or increased compensation".[8]

## 3. Compulsory Purchase Compensation

The detailed rules of compensation are discussed in Chapters 26–28. It will, however, be convenient to consider them briefly in outline here in order to demonstrate the legal basis on which they rest.

The acquiring authority must compensate the expropriated owner for the land taken, by way of purchase price, and for any depreciation of land retained by him, as well as for "all damage directly consequent on the taking".[9]

The basis of compensation for the taking or depreciation of land is "market value", namely "the amount which the land if sold in the open market by a willing seller might be expected to realise". "Special suitability or adaptability" of the land which depends solely on "a purpose to which it could be applied only in pursuance of statutory powers, or for which there is no market apart from the special needs of the requirements of any authority possessing compulsory purchase powers", must be disregarded. There must be no addition to nor deduction from market value purely on the

---

[7] Compulsory Purchase Act 1965, Section 11.
[8] Acquisition of Land Act 1981, Section 4.
[9] *Harvey* v *Crawley Development Corporation* [1957] 1 All ER 504, *per* Denning LJ.

ground that the purchase is compulsory, nor any addition or reduction solely and specifically on account of the project to be carried out (on the claimant's land or any other land) by the acquiring authority. An increase in the value of adjoining land of the owner not taken by the authority, if it is attributable solely to the acquiring authority's project, must be "set off" against compensation.[10]

If the property has been developed and used for a purpose which has no effective market value, such as a church, then the Lands Tribunal may order that compensation "be assessed on the basis of the reasonable cost of equivalent reinstatement", if "satisfied that reinstatement in some other place is bona fide intended".[11]

These intricate legal rules are intended for the guidance of valuers rather than lawyers. Valuers engaged in the assessment of the compensation are required, subject to such guidance, to reach a figure which will put the expropriated owner in a position as near as reasonably possible to that in which he would find himself if there had been no compulsory acquisition and he had sold his land in an ordinary private sale.

Market value, however, has in any case two distinct main elements: "existing use value" and "prospective development value". Since development is not lawful without planning permission, the absence of permission will inhibit purchasers from paying any amount over and above "existing use" value, whether the land is built on or vacant in its present state of development. Before the days of planning control, "prospective development value" over and above "existing use value" depended on market demand. It must not be forgotten that this is still true. "It is not planning permission by itself which increases value. It is planning permission coupled with demand."[12]

Assessing the existence of demand is essentially a question of valuers' expert evidence; though of course the Lands Tribunal is better qualified than a court to pronounce on such evidence.

Assessing the availability of planning permission, however, calls for special statutory rules. This is because there are many cases

---

[10] Land Compensation Act 1961, Sections 5–9 and Schedule 1, as amended by the Planning and Compensation Act 1991.
[11] Land Compensation Act 1961, Section 5 (Rule 5).
[12] *Viscount Camrose* v *Basingstoke Corporation* [1966] 3 All ER 161, *per* Denning MR.

where planning permission is refused purely because some proposed private development, which might otherwise be acceptable and if so would be capable of giving rise to appreciable development value, is ruled out by the impending compulsory purchase notwithstanding that this purchase will often be for the purpose of a public works project giving rise to little or no development value.

"Assumptions as to planning permission" are therefore authorised by statute. The most useful of these turn on the allocation or "zoning" in the current development plan of areas of land which include the owner's property for uses which command a lucrative development value: residential, commercial or industrial. There may be a range of such uses. But permission can only be assumed it is also reasonable to do so in relation to the particular circumstances of the land itself. If the development plan does not "zone" the land in this way, the owner (or the authority) can apply to the local planning authority for a "certificate of appropriate alternative development" in relation to the particular circumstances of the land to show what planning permission would have been granted if the acquisition had not occurred. A "nil certificate" will, however, state that planning permission would not have been granted for any other development. Appeal lies to the Secretary of State; and from him in turn lies the usual limited right of appeal within six weeks to the High Court (Land Compensation Act 1961, Sections 17–22, as amended) on points of law.[13]

If, within 10 years after a compulsory acquisition, any subsequent planning permission is granted which adds to the development value of the land acquired, that additional value may be claimed by the expropriated owner from the acquiring authority, calculated on the basis of values at the time of the acquisition from him.[14]

---

[13] In *Porter* v *Secretary of State for Transport* [1996] 2 EGLR 10, the Court of Appeal held that a "certificate of appropriate alternative development" does not "estop" (i.e. prevent) the Lands Tribunal from making its own assessment of the factors likely to affect development value, including, if possible, alternative schemes to that of the acquiring authority.

[14] Land Compensation Act 1961, Part IV, inserted by the Planning and Compensation Act 1991, Section 66 and Schedule 14. There are various special cases specified in these provisions where the right to additional compensation is excluded. The time of the acquisition is in normal cases to be taken as the date of the notice to treat.

In addition to purchase price compensation there is compensation for "injurious affection", i.e. depreciation of land retained. This is "severance" if caused by a reduction in the value of the land retained greater than that caused by the reduction in size. Depreciation caused by what is done on the land taken by the acquiring authority is closely analogous to damages in tort for private nuisance, though it may well include loss not compensatable in tort.[15] But if the harm done goes beyond what is authorised by the statutory powers of the acquiring authority, then it will in any case be unlawful and so compensatable (if at all) in tort and not as "injurious affection".

It is also possible to obtain compensation for "injurious affection" when no land has been acquired from the claimant. Here it is necessary to prove four things:

(a)  the loss is caused by activity authorised by statute;
(b)  it would be actionable at common law or in equity if it were not so authorised;
(c)  it is strictly a depreciation in land value; and
(d)  it arises from the carrying out of works on the compulsorily acquired land and not from its subsequent use.

These four rules are customarily attributed to the decision of the House of Lords in *Metropolitan Board of Works* v *McCarthy*.[16] But depreciation caused by the use of public works, including highways and aerodromes, is in many cases now compensatable under Part I of the Land Compensation Act 1973, if attributable to "physical factors". The claim period runs for six years starting one year after the use begins.

Another head of compensation is "disturbance", which is not strictly land value but "must . . . refer to the fact of having to vacate the premises".[17] Thus it may include the loss of business profits and goodwill, removal expenses and unavoidable loss incurred in acquiring new premises. It has been held that to claim for "disturbance" an owner must forego "prospective development

---

[15] *Buccleuch (Duke of)* v *Metropolitan Board of Works* (1872) LR 5 HL 418. On this, see Chapter 27.
[16] (1874) LR 7 HL 243. The claimant's business premises in London became less valuable when an adjoining public dock (a public right of way) was destroyed by the construction of the Victoria Embankment at Blackfriars.
[17] *Lee* v *Minister of Transport* [1965] 2 All ER 986, *per* Davies LJ.

value" in his purchase price compensation; that is to say his "true loss" is whichever is the higher: "existing use" value plus "prospective development" value, or "existing use" value plus "disturbance".[18]

Disturbance compensation is (illogically) regarded in law as an integral part of land value. It is therefore not payable where the acquiring body, having expropriated the landlord, do not expropriate a short-term tenant but displace him by notice to quit. In such cases the Land Compensation Act 1973 (Sections 37–8) provides for "disturbance payments" by the acquiring body to the tenant (unless he is an agricultural tenant, for whom separate compensation provisions exist). The 1973 Act also authorises the payment of "farm loss payments" to displaced owner-occupiers of farms (Sections 34–6) and "home loss payments" to displaced occupants of dwellings who have lived there for five years (Sections 29–33).

A claimant "must once and for all make one claim for all damages which can be reasonably foreseen".[19] The date of the notice to treat fixes the interests which may be acquired, but does not govern compensation which, as the House of Lords held in the case of *Birmingham Corporation* v *West Midland Baptist (Trust) Association (Inc)*,[20] must be assessed as at the time of making the assessment, or of taking possession of the land (if earlier), or of the beginning of "equivalent reinstatement".

## 4. Compulsory Purchase in Planning

The Town and Country Planning Act 1990 is itself the authorising Act for certain kinds of compulsory purchase of land; and Part IX

---

[18] *Horn* v *Sunderland Corporation* [1941] 1 All ER 480. In this case Scott LJ referred to the claimant's "right to receive a money payment not less than the loss imposed on him in the public interest but on the other hand no greater." Lord Nicholls in the Privy Council in *Director of Buildings and Lands (Hong Kong)* v *Shun Fung Ironworks Ltd* [1995] 1 All ER 846, emphasised that this right justifies including in the compensation any "personal losses" in the nature of disturbance where "the general principle of fair and adequate compensation" so requires. This may extend to relocation of a business, but only in appropriate circumstances which demonstrate that the claim is genuine and reasonable.

[19] *Chamberlain* v *West End of London etc. Rail Co* (1863) 2 B&S 617, *per* Erle CJ.

[20] [1970] AC 874; [1969] 3 All ER 172; (1969) RVR 484.

of the Act authorises acquisition of land "for planning purposes." This means land which "is suitable for, and is required in order to secure the carrying out of one or more of the following activities, namely "development, redevelopment or improvement", or "for a purpose which it is necessary to achieve, in the interests of the proper planning of an area in which the land is situated" (Section 226). Local authorities in general have this power, subject to the standard compulsory purchase procedure. They can themselves develop land so acquired, but not without the Secretary of State's consent. More usually they dispose of the land to private developers, "in such manner and subject to such conditions as may appear to them to be expedient" (Section 233). Other planning purposes for which land can be compulsorily purchased include under Section 228 of the 1990 Act, "the public service", "proper planning" or "the best or most economic use of the land" (the acquisition must be by the Secretary of State) and also the preservation of listed buildings, under the Planning (Listed Buildings and Conservation Areas) Act 1990, Sections 47–50.

Another aspect of compulsory purchase in planning, under the 1990 Act (Part VI), is "inverse" compulsory purchase, of which there are two species: "purchase notices" and "blight notices". The owners supply the compulsion in these cases, not the acquiring authorities. A purchase notice (Sections 137–148) is served in consequence of an adverse planning decision; but a blight notice (see below) is served in consequence of adverse planning proposals.

If planning permission is in a particular case refused, or granted subject to conditions, so that as a result "the land has become incapable of reasonably beneficial use in its existing state", then an owner may serve a purchase notice on the local borough or district council. If the council are unwilling to accept it they must normally refer it to the Secretary of State who must then exercise his own judgment as to whether the notice is justifiable and ought to be upheld. He must not uphold it merely on the ground that "the land in its existing state and with its existing permissions is substantially less useful to the server", since that is true of nearly all planning refusals. The land must in fact be virtually useless. There are other adverse decisions which justify a purchase notice, e.g. revocation orders.

A "blight notice" (1990 Act, Sections 149–171) is served on a prospective acquiring authority. There are four principal requirements: (1) the land must be "blighted land" within one of the categories set out in Schedule 13 to the 1990 Act; (2) the server must be an owner-occupier (or his mortgagee) with an interest "qualifying for protection"; (3) the server must have made genuine but unsuccessful attempts to sell for a reasonable price on the open market; and (4) the authority must in fact intend to acquire the land. Within two months the authority concerned may serve a counter-notice alleging that any of the above requirements has not been met. The claimant then has two more months in which to refer the dispute to the Lands Tribunal, before whom the burden of proof is on the authority if they deny an intention to acquire any or all of the land but on the claimant in all other cases.

The "blighted land" includes the following situations, all of which involve the threat, though not the certainty, of compulsory purchase of the land itself (not neighbouring land as such), in the foreseeble future. The land must be indicated as being required for the functions of a public body in a local plan or, failing that, in a structure plan or, failing that, indicated in a development plan as required for a highway; or as land in or beside the line of a trunk or special road, or sufficiently indicated in writing by the Secretary of State to the local planning authority as required for such a road or selected for a highway by a resolution of a local highway authority; or as land covered by a CPO which has not yet been acted upon, or else subject to compulsory purchase by virtue of a special enactment. Inclusion of land in a slum clearance area, or in road-widening proposals, is also within the "specified descriptions".

An interest in land "qualifying for protection" is that of: (a) a freeholder or a leaseholder with over three years to run, who is (b) a resident owner-occupier of a dwelling or the owner-occupier of either (i) all or part of an "agricultural unit" or (ii) other (non-residential and non-agricultural) premises with a rateable value currently (1990) not exceeding £18,000 (under the Town and Country Planning (Blight Provisions) Order, SI 1990 No 465). The period of actual occupation must have been six months either (a) immediately before serving the blight notice or (b) immediately before leaving the premises unoccupied for not more than 12 months before serving the notice. A mortgagee of a person "qualified for protection" may also serve a blight notice provided that his power of sale has arisen, and he is given an extra six months in which to do so. Special provision is made for cases where land is held by a partnership or has passed to personal representatives.

# Chapter 16

# Development Properties

## 1. Generally

The term "Development Properties" is used here to indicate the type of property the value of which can be increased by capital expenditure, by a change in the use to which the property is put, or possibly by a combination of capital expenditure and change of use. It has commonly been applied to areas of undeveloped land likely to be in future demand for building purposes; to individual sites in towns, at present unbuilt on; and to other urban sites occupied by buildings which have become obsolescent or which do not utilise the site to the best advantage. The value, which in these cases is latent in the property, can only be released by development or refurbishment and in all cases is subject to any necessary planning permission being granted.

Because of the pace of modern technology the valuation approach can also be applied to existing property which is no longer suitable to today's needs but which can be refurbished and altered to provide once again desirable space.

## 2. Valuation Approach

The method of valuation most commonly applied to development properties is the residual method described in Chapter 11. The alternative method is the comparison method where there is evidence of actual sales for development purposes.

In making the valuation it is necessary to determine several factors. It is first necessary to decide the type of development or redevelopment for which the land or building is best suited, due regard being had to the planning permission likely to be granted. It is frequently necessary to work on an assumption as regards what consent will be forthcoming and the conditions attached to it, particularly in relation to density and car parking. Where this is done the valuer must state clearly what consent is assumed and what the conditions are, and also state that this valuation is conditional upon that consent being granted. The valuer must then estimate the market value of the developed building or land when put to the proposed use; to consider the time which must elapse before the property can be so used; to estimate the cost of carrying out the works required to put the property to the proposed use, together with such other items involved as legal costs and agent's commission on sales and purchases and fees for planning applications; and to assess the cost of financing the project. Where the development is forecast to take a considerable time, or where the development is to be phased, the valuer must also consider the effect of future changes in realisation and costs on the land value.

Where data of recent similar transactions exist, the valuer may be able to use the comparison method of valuation. He may do so even if he has to look outside the area in which the property is situated, for example sales of office sites in other towns with the same level of office rental values or residential sites with closely similar house prices. The above factors will then bear on his consideration somewhat indirectly. Even when he can value by comparison he would be wise to make an alternative valuation by the Residual Method. Since this is essentially a forecast of sales and expenditure, it is possible to set out the figures in a different form; such a statement is known as a Viability or Feasibility Report or Study. Examples are given below but such a Report or Study contains the same figures as are employed in a Residual Valuation.

When the residual method is employed it is obvious that any errors made in the estimates of completed value, cost of development, etc., will be reflected in the valuation arrived at, so that considerable skill is required when applying this method if consistent and accurate results are to be obtained. Moreover, it frequently happens that the amount of the estimates of completed value and cost are very large compared to the value of the property in its present state; a small error in the estimates will entail a large error in the residue in such cases.

Although at one time the residual method was frequently used as the only method of valuation applied in any particular case, direct comparison of the property being valued with other similar properties may be a more reliable method, and is certainly preferred by the Lands Tribunal[1] although the residual method has

[1] See, for example, *Fairbairn Lawson Ltd* v *Leeds County Borough Council* (1972) 222 EG 566; *South Coast Furnishing Co Ltd* v *Fareham Borough Council*

been accepted by the Tribunal in some cases,[2] particularly where no comparables are available, and was accepted by the Court of Appeal when six residual valuations were submitted.[3]

The Tribunal's criticism of the residual method is essentially that of the inherent weakness referred to in the preceding paragraphs coupled with the fact that valuations for the Tribunal are not "tested" in the open market. The quality of the ingredients is probed by both sides in a willing buyer/willing seller scenario, but in the end event both want a completed transaction; this probing is not possible in the non-market scenario of a reference to the Lands Tribunal.[4] If an optimistic view of values and costs is adopted (high values and low costs) a high land value will emerge, whilst a pessimistic view will produce a low value. If such opposing but genuinely held views are adopted by the parties to a dispute then the arbitrator will be faced with markedly different values. This is the common experience of the Tribunal and it results from the fact

---

[1] *(continued)*
[1977] 1 EGLR 167; *Essex Incorporated Congregational Union* v *Colchester Borough Council* [1982] 2 EGLR 178.

[2] See *Baylis's Trustees* v *Droitwich Borough Council* (1966) RVR 158; *St Clement's Danes Holborn Estate Charity* v *Greater London Council* (1966) RVR 333; *Clinker & Ash Ltd* v *Southern Gas Board* (1967) RVR 477; *Trocette Property Co Ltd* v *Greater London Council* (1974) 28 P&CR 256.

[3] See *Nykredit Mortgage Bank plc* v *Edward Erdman Group Ltd* [1996] 1 EGLR 119. This was a case claiming negligence concerning a valuation of a development site when the valuations of six experts were submitted. As to the method of valuation to be adopted, it was said that "All are agreed that a residual valuation is the appropriate method for a site on which development is to be carried out by the purchaser."

[4] See *Liverpool and Birkenhead House Property Investment Co* v *Liverpool Corporation* (1962) RVR 162

In *Wood Investments Ltd* v *Birkenhead Corporation* (1969) RVR 137, Mr John Watson, a member of the Lands Tribunal said, with regard to a witness's residual valuation, that it "provides a telling illustration of its [the residual method's] uncertainties. The key figures are (a) the value of the completed buildings estimated at £265,537 and (b) the cost of providing it estimated at £230,210. £35,327, which is (a) less (b) is the land, but (a) and (b) are necessarily rough estimates and there must be some margin of error. If (a) turned out to be only 5% too high and (b) 5% too low the residual value of the land would be approximately £10,500 instead of approximately £30,000 and if the 5% errors happen to be the other way round it would be over £60,000."

that no actual transaction is contemplated; it is only hypothesised. The residual method is thus mistrusted in such cases. However, in commercial situations, where there is a willing purchaser and a willing seller, a realistic view must be adopted and the residual method is accepted as a proper approach. In effect, the various participating parts of the valuation are "tested" by both sides and a compromise value for the land is usually reached.

## 3. Types of Development

It can be said that there is at any given time a general demand for land for building purposes dependent upon factors applicable to the country as a whole, and that the extent to which this demand is localised in certain areas or in the neighbourhood of certain towns will depend upon local factors.

General factors affecting the demand for land include the state of prosperity of the country and population trends.

It is obvious that in prosperous times there will be an increasing demand for sites for such buildings as offices, shops and the like and for an improved standard of housing, both in quality and quantity. When the population is increasing there will be a larger demand for houses.

It does not, of course, follow that, because a certain type of development is provided for in the development plan for the area, the land can profitably be developed for such a use immediately or even within a reasonable period in the future. For instance, with a view to increasing employment in run-down locations such as inner cities, areas may be allocated for industrial use, but unless industrialists are willing to set up businesses in these areas the development of the land for this purpose will be unprofitable.

The prospect of profitable development of any particular piece of land within the conditions imposed, or likely to be imposed, by the planning authority will depend largely on local circumstances, and past evidence of trends of development in the neighbourhood will have to be taken into account. The valuer has also to consider general trends affecting development in the country as a whole.

Unless there is a valid planning consent in existence it will always be necessary to obtain an indication of the sort of planning consent likely to be forthcoming by informal discussion with the local planning authority before putting forward even a tentative valuation. If the existing consent is approaching the end of its life (it will only normally be valid for five years and sometimes less) then discussions with the local planners will also be necessary in order to assess whether the consent will be renewed. In the course of such discussion any requirements for "planning gain" will become apparent. For example, the planning authority may require the provision by the developer of children's play areas in residential developments or a community centre in a district shopping centre. The financial implications of such arrangements must be established and allowance made for any additional costs which may fall on the developer. There is a difference between a valuation for the purposes of a sale and a valuation for mortgage purposes. The former can be made conditionally upon the assumed planning consent being granted, whereas the latter can only be based on a planning consent which exists, or is so certain to be granted that there is no doubt that it will exist (this of course only applies where the loan is being made unconditionally; a mortgage valuation conditional upon a consent being granted can of course be given).

The nature of the development likely to be permitted having been ascertained or assumed, it is also necessary to assess its commercial possibilities. It does not follow that, because planning consent is likely to be obtained, there is necessarily a market for that particular form of development.

## 4. Viability Statements (Studies)

As explained above these are a forecast of the profits which will be earned from the developments. They are based on forecasts of sales and costs of development. They can be presented in a number of different forms including that of profit and loss accounts. The form used herein is one frequently employed where the land cost is known and the profitability of the scheme is being tested.

### Example 16–1

A developer has been negotiating for the purchase of a freehold site which could accommodate a small block of four self-contained flats each with garage. He can buy it for £80,000 subject to contract and to outline planning consent for the development.

Sketch plans indicate that the flats will have gross floor areas of 70 m² (including common parts) each and investigation of sales of flats in the same area indicate that selling prices will be £63,500, including garage, on a leasehold basis – 99 years at a ground rent of £100 per annum per flat.

You are instructed to advise a finance house on the proposal generally and your report is to include a Viability Statement.

*Viability Statement*

Sales:

| | | | |
|---|---|---|---|
| 4 flats at £63,500 each | | | £254,000 |
| Ground Rents: 4 × £100 p.a. | | | |
| at 7 YP | | | £2,800 |
| | | | £256,800 |

Costs:

(a) Land Costs:

| | | | |
|---|---|---|---|
| Land | | £80,000 | |
| Stamp Duty on land | | £800 | |
| Legal Costs on purchase | | | |
| of site | | £800 | £81,600 |

(b) Costs of Development:

Cost of building:

| | | | |
|---|---|---|---|
| 280 m² at £450 per m² | £126,000 | | |
| 4 garages at £2,000 each | £8,000 | | |
| Site preparation, approach | | | |
| road and gardens | £4,000 | £138,000 | |
| Architect's fee for plans | | | |
| (agreed) | | £3,000 | £141,000 |

(c) Finance Costs:

| | | | |
|---|---|---|---|
| Land Costs £81,600 for say | | | |
| 12 months at 9% | | £7,344 | |
| Building Costs £141,000 for | | | |
| say 6 months at 9% | | £6,345 | £13,689 |

(d) Sale Fees:

| | | | |
|---|---|---|---|
| Agent's Commission on sale | | | |
| of flats (agreed) | | £4,000 | |
| Legal Costs on sale of | | | |
| flats (agreed) | | £2,000 | £6,000 |
| | | | £242,289 |
| Estimated profit | | | £14,511 |

Many developers judge whether the profit level is sufficiently worthwhile to warrant undertaking the development by expressing the estimated profit as a percentage of the total costs of development or of the total estimated realisation (known as the Gross Development Value [GDV]). In the above case the return is 6% since £14,511 represents 6% of £242,289. This percentage can be used as a basis for comparison between one development and another of the same type. Alternatively, the estimated profit can be expressed as a percentage of the total sales, £14,511 on £256,800 producing 5.7%, or the amount of profit per flat, £3,628 per flat in the example.

In this example the level of profit is too low no matter how it is looked at. Developers usually look for a profit of 12.5% to 15% of total sales or 15% to 20% on total costs. In order to achieve this level of profit the study must be turned into a residual valuation as follows:

Sales:
– as before                                                                  £256,800

Costs:

| | | |
|---|---|---|
| Cost of building – as before | £141,000 | |
| Finance on building cost | £6,345 | |
| Agent commission | £4,000 | |
| Legal costs | £2,000 | |
| Profit – say 12.5% of sales | £32,100 | £185,445 |
| | | £71,355 |

Interest – 12 months @ 9%
$(71,355 - (71,335 \div 1.09))$                                              £5,892
                                                                             £65,463

*Less*
Stamp Duty[5] and legal costs
on purchase @ 2%
$(65,463 - (65,462 \div 1.02))$                                              £1,283

Value of Land                                                                £64,180

---

[5] Stamp duty at this level of value is 1%. The allowance would need to be higher at higher level of value when stamp duty increases – see Chapter 22.

## 5. Factors Affecting Value

In dealing with a particular area of land the local circumstances must be carefully considered, e.g. the prosperity of the town, the existing supply of houses, factories or other buildings, and the amenities of the particular property under consideration.

It is, of course, essential that proper access to the land is available. The proximity and availability of public services such as public sewers, gas, electricity, water and telephone supplies, is also of major importance. The existence of restrictions and easements must be checked.

These factors vary according to the type of development for which it is considered the land is most suitable, due regard being had to town planning consent likely to be forthcoming or assumed.

In the case of land to be used for houses or flats, these will include the proximity of good travel facilities, and also the existence and proximity of shops, schools, places of worship, and the like or the possibility of the provision thereof in the future. Local employment conditions also affect demand. Regard must also be had to the presence of open spaces, parks, golf courses and the like and the reservation of land under planning control for similar leisure purposes. The character of the neighbourhood must be considered to determine the most suitable type or types of development; also whether the character is changing.

The principal factor in the case of retail development is the location of the land. Where the land is in an established prime shopping location it is important to determine whether there are proposals such as road schemes or major retail developments nearby which might draw shoppers away from that location. Edge of town developments, normally large retail warehouse schemes or new shopping centres, depend on access to major roads in the area and the site being large enough to provide extensive customer car parking and a sufficiently large scheme to compete with existing centres.

Offices depend on several factors, such as access by road and public transport generally and an adequate supply of suitable labour. Many occupiers need to be close to other companies in the same area of business, as witnessed by the grouping of professional firms, or those engaged in financial services. The success of an office scheme also depends on proximity to the established town centre. Office workers usually like to visit shops, banks etc., and they like to travel on foot to these from their place of work. Many office workers have to fit their house-keeping requirements into their work regime and they wish to minimise the inconvenience of so doing. Non-centralised office developments have often failed because they have not taken these requirements into consideration, although offices set in attractive landscaped areas have met with success.

The most important factors in the case of warehouse and industrial developments are access by road and an adequate supply of suitable labour. Proximity to ports and markets and to sources of power and materials are now less important than they were due to the improvement of road transport facilities.

Whether development is likely to be profitable or not is dependent upon the demand for the property when it has been completed. The physical state of the land, availability of services, etc., are also important factors to be taken into account. Until 1947 these were the only factors to be considered, but since then the town planning provisions affecting the land are the overriding factor which the valuer has to consider.

## 6. Residential Development Schemes

### (a) Generally

In relation to small areas of land it may be a comparatively simple problem to determine the best use to which the land can be put, having regard to general trends of development in the neighbour-hood, the factors affecting the particular piece of land under consideration and the provisions of the relevant Development Plan.

### Example 16–2

You are asked to advise on the value of land fronting a residential road which has been made up and taken over by the local authority. There are foul and surface water sewers in the road and gas, water and electric services are available. The property has a total frontage of 146 metres and a depth of 36 metres. This is the remaining vacant land in this road; the rest of the frontage has been developed for houses currently selling at about £100,000 with frontages of about 9 metres each. From the enquiries you have made you find that similar plots in the area have sold at prices ranging from £35,000 to £40,000 each. There is a demand for houses of the character already erected; it is expected that all of the plots could be built on and the houses disposed of within a year.

You have ascertained from the local planning authority that they would permit the development of this land for residential purposes. The land is in an area allocated for residential development.

## Valuation

In this instance there cannot be very much doubt as to the type of development that should take place, i.e. the construction of 16 semi-detached houses.

After inspecting the land and the area and taking into account the prices realised and the upward trend in values, it might be considered that a fair value per plot would be £40,000, giving a total valuation of £640,000 on the assumption that planning consent for this form of development would be granted subject only to usual conditions.

## Note

In this instance the procedure has been a direct comparison with the sale prices of other similar properties in the vicinity. A speculative builder, before buying, even at a figure which he feels satisfied is a fair market value, would be wise to prepare a viability report, and a bank lending on the scheme should insist on such a report on the following lines to forecast the probable rate of profit. A valuer advising on the sale or purchase price of the land would be wise to do the same since the forecast may bring to light some factor which might otherwise be overlooked or given too little weight.

## Viability Statement

Sales:

| | | |
|---|---|---|
| 16 houses @ £100,000 each | | £1,600,000 |

Costs:

| | | |
|---|---|---|
| (a) Land | £640,000 | |
| Stamp Duty on land | £12,800 | |
| Legal costs on purchase of land | £6,400 | £659,200 |

(b) Costs of Development:
Building Costs - 16 houses of
90m² each at £400 per m²

| | | |
|---|---|---|
| including garages, say | £576,000 | |
| Plus plans (see note 3) | £2,500 | £578,500 |

(c) Finance Costs:

| | | |
|---|---|---|
| Land Costs £659,200 for 12 months @ say 9% | £59,330 | |
| Building Costs £578,500 for say 6 months @ 9% | £26,030 | |
| | £85,360 | |

But say 50% (see Note 5 below)

| | |
|---|---|
| Say | £43,000 |

(d) Sale Fees:

| | | |
|---|---|---|
| Agents commission @ £1,000 per house | £16,000 | |
| Legal Costs on Sale, say | £8,000 | £24,000 |
| | | £1,304,700 |

| | |
|---|---|
| Say | £1,300,000 |

| | |
|---|---|
| Estimated Profit | £300,000 |

£300,000 represents 23.1% of the total costs of £1,300,000, or £18,750 per house, or 18.75% of the sale price. This is a high level of profit which would indicate that the land was worth more than £40,000 per plot or the costs of development have been under-estimated.

*Notes*

1. The sale price of £100,000 is substantiated by reference to sales of similar houses in the same area. If there has been a recent trend of rising prices which is anticipated to continue for such houses in the area, the developer may deliberately increase his estimate to allow for a continuation of increase, or reduce his profit requirement which he would make up from the rising prices. This is the classic developer's predicament; if he adhered to prices current at the date of the purchase of the land he might find that he was constantly losing opportunities of purchase due to higher bids from others, if he worked on the same level of profit as them. He might also increase his estimated realisation to reflect the fact that he will be offering new houses and not second-hand ones in an area where new

houses command a premium. The developer would probably do a number of such calculations within a price bracket. It is preferable to work on prices current at the date of purchase and a lower profit margin and to regard any increase in prices obtained in the event as "super-profits". If he had regard to the possibility of rising sale prices, he should also consider the effect of rising costs. To be too optimistic is likely to bring disaster as was seen following the 1989 collapse in prices.

2. There is an obvious interdependence between the figures of selling prices (£100,000 above) and building costs (£400 per m≤ above). The better the quality of the house offered the higher will be the price realised and the higher the cost of building. The object of the developer is to maximise the difference between the two figures by the use of his experience, expertise and skill in design, and efficiency in sales organisation and building cost control.

3. Most developers use stock plans or adapt plans they have used before and a full-scale architect's fee would not be payable in most cases. Where fresh plans have to be prepared, a much larger figure would have to be allowed; if the developer offers the designer too small a fee, the design and hence the saleability of the houses is likely to suffer. Many developers have their own in-house architects/designers so that the costs would be reflected in their profit targets with no specific allowance for architects' fees or plans. In deciding on the likely level of architects' fees, regard should be had to the number of times the design will be repeated on the site.

4. There is a similar interdependence between the amount of the selling agents' commission and the prices obtained. If selling agents are paid too little they will not use maximum efforts to sell. On the other hand a developer is able to negotiate better terms for multiple instructions, particularly if the market is buoyant and the houses will sell quickly.

5. Since proceeds of sales will start to be received in say 6 months because of a phased building scheme it will not actually be necessary for the developer to borrow the whole of the estimated land and building costs (£1,300,000) for the whole period of the development (12 months). Under half the total figure for something over half the total period will give the cost of finance in this case.

In order to substantiate this figure, a "cash flow" forecast should be prepared. The Table on p. 272 assumes the sale of two houses per month starting in the sixth month and three houses per month in the last two months. Such a forecast would be useful to (and probably required by) any finance house providing the money for the development, and can easily be produced utilising a simple computer programme.

Most valuations and appraisals of this kind are now prepared on computer which allows sophisticated cash flows to be produced. Modern programmes also allow for increases in selling prices and costs to be incorporated within the calculations. They can also cater for a more accurate spread of expenditure. Building cost expenditure is not usually evenly spread over the construction period but is represented by an S curve.

It may be necessary in the case of larger areas to prepare sketch plans of a suitable development, making provision, for instance, for certain portions of the estate to be developed for shops, others for flats, and others for houses, probably with varying densities. In such cases, the problem is more complicated but the principle is the same.

It is of course simple to set out the above figures in the form of a residual valuation.

| | | |
|---|---|---|
| Selling prices: | | £1,600,000 |
| *Less* | | |
| Building costs – 16 houses at 90 m$^2$ | | |
| each at £400 per m$^2$ | £576,000 | |
| Plans | £2,500 | |
| Legal costs | £8,000 | |
| Commission | £16,000 | |
| Finance | £13,000 | |
| | £615,500 | |
| Developer's profit – say 18.75% × GDV say | £300,000 | £915,500 |
| Land balance | | £684,500 |
| Land price and interest + stamp | | |
| duty/legal costs | 1.03 × = | £664,563 |
| ∴Value of Land + interest | | ÷ 1.045 |
| ∴Value of Land | | £635,946 |
| say | | £640,000 |

*Note*: If the profit target were 20% then the land value would be greater, as indicated before.

| Month | 1 £ | 2 £ | 3 £ | 4 £ | 5 £ | 6 £ | 7 £ | 8 £ | 9 £ | 10 £ | 11 £ | 12 £ |
|---|---|---|---|---|---|---|---|---|---|---|---|---|
| Items | | | | | | | | | | | | |
| Land | 640,000 | | | | | | | | | | | |
| Stamp duty | 12,800 | | | | | | | | | | | |
| Building costs | | 52,360 | 52,360 | 52,360 | 52,360 | 52,360 | 52,360 | 52,360 | 52,360 | 52,360 | 52,360 | 52,500 |
| Plans | | | 2,500 | | | | | | | | | |
| Legal fees | 6,400 | | | | | | | | | | | |
| Interest at 0.75% per month | | 4,944 | 5,374 | 5,826 | 6,262 | 6,702 | 5,667 | 4,625 | 3,575 | 2,517 | 1,451 | CR414 |
| | 659,200 | 716,504 | 776,738 | 834,924 | 893,546 | 952,608 | 813,635 | 673,620 | 532,555 | 390,432 | 247,243 | 3,829 |
| Net proceeds of sales i.e. 2 or 3 houses at £100,000 each less £1,500 per house (legal costs and sale commissions) | — | — | — | — | — | 197,000 | 197,000 | 197,000 | 197,000 | 197,000 | 295,500 | 295,500 |
| Net balance of drawings | 659,200 | 716,504 | 776,738 | 834,924 | 893,546 | 755,608 | 616,635 | 476,620 | 335,555 | 193,432 | CR48,257 | CR291,671 |
| Total Interest = £46,424 | | | | | | | | | | | | |

Note how the amount drawn builds up from the initial payment for the land to the point in time when the first sales are made from when it decreases until it is all re-paid. The profit in cash all accrues to the developer in the last two months.

In the above table it is assumed that the interest is added to the loan, i.e. it is "rolled up", as distinct from being paid regularly by the borrower from his own resources.

This table could be amended to reflect a developer's deliberate policy of pricing the first houses lower and increasing them progressively. It could also reflect a different pattern for paying interest.

It will be seen that the amount of interest was under-estimated (£13,000 + £28,617 (664,563 − 635,946) instead of £46,529) resulting in a lower profit of £291,671 instead of £300,000. This gives a return on cost of 22.36% or 18.23% on selling price, both of which are well within the acceptable range

## (b) User of Land

When preparing a scheme in outline, the extent of the various uses to which the land is to be put must be determined, and a lay-out the provision of suitable size plots and to any restrictions in force relating to the density of buildings and to the proportion of site that may be covered.

In the case of large estates it may be necessary to make reservations for special plots for open spaces, for a school or a church or community centre, and for similar local amenities.

It will also be necessary to phase the development over a number of years based on a realistic sales programme. This will be considered further in Section 8.

In relation to those parts allocated for commercial purposes, it may be considered that a site should be set aside for a petrol-filling station, public-house or neighbourhood shopping centre.

Even though a scheme is prepared primarily for the purpose of arriving at the value of the land it is usual to discuss the scheme informally with the local planning authority. Such discussions will identify whether different densities should be allowed for in different parts of the land and what provision is to be made for roads, open spaces, etc. Indeed as the scheme is worked up, further valuations will be required to monitor the value implications.

## (c) Infrastructure

In a scheme of even a modest size, close attention must be given to the provision of infrastructure such as roads and services since these will represent a significant cost. This applies not only within the area of the scheme but also to off-site works.

The internal road layout should be such as to provide good and easy access to all parts of the estate. If there are any changes of level, regard must be had to the provision of easy gradients. So far as is practicable, plots of regular shape and of suitable size must be produced by the road layout, at the same time avoiding undue monotony and lack of amenity. Roads producing no building frontage should be kept to a minimum.

The layout of soil and surface water sewers must be determined in relation to the available outfall. Both roads and sewers will need to be constructed to the requirements of the local authorities prior to them being taken over (or "adopted") by the highway authority or statutory undertaker.

In addition to agreeing the internal road and sewer provisions with the appropriate authorities, it may well be necessary to agree off-site works. For example, the local road serving the site may need to be widened or a roundabout or traffic lights provided. Similarly, the existing sewerage system may be inadequate and need to be upgraded with larger pipes or by enlarging the sewage works. Such works will be carried out by the authorities with the developer bearing all or part of the costs.

It may be found that the land is so situated that it would not be possible to connect to a public sewer by gravity and allowance would have to be made for the cost of constructing a pumping station. Surface water disposal might require the provision of a balancing pond, although these can often provide an attractive amenity feature within the overall design.

Enquiries must also be made as to the terms upon which supplies of gas and electricity and the installation of telephone cables can be obtained, as the nearest mains or cables may be at some considerable distance and supplies inadequate.

It will also be necessary to consider the nature of the ground and its effects on building costs. Piling may be necessary to overcome poor land bearing capability. Regard must also be had to the possibility that some or all of the land may be contaminated or that the site may be of archaeological significance. The valuer must make adequate enquiries before making his assumptions as to these factors and then he must state his assumptions.

## 7. Commercial Development Schemes

The same principles apply to commercial development schemes as apply to residential schemes, save that the unit of supply is different; it may be an office block or a factory estate or a mixed development of shops, offices and cinemas.

The development period may be much longer than a year so that compound interest becomes very relevant and likewise the effect of changing rents/yields and rising costs. Some developments will be phased and others will be single phase.

As stated above, most residual development calculations are prepared on computers. The programmes used are very sophisticated and must allow for interlocking phasing as well as overlapping phases. A consideration of these programmes is beyond the scope of this book, but generally they rely on DCF techniques and use cash flow as the basis of calculation rather than the simple residuals considered here.[6]

## 8. Period of Development

There are two main factors that will determine the speed of development. The first is the physical factor of how quickly the actual construction work can be carried out and the second is the rate at which the completed buildings can be sold or let.

In the case of a small housing estate, it may be reasonable to assume that each house will be sold immediately it is completed, so that no deferment of development costs and sale proceeds are necessary.

However, in the case of an estate of several hundred houses, it will be necessary to estimate the rate at which the market will take up the houses, which will depend on a number of factors including the strength of demand in the area and any competing developments being carried out, and to assume that the development will be appropriately phased over a period of years. The easiest way to value a large area of land of this kind is to consider first the value of the land for an average phase, say of 50 houses. If the total scheme envisaged 300 houses there would be six phases, but, because of the holding cost of the land, its current value could not be six times the first phase value. The value (in the absence of any assumptions as to increasing sales prices and costs) would have to be reduced by the compound interest cost of each phase as shown below.

Assume that the land value per phase is £1m, each phase lasts one year, and the holding cost is 10% p.a:

| | | |
|---|---|---|
| Phase 1 land | | £1,000,000 |
| Phase 2 land | £1m | |
| × PV of £1 in 1 yr @ 10% | 0.91 | £910,000 |
| Phase 3 land | £1m | |
| × PV of £1 in 2 yrs @ 10% | 0.83 | £830,000 |

---

[6] A more detailed consideration of Development Practice is available in Chapter 13 (4th Edition) of *Valuation: Principles into Practice* Ed. W. H. Rees (Estates Gazette).

| | | |
|---|---|---|
| Phase 4 land | £1m | |
| × PV of £1 in 3 yrs @ 10% | 0.75 | £750,000 |
| Phase 5 land | £1m | |
| × PV of £1 in 4 yrs @ 10% | 0.68 | £680,000 |
| Phase 6 land | £1m | |
| × PV of £1 in 5 yrs @ 10% | 0.62 | £620,000 |
| | | £4,790,000 |

It may not be possible to purchase at this price because it assumes that land values will remain constant for six years. It may be necessary to build in an inflation index.

Quite often the original purchaser of a large scheme will sell off parcels of serviced land to other developers. One reason is to recoup some of the initial heavy costs of providing the basic infrastructure. However, the introduction of a different developer with his own design standards can add variety to the house types and layouts on offer and so give an impetus to interest in the whole scheme with benefits to the vendor developer through increased take-up of houses. Developers might even join together from the very start and form a joint venture consortium to carry out the scheme.

## 9. Incidental Costs

In most cases certain additional expenses will be incurred beyond the bare constructional costs of roads and sewers.

If open spaces or other amenities such as belts of trees are provided by the developer the cost must be taken into account.

Professional services are required for the preparation of detailed layout schemes and the drawings and specifications for constructional works; fees for this may be taken from 4% to 12.5% on the cost of the works, depending on the amount of professional input required. Even where a developer has an in-house design department some independent professional advice may still be required, particularly from highway and civil engineers. In the case of large and contentious schemes where the planning permission must be pursued through the appeals procedure with a public hearing, the fees of planning lawyers, town planners and other professional advisers can amount to very large sums.

Legal charges may be incurred both in respect of the purchase of the land (and stamp duty) and of the sales. The amount to be allowed will depend on whether the land is registered or unregistered. Further legal charges may be incurred where planning, highway and other legal agreements are required, including financing agreements.

A detailed estimate of the cost of advertisements and commission on sales payable to agents must also be made. This should include the costs, where appropriate, of furnishing and staffing a show house.

## 10. Site Assembly

The acquisition of land for large schemes almost invariably involves the purchase of several land holdings. The problem for the developer is that he is usually buying land without planning permission. He will be unwilling, and probably unable, to commit large sums of money without the certainty of being able to carry out development and so justify the expenditure.

The normal solution to this problem is for the parties to enter into option agreements. These may be call options, whereby the developer can require the owner to sell to him on the grant of planning permission, or put and call options, whereby either party can require the other to carry out the sale/purchase of the land.

The option agreement must cover many matters, including the means of determining the price payable. Sometimes this will be at a stated figure, although when land values are rising this is unfavourable to the landowner. Alternatively, the price will be market value, to be determined at the time of the exercise of the option. Commonly the price will be a percentage of market value, say 70% to 90%, to reflect the costs of assembly and of obtaining planning permission which will be borne by the developer.

It is essential that the provisions for determining the price should be set out to enable the valuers for the parties to have a clear understanding of the approach to the valuation. For example, where the price is to be assessed at an agreed price per hectare, it must be clear whether this is the gross area or some lesser area such as the net developable area which needs to be defined. Where it is price per plot, how the valuer should deal with plots which straddle the boundaries of land ownership needs to be stated. If a landholding controls the access to other land, the valuer needs to know whether the special value of this strategic land must be reflected – "ransom value". In options where a developer may exercise the option without planning permission having been granted, the planning assumptions for the valuation should be present; if, as is usual, the developer has to bear the cost of extensive and expensive infrastructure, the valuer needs to know how this factor should be reflected in the valuation.

It is clear that considerable care is required in the preparation of an option if extra complications in what is in any event a difficult and contentious area of valuation are to be avoided.

In urban areas many sites are assembled by a developer buying properties at existing use value and then letting them on short-term leases with rebuilding clauses. In these cases the holding costs are met wholly or partly out of the rents.

## 11. Hope Value

The development value of land can be identified by the existence of a planning permission. It is obvious, however, that planning permissions do not appear out of the blue so that, on one day, a planning permission for residential development is given when the land acquires a residential development value. What happens in practice is that a view is taken as to the likelihood of planning permission being given, either now or at some time in the future. If the likelihood is nil, such as prime agricultural land in the Green Belt in an Area of Outstanding Natural Beauty and with rare wild flowers found nowhere else, it is unlikely in the extreme that anyone would pay more than agricultural value.

In other cases, whilst it is agricultural land with no immediate prospect of development, it may be felt that in a few years, when say a proposed new road has been built nearby, some form of development might be allowed. In such cases purchasers might be found who will pay above agricultural value in the hope that, after a few years, they will realise development value and make a large profit. This price above value reflecting only the existing use but below full development value, is termed hope value.

A valuation to determine hope value is often impossible other than by adopting an instinctive approach, particularly in the stages when the hope of permission is remote; it can only be a guesstimate of the money a speculator would be prepared to pay. As the hope crystallises into reasonable certainty of a permission at some stage, a valuation can be attempted based on the potential development value deferred for the anticipated period until permission will be forthcoming, but with some end deduction to reflect the lack of certainty. Indeed, since most developers will buy only when permission is certain (preferring an option to buy or a contract conditional on the grant of permission before certainty has been reached) any sale in the period of uncertainty will probably require a significant discount on what might otherwise appear to be the full hope value.

In some cases there may be little risk that planning consent will be refused, e.g. with an infill site for one house in a whole road of houses and where there is no view which will be obstructed by the development. In such a case the market may assume that consent will undoubtedly be granted and there will be little if any discount from the development value.

## 12. Urban Sites

### (a) New Development

The previous sections of this chapter have been concerned mainly with land, not built upon, in the vicinity of existing development, commonly referred to as greenfield sites.

In the centre of towns, in built-up areas, it is often necessary to consider the value of a site which has become vacant through buildings having been pulled down, sometimes termed a brownfield site, or which is occupied by buildings which are obsolete or do not utilise the site to the full.

The general method of approach suggested in Section 2 can be used, i.e. to value by comparison with sales of other sites by reference to an appropriate unit, e.g. per square metre or metre frontage and to check the result by drawing up a viability statement or by a valuation by the residual method. For example, with office sites, there is a correlation between rental value, building cost and site value. In areas with closely similar rental values and site conditions, an office site might be valued at £x per gross square metre of offices to be built. Thus in an area where offices are letting at £160 per m$^2$ analysis of site purchase prices might show that prices represent £1,000 per m$^2$ of the gross floor area of the offices to be built. So a site with permission to build 5,000 m$^2$ gross of offices would be valued at £5 million.

As in the case of building estates, the type of property to be erected and the use to which it can be put when completed is entirely dependent upon the planning permission which will be granted. An indication will probably be obtained by an inspection of the development plan and discussion with the local planning authority.

Sites are often restricted as to user, height of buildings, the percentage of the site area that may be covered at ground floor and above, and by conditions imposed when planning permission is granted. Car parking standards can have a significant effect on the size of building that can be erected.

There may be further restrictions on user, owing to the existence of easements of light and air or rights of way.

A site may be bare, but may contain old foundations or be contaminated and the cost of clearing may be considerable, whilst on the other hand advantage may be taken of existing runs of drains.

In many areas the cost of development may be significantly increased by the presence of underground water or the difficulty of access with building materials in a crowded and busy thoroughfare.

Factors of this kind must be carefully considered both in relation to any suggested scheme of development and also when comparing two sites which apparently are very similar but which are subject to different restrictions or conditions.

The existence of restrictive covenants may reduce the value of a site, although the possibility of an application for their modification or removal under Section 84 of the Law of Property Act 1925[7] must not be lost sight of. It is frequently possible to insure against restrictions being enforced where these are contained in old documents and where there is considerable doubt as to whether they are any longer extant. Where applications are made to the Lands Tribunal it must be remembered that the Tribunal can award compensation to the person having the benefit of the covenant. In the case of restrictions preventing the conversion of houses it may be possible to take similar action to that under Section 84 of the Law of Property Act 1925, under the Housing Act 1985.[8] The alternative to a Section 84 application is to buy out the restriction by agreement with those who have the benefit of the covenants or rights.

---

[7] As amended by the Landlord and Tenant Act 1954, Section 52 and by the Law of Property Act 1969.
[8] Section 610.

Where possession of business premises is obtained against tenants in occupation at the termination of their leases for purposes of redevelopment under Section 30 of the Landlord and Tenant Act 1954,[9] the amount of compensation to be paid must be deducted as part of the costs of development.

As in the case of a building estate, the value of urban sites is best found by direct comparison using an appropriate unit. However, urban sites are not generally susceptible to direct comparison since they commonly provide a mix of uses. The residual method is therefore usually adopted.

### (b) Refurbishment

In urban areas, development can take the form of work to existing buildings rather than the erection of new buildings. In conservation areas or with listed buildings, this may be the only possible form of development. Even in other cases it may be found that it is more profitable to modernise a property rather than to demolish and replace it. This approach is known as refurbishment.

When considering refurbishment it is necessary to consider whether any planning permission will be required. This may be so because alterations will be made to the external appearance such as inserting new windows or making additions such as plant rooms on the roof or because the building is listed as being of architectural importance or is in a conservation area. Equally, where a change of use is contemplated then the need for planning permission must be considered.

The other important factor to be considered is the estimated cost of the works. A typical refurbishment scheme of an office building would include installation of suspended ceilings and raised floors where possible, upgrading of all services including heating, toilets and lifts, re-design of the entrance hall and common parts, and even the re-cladding of external parts. Similar considerations apply to blocks of flats, shopping centres and other buildings. Unlike new development, where building costs may be available by comparison with other new buildings, each refurbishment is unique and needs the services of a building cost surveyor to estimate the costs involved.

---

[9] See Chapter 18.

A common practice is for planning authorities to allow new buildings subject to the retention of existing façades. This film set architecture produces a hybrid new development/refurbishment scheme which draws upon the elements of both.

Since refurbishment is a form of development the residual method of valuation is appropriate in valuing such property. Indeed, the valuer will often need to prepare a residual valuation assuming refurbishment and also one assuming redevelopment. A comparison of the valuations will provide a strong indication of which approach is to be favoured.

*Example 16–3*

A freehold office building erected 30 years ago is on 5 floors with 400 m$^2$ net on each floor. There is surface car parking for 40 cars. The site area is 1,200 m$^2$.

The property is occupied by a tenant whose lease has 2 years to run at a current rent of £30,000 p.a. The rateable value is £30,000.

The property is lacking in modern amenities. It would cost £700,000 to bring the property up to current standards when the rental value would be £120 per m$^2$. The planning authority has indicated that a new office building would be permitted at a plot ratio of 25:1 producing 3,000 m$^2$ gross, 2,300 m$^2$ net with 15 car spaces at basement level. The rental value would be £150 per m$^2$.

What is the value of the freehold interest?

*Valuations*

The first step is to establish the value assuming refurbishment and the second the value assuming redevelopment.

*A.  Refurbishment*

GDV:
Offices 2,000 m$^2$ at
£120 per m$^2$                £240,000  p.a.
Car Spaces 40 spaces at
£200 per space p.a.            __£8,000__
                                £248,000  p.a.
YP perp at 6.75%              __14.81__  say    £3,673,000

*Less*
Building Costs
Estimated Costs          £700,000
*Add*
Fees at say 10%           £70,000        £770,000
Finance:
Assume 8-month scheme
£770,000 for 4 months at
9% say                                   £23,100
Letting Costs say 15%
of £248,000                              £37,200
                                         £830,300
Developer's Profit at
20% of £830,30                           £166,060          £996,360

Land Balance including acquisition
costs and interest and profit                              £2,676,640
*Less*
Interest for 8 months @ 9%
Land balance including acquisition
costs (3%) and profit (20%)                                £2,525,132 (a)
*Less*
Acquisition costs @ 3% and
profit 20%                                                 £472,179
Land Value and Profit                                      £2,052,953 (b)
Land Value                                                 £1,710,794 (c)
Say                                                        £1,700,000

(a)  This figure can be calculated direct by multiplying the land
     value including interest by the PV of £1 for the development
     period and this will automatically allow for compounding of
     interest.
(b)  This figure is calculated direct by dividing the land balance
     including acquisition cost by 1.03.
(c)  This figure is calculated direct by dividing the land balance
     and profits by 1.2. Alternatively, developer's profit can be
     taken as a percentage of GDV (in this instance 13.84%) which
     allows for only one deduction to appear in the calculation.

## B. New Development

GDV:
Offices 2,300 m²

| | | | |
|---|---|---|---|
| at £150 per m² | | £345,000 p.a. | |
| Car Spaces 15 spaces at | | | |
| £300 per space p.a. | | £4,500 | |
| Say | | £350,000 p.a. | |
| YP perp at 6.25% | | 16.0 | £5,600,000 |
| *Less* | | | |
| Building Costs | | | |
| 3,000 m² at | | | |
| £500 per m² | £1,500,000 | | |
| Basement say 400 m² | | | |
| at £400 per m² | £160,000 | | |
| | £1,660,000 | | |
| *Add* | | | |
| Fees at 12% | £199,200 | £1,859,200 | |
| | | | |
| *Finance* | | | |
| Assume 18-month | | | |
| scheme £1,859,200 for | | | |
| 9 months at 9% | | £125,496 | |
| Letting Costs say 15% | | | |
| of £350,000 | | £52,500 | |
| Developer's Profit at | | | |
| 20% of £2,037,196 | | £407,439 | £2,444,635 |
| Land Balance including | | | |
| acquisition costs and | | | |
| interest and profit | | | £3,155,365 |
| *Less* | | | |
| Interest for 18 months | | | |
| @ 9% p.a. | | | £382,620 |
| Land balance including | | | |
| acquisition costs (3%) | | | |
| and profits (20%) | | | £2,772,746 |
| *Less* | | | |
| Acquisition costs | ÷ 1.23 | | £518,481 |
| Land Value | | | £2,254,265 |
| Say | | | £2,250,000 |

*Value*

Hence it appears that redevelopment and refurbishment are of similar value. The decision on what course to follow would depend on resources available and the state of the market at the end of the lease.

*Valuation*

| | | |
|---|---|---|
| *Term* – Rent reserved | £40,000 p.a. | £72,000 |
| YP 2 yrs at 7.5% | 1.80 | |
| *Reversion* – to Development | | |
| Value say | £2,200,000 | |
| *Less* | | |
| Compensation to Tenants say | | |
| 2 × RV £30,000 | £60,000 | |
| | £2,140,000 | |
| PV £1 in 2 yrs at 6.75% | 0.88 | £1,883,200 |
| | | £1,955,200 |
| Value say | | £1,950,000 |

*Note*: In practice, valuations should also be made on the basis that the building would be redeveloped, or be modernised to lesser standards at reduced costs of improvement.

If the work could be carried out with the tenants remaining, this might also be a significant factor.

## 13. Ground Rents

A common feature in the development process is the release of land to developers by landowners who wish to retain an interest in, and some control over the future use of, the land to be developed. This is achieved by the grant of a ground lease to the developer.

Ground leases have been a common feature of the development process for several centuries. The best known examples from the past are the large estates controlled by families or charities whereby the overall estate remained in the ownership of the estate owner who controlled the development of the estate in accordance with a general estate plan. The Grosvenor and Cadogan estates in central London, Alleyn's estate in Dulwich, and the Calthorpe estate in Birmingham and the Crown Estate generally are typical examples.

Control was, and still is, exercised by the granting of ground leases which impose restraints on the manner in which any parcel of land is to be developed for the benefit of the estate at large.

In recent times the ground lease has been adopted by local authorities and new town corporations whereby areas of land can be developed by the private sector whilst control of such development can readily be exercised through the landlord's powers under the ground lease, notwithstanding the powers they may have in addition as planning authorities.

Ground leases, however, are not confined to large-scale, estate development. Many individual sites may be offered on a ground lease basis since such an approach may be attractive to others who are taking a long-term view. This is particularly so with the advent of rent review clauses. A further significant area where ground leases have become common is that of developments of flats and maisonettes. The legal system in the United Kingdom, apart from Scotland, is so structured that a freehold interest in a flat which is part only of a property raises considerable problems. These are readily solved by the grant of a ground lease of the flat.

In earlier times, ground leases were usually granted for 99 years at a fixed rent, although other terms were sometimes adopted. For example, there are in existence many leases granted for 999 years at a peppercorn ground rent which are effectively freehold interests.

In recent times, two important changes have taken place. First, there has been a movement towards leases of 125 years' duration. The pressure for this appears to be two-fold. One is that the pace of change is such that it is felt that the life of buildings is shortened. Hence if a building will last around 60 years then a 99 year-lease does not allow sufficient time to justify the redevelopment of the building after 60 years. The other is that the major funding institutions, the pension funds and insurance companies, have argued forcefully that a 125-year lease is the minimum period to justify their investing in such interests. Hence, building leases for 125 years have become more common. However, the power of local authorities to grant leases of more than 99 years was for some time tightly controlled by central government so that they normally granted leases for the traditional term of 99 years.

The other change which has occurred is far more significant. This is the adoption of rent review clauses. At first, rent reviews were introduced into 99-year leases after 33 years and 66 years, but over time these intervals have shortened and it is not now uncommon for a rent review to operate each 5 years, as in occupation leases.

The valuer therefore is likely to be faced with a variety of ground leases. The general principles to be adopted in valuing a freehold interest subject to a ground lease are set out in Part 1. In this chapter the current forms of ground lease are considered since these are the ones to be found in properties suitable for development. They are considered in two broad categories, residential and commercial.

## (a) Residential Ground Leases

As has been explained, developers of flats have overcome the problems of divided freehold interests by the grant of long leases. Typically, such flats will be offered for sale at a stated price plus a ground rent. The ground rents are commonly quite small with few reviews. Technically what is offered is a lease at a lower rent and a premium, but such terminology is rarely, if ever, used when they are marketed.

The market is likely to change considerably over the next few years following the introduction of the Leasehold Reform, Housing and Urban Development Act 1993, where qualifying lessees can require their landlords to take a surrender of the existing lease and grant a new lease for the original term plus 90 years or can collectively buy the freehold (see Chapter 18). This is likely to lead to leases being sold once again on the basis of 999 years at a peppercorn.

The determination of the original ground rent bears no relationship to the rental value of the land but tends to be derived from prevailing levels of such rents charged in the area. For this reason they are not truly ground rents though so described. This may not be the case on review.

Many ground rents have fixed rent reviews; thus a typical ground rent may be £30, doubling every 25 years. Where the ground rent is subject to an open market review the critical factor is whether regard should be had to the premium payable for the lease, if not the ground rent payable on review will be a true rent, being effectively the reciprocal of the capital value in exactly the same way as a Section 15 rent under the Leasehold Reform Act 1967 is a true ground rent (see Chapter 18).

In valuing such ground rents the principal characteristics to note are that the sums tend to be small with a consequently disproportionately high management cost and that the income, though secure, is fixed for a long period. Returns tend to equate with those on medium dated government stock or with PIBS (see Chapter 7).

*Example 16–4*

Value the freehold ground rents derived from a block of 24 flats erected two years ago. Each flat owner pays a ground rent of £30 p.a. rising to £60 p.a. in 32 years' time and £90 p.a. in 65 years' time.

| Income | | | |
|---|---|---|---|
| 24 flats at £30 p.a. | | £720 p.a. | |
| YP 32 years @ 9% | | 10.41 | |
| | | | £7,495 |
| Reversion to | | | |
| | | £1,440 p.a. | |
| YP in say perp @ 9% | 11.11 | | |
| Deferred 32 years | 0.7 | 0.77 | £1,120 |
| | | | £8,615 |

Where the unexpired period of the ground lease is shortening to the point where purchasers of the ground lease have difficulties in raising mortgages, the ground lessee may be prepared to offer a sum significantly in excess of the normal investment value to the freeholder in order to overcome the problem. Alternatively, the ground lessee may be prepared to pay a premium for a new lease or an extended lease. This extra payment is generated because of the marriage value which will exist. A lower yield may also be achieved because some purchasers will acquire the "estate" in order to manage it and derive management fees and insurance premiums. Thus where these are available the value in the above example may easily rise to £10,000.

### (b) Commercial Ground Leases

Where commercial sites are offered on ground leases, the ground rent will tend to represent the annual rental value of the site. The rental value clearly depends on the surrounding conditions under the lease.

A typical approach to the matter is that the freeholder offers the site on ground lease. Initially, the developer/lessee will be granted a building agreement which requires him to carry out the proposed development. The agreement provides that, on satisfactory completion of the development, a building lease will be granted in the form of the draft lease attached to the agreement. The lease will provide that the lessee is responsible for the property under the normal repairing and insuring obligations. As to the ground rent, the initial rent will be agreed as the prevailing ground rental value. Current practice is to provide for the ground rent to be reviewed at frequent intervals, commonly five years, and for the basis of review to be a geared rent. This means that, at the grant of the lease, the parties will agree what is the estimated rental value of the property to be built. This establishes a relationship between initial ground rent and full rental value, normally expressed as a percentage. Thereafter, at each review, the ground rent will rise to that same percentage of the then prevailing rental value.

For example, on the grant of a ground lease of an office site, the ground rent may be fixed at £20,000 p.a., and it is agreed that the rental value of the offices, if they were already built, would be £80,000 p.a. It is clear that the initial ground rent is 25% of the rental value of the offices. At the first review, it may be agreed that the rental value of the offices is £140,000 p.a. If so, the ground rent will become 25% of £140,000 = £35,000 p.a.

In this way the ground rent will follow the rental pattern of the finished building rather than movements in land values. This tends to be preferred since it brings greater certainty in the rental value pattern, links movements in income to larger and more acceptable types of investment, and obviates arguments over land values at each review which tend to be less easily determined and thus less certain of agreement.

There are therefore two principal matters which demand the valuer's skills. One is determination of the initial ground rent, the other is the valuation of the freehold ground rent once in being.

### (c) Ground Rental Value

Since the ground rental value is the annual equivalent of the site capital value, ground rents may be derived from the capital value of the site which can be determined in the manner described in this chapter.

From the freeholder's point of view, the ground rent should be at a level which, when capitalised, will produce the capital value of the site which it has before the grant of the lease. This level is therefore the level derived from applying the prevailing yield for such ground rents. Hence if a site has a capital value of £300,000, and ground rents are valued at 7% at full rental value, then the freeholder will require a ground rent of 7% of £300,000 = £21,000 p.a.

Such a result might not emerge in practice. A developer offered a ground lease might prefer such an arrangement, as he has to raise less building finance, and might therefore go above this level. On the other hand, he might find it more difficult to raise development funds if he can offer only a lease as security and so might reduce his offer to encourage an outright sale of the freehold. These factors depend on the state of the market and the nature of development, but the general rental value can be derived from such a straightforward approach, which at least provides a starting point.

The developer/lessee's approach to ground rents is different from a landlord's. To him a ground rent is an outgoing, and he may choose to determine the ground rent he can afford by a residual approach. The same principles will apply as to the residual method adopted to determine capital value, as described earlier, save that the annual cost of the development coupled with an annual profit will be deducted from the prospective annual income, any difference representing the "surplus" he can offer on a ground rent.

*Example 16–5*

X has been offered a ground lease of a site for which planning permission exists to erect a warehouse of 4,500 m≤ gross internal area. The lease is to be for 99 years from completion of the building. What rent can X afford to pay for the ground lease?

| | | |
|---|---|---|
| Rental Value of completed building 4,500 m$^2$ gross internal at £60 per m$^1$ = | | £270,000 p.a. |
| Development Costs Building costs 4,500 m$^2$ at £330 per m$^2$ say | £1,500,000 | |
| Professional fees at 10% | £150,000 | |
| Finance say 6 months at 9% | £74,250 | |

Acquisition Costs
Legal and agent's fees and
stamp duty say                          £40,000

Letting Costs
Legal and agent's fees say              <u>£30,000</u>
                                        £1,794,250

Say                                     £1,800,000

Annual Equivalent of Costs
Interest at 8%                  0.080
ASF 50 yrs at 3% adj.
 for Tax@ 35%                   0.014
Profits at 1.5%                 <u>0.015</u>       <u>0.109</u>        <u>£196,200</u>  p.a.
Surplus for Ground Rent                                  <u>£73,800</u>  p.a.

A comparison can be made with a capital value approach to assess
whether the ground rent is realistic.

GDV:
Rental Value                            £270,000  p.a
YP in perp @ 8%                         <u>12.5</u>        £3,375,000
*Less*
Sale costs @ say 3%                                  <u>£101,250</u>
                                                    £3,273,750

*Less*
Costs of Development                    £1,794,250

*Deduct*
Acquisition costs                                   <u>£40,000</u>       <u>£1,754,250</u>
Land Balance                                                £1,519,500
Land Value                  1.000 x
Stamp Duty 2%               <u>0.020 x</u>
                            1.020 x

Funding 1 year @ 9%        <u>0.092 x</u>
                            1.112 x

Profits @ 20%              <u>0.222 x</u>          1.334  x =  <u>£1,519,500</u>
Land Value (x) =                                            £1,139,055

Say                                                        <u>£1,140,000</u>

A ground rent of £73,800 p.a. represents a return of close to 6.5%
which can be regarded as realistic.

## (d) Capital Value of Ground Rents

Once a building lease is in operation the valuation of the ground rents produced follows the general principles in determining the capital value of any investment, as described in Part 1.

However, a ground rent is generally less than the full rental of the property so that, in the case of a ground rent with geared reviews, apart from having the same qualities as the investment from which it derives, the rent is that much more certain of receipt, and has no risk of income interruption because of voids. For these reasons a ground rent will generally be valued at below the prevailing yields for the type of property from which it derives.

On the other hand, it must be stressed that the general content of the ground lease will determine the yield. If rent reviews are widely spaced, so that the rent is fixed for longer than acceptable periods, then the yield will rise to reflect this. Similarly, at one time developers commonly entered into leasing arrangements whereby the landlord received a certain minimum proportion of the rent of the buildings, leaving the developer with a share of the marginal rent (a "top slice" arrangement). Clearly the valuation of the landlord's interest would reflect his added security, whereas the valuation of the ground lessee's interest would need to reflect his exposure to the changing fortunes of the market place.

The foregoing comments have concentrated on the traditional situation where the freeholder grants a ground lease to the developer. Recent times have seen the use of ground leases where several parties come together to carry out large-scale schemes, typically town centre redevelopments. The parties will include a developer, who will carry out the development and manage it thereafter, a funding institution which will put up the development funds, and the local authority which will provide planning and other support including the use of compulsory purchase powers to ensure site assembly. In these cases the agreement might be that the freehold interest will vest in the local authority who will grant a head lease to the funding institution who in turn will grant an under ground lease to the developer. The terms of the ground leases will reflect the agreement between the parties and their financial involvement. In these cases the ground lease terms will be arrived at in a different manner from that previously described. For example, the under ground lease might well provide for a rent which is a percentage of money given to the developer and, as such, the agreement is more of a funding document than a traditional ground lease. Even so, the agreements must be related to the development and "side by side" agreements are common whereby all the parties share in the growth in rental value whilst sharing the downside risks of a fall.

Ground rents and ground leases have seen a fairly rapid development in their nature and make-up over recent years. At one time the valuer was concerned with whether they were "well secured" or not and, if so, they attracted minimum yields. Today the ground lease has evolved into the expression of a commercial arrangement and the valuation of ground lease investments has changed accordingly. They are now seen as an investment to be judged critically along with the other investment opportunities available.

# Chapter 17

# Residential Properties

## 1. Generally

This is the area of property which touches on most people in their everyday lives. The range of properties is vast, with tenements and cottages at one extreme and country estates at the other. It is also the area of property which is most affected by legislation because of the perceived need to protect the individual against exploitation.

The residential property market is dominated by the individual who purchases for owner occupation, as compared to the commercial property market which is dominated by the investor. This is not to say that many properties are not rented, but the great majority of rented property is owned by the public sector (Local Authorities and Housing Associations) with only a comparatively small percentage being privately owned. Approximately 65% of all United Kingdom residential properties are owner-occupied and only approximately 10% are privately rented. This difference between residential property and commercial properties is important because the direct correlation between rent and capital value which exists in the commercial market does not usually apply in the residential market.

The main statutory areas of law which will be looked at are those concerned with:

(i) the protection of residential tenants in terms of security of tenure and rent control;
(ii) the availability of financial assistance for the acquisition, provision and improvement of dwellings; and
(iii) the protection of owner occupier tenants in connection with the management of their property, and the rights of these residential tenants to extend their leases or acquire the freehold of the property which they occupy.

It is necessary to distinguish between properties which are let under existing tenancies enjoying security of tenure or which are

295

only suitable for lettings (with these being bought and sold for investment), and the great bulk of properties which are usually on the market with vacant possession. The two markets are, however, not independent, as the former category is linked to the latter category because of the potential for capital gain if vacant possession is achieved.

The properties offered on the market with vacant possession are best valued by direct comparison with achieved sales of similar property with the same accommodation in comparable locations. Where no direct comparable exists then recourse should be had to the best comparable available, suitably adjusted for differences in accommodation and location. Except in the most unusual of situations the value of a let property cannot be the same as the value of a vacant property since the latter represents the highest value achievable. This is because every vacant property can be let, but vacant possession is rarely immediately available from a let property.

There are a number of factors which influence the value of any residential property. The following list is not exhaustive but encompasses the principal factors, not necessarily in order of importance:

(i)     location;
(ii)    house or flat type. In the case of houses, whether they are detached, semi-detached or terraced. In the case of flats, whether they are standard, penthouses, duplexes or are maisonettes (i.e. flats with their own independent street access);
(iii)   accommodation; including the number of bathrooms and the extent of ancillary areas such as utility rooms, play rooms, swimming pools etc;
(iv)    design and layout;
(v)     the number of storeys;
(vi)    the extent of the grounds, gardens, etc;
(vii)   the topography of the site;
(viii)  the state of repair;
(ix)    the standard of finish;
(x)     historical associations.

Most of the above are self-explanatory and all could be the subject of significant amplification. It will be appreciated that many of these points are subjective and what one owner might think of as an improvement could be regarded as the opposite by a prospective purchaser; in some areas the proliferation of gilt and marble might detract from rather than add to the value of a property.

Of all the items on the list, the most important is location, by which is meant the reputation of the area as well as its locational advantages and disadvantages. The latter relates to proximity to work, schools and shopping centres, as well as to communication, places of worship, golf courses, etc. The factors which determine the quality of a location are not always susceptible to definition since there may be historical reasons why a given area is regarded as "upmarket" whilst other areas are considered to be "downmarket". Historically, much had to do with the wind direction (i.e. being on the freshest side of the town; and that in many towns West tends to be more fashionable than East) or due to the decision of a particular person to settle in an area which then attracted friends to move close by. The quality of an area can also change and in valuing a property regard must be had to changes in fashion which might be caused by a change in infrastructure or even by a change in the political complexion of the local authority.

To summarise the effect of position and the other factors, it can be said that the general level of values of a neighbourhood is determined by locational factors, with differences in value between individual properties being determined by the nature and extent of the accommodation offered.

## 2. The Rent Acts

Until the passing of the Housing Act 1988, the main body of legislation concerned with rent control of dwelling-houses was contained in the Rent Act 1977. This control is being "phased out" gradually in consequence of the 1988 Act but will, though diminishing, continue to exist.

The Rent Act 1977, where it continues to apply, contains the law enacted in the Rent Act 1965 which first brought in the concept of "Fair Rent". It applies to all tenancies granted before 15 January 1989 where the rent passing was more than two-thirds of the Rateable Value and where the Rateable Value was less than £1,500 in London and £750 elsewhere. It will be referred to in the remainder of this account of the law, together with later amendments as the 1977 Act. It should be noted that certain kinds of tenancy are excluded from Rent Act protection even if they are not at "low rents" and do not exceed the rateable value limits. Examples are holiday lettings, lettings by certain educational institutions to their students, lettings where the rent includes payment for board or attendance, lettings by the Crown or other public bodies, and lettings of parsonage houses, public houses or farms.

The 1977 Act also prohibits the evasion of new control by charging the tenant a "premium" (i.e. a lump sum over and above the rent). This is a criminal offence and the tenant can claim repayment.

Tenancies at low rents (less than two-thirds of Rateable Value) granted for terms in excess of 21 years come within the provisions of Part I of the Landlord and Tenant Act 1954 and also within the Leasehold Reform Act 1967 and the Leasehold Reform, Housing and Urban Development Act 1993 (where the lease was granted for more than 35 years the low rent test does not apply).

Tenancies within the ambit of the 1977 Act are known as "Regulated Tenancies". The tenants have security of tenure and the right to have their rents fixed by an independent local official known as a "Rent Officer", although there is a right of appeal to a Rent Assessment Committee. Where a rent is fixed it is known as a "Registered Rent" and in the absence of a major change in circumstances the rent is fixed for two years. At the end of two years when the fair rent is varied it is known as a re-registration. Under the Rent Acts (Maximum Fair Rent) Order 1999 from 1 February 1999 the maximum increase on the first re-registration is 7.5% above inflation and on subsequent re-registrations 5% above inflation. Inflation is measured by the change in the Retail Price Index between the month before the date of the earlier registration and the month before the re-registration.

### Rents under Regulated Tenancies (1977 Act, Part III)

Under the regulated tenancy system, the landlord or tenant or both can apply to the Rent Officer for the registration of a Fair Rent (subject to a right of appeal to the local Rent Assessment Committee). Once such a rent has been determined and registered it is the maximum rent which can be charged for the property during the continuance of a regulated tenancy. Mixed business and residential premises are protected by Part II of the Landlord and Tenant Act 1954 (see Chapter 18).

A Fair Rent is an open market rent subject to certain special rules laid down in Section 70 of the 1977 Act, which states that in determining a fair rent, regard must be had to all circumstances (other than personal circumstances) and in particular to the age, character and locality of the dwelling-house and to the state of repair; it must be assumed that the number of persons seeking to become tenants of similar dwelling-houses in the locality on the terms (other than those relating to rent) of the tenancy is not substantially greater than the number of such dwelling-houses in the locality available for letting on such terms. To be disregarded are disrepair or other defects attributable to the tenant and any improvements carried out, otherwise than in pursuance of the terms of the tenancy, by the tenant or his predecessors. Where a landlord can show that there is no scarcity of housing for rental in the locality where the tenant can obtain security of tenure then no discount from the open market rental will be made.[1]

## Security of Tenure (1977 Act, Part VII)

Limitation of rents would be of little use to protected tenants if they did not also have security of tenure, because otherwise their landlords could at common law terminate their tenancies by notice to quit (in the case of periodic tenancies) or refuse to renew them (in the case of fixed-term tenancies). The 1977 Act therefore also provides that when the contractual ("protected") tenancy comes to an end in this way it is prolonged indefinitely in the form of a "statutory tenancy", which is transmissible on the tenant's death (though not otherwise except by agreement) to a surviving spouse or other relative (if resident), and similarly a second time on that person's death (there can only be two successions). However, where the tenancy is transmitted to a non-spouse the rent ceases to be a Fair Rent but becomes an Assured Rent.

This security of tenure, however, can be terminated if the county court grants possession to the landlord; though the court can only do this in certain specified cases, some of which arise because of the requirements of the landlord and others because of some default by the tenant. These various grounds will be found in Schedule 15 to the 1977 Act. Some of these grounds are discretionary and some are

---

[1] *BTE Ltd* v *Merseyside and Cheshire Rent Assessment Committee* [1992] 1 EGLR 116.

mandatory. In the case of the discretionary grounds the landlord will only succeed if he can satisfy the court that:

(a)   it would be reasonable for him to be awarded possession;
(b)   there is suitable alternative accommodation; and
(c)   there exists no special reason for withholding the award of possession.

Alternative accommodation is not a requirement of the Schedule for obtaining possession except where this is the sole ground, but in practice it will be required. The Act defines what is "suitable" as being suitable to the means as well as the needs of the tenant with particular reference to his place of work. The alternative accommodation need not be the same size as the existing accommodation and it could, for example, comprise part of the existing premises.

As stated above, security may only be transmitted twice following the death of the original tenant. However, the rent will only be a Regulated or Fair Rent if the transferee is the spouse of the original tenant. Where the transferee is not the spouse, he or she only acquires an Assured Tenancy (see later) where there is security of tenure but the rent is the open market rent. County courts also have jurisdiction over "rental purchase agreements", which were previously outside the scope of statutory protection (Housing Act 1980, Part IV).

### 3.  The Housing Act 1988

Under this Act no new Regulated Tenancies can come into existence after 15 January 1989 except by Order of the County Court (e.g. on the grant of a tenancy of suitable alternative accommodation). Two new forms of tenancy were created: (i) Assured; and (ii) Assured shorthold.

Where an Assured Tenancy is granted, the tenant has security of tenure but pays an open market rent. The tenancy may be either periodic or for a fixed term. At the end of the contractual period the parties can agree to a revised rent by the landlord serving a notice with a proposal which the tenant does not oppose. If the parties do not agree a revised rent, the tenant makes an application to the Rent Assessment Committee which fixes a market rent. Possession can only be obtained on specified grounds which follow Schedule 15 to the Rent Act 1977. The letting can provide for rent reviews which can be fixed to an actual assessment or to a formula.

Where an Assured Shorthold Tenancy is granted, the tenant does not have security of tenure but, at the commencement of the tenancy, the tenant can challenge the rent as being higher than comparable rents in the locality; the Rent Assessment Committee has powers to lower the rent. If the tenancy was created prior to the coming into effect of the Housing Act 1996, great care had to be exercised in complying with the formalities pertaining to the grant of the tenancy, as otherwise the tenancy would become an Assured Tenancy. The term had to be a term certain of not less than six months, determinable on not less than two months' notice and the tenant had to be notified in advance of signing the tenancy of its nature and the fact that he would not have security of tenure. Best practice dictated that this notice was given more than 24 hours before the tenancy was signed and that the tenant acknowledged receipt of the notice by signing and returning a duplicate copy stating the time and date received.

Under the Housing Act 1996 all tenancies created after the coming into effect of the Act (i.e. after 1 March 1997) are automatically Assured Shorthold unless an Assured Tenancy is specifically created by agreement.

## 4. The Landlord and Tenant Act 1954 (Part I), the Leasehold Reform Act 1967 and the Leasehold Reform, Housing and Urban Development Act 1993 (Part I)

The protection of tenants occupying dwelling-houses on ground leases has caused concern for a number of years and these three pieces of legislation represent the steps taken to ensure that such tenants are not automatically dispossessed when their contractual right to remain in occupation has expired.

The Landlord and Tenant Act 1954, (Part I), which came into effect on 1 October 1954, applies to houses let on "long tenancies" (i.e. for more than 21 years) which, on account of the rent being a "low rent" (i.e. less than two-thirds of the rateable value on 23 March 1965, or when first rated thereafter) are outside the protection of the Rent Acts. The limits of rateable value within which Part I of the 1954 Act applies are those applicable under the Rent Act 1977, referred to above. A tenant is not protected if the landlord is the Crown or a local authority, the Development Corporation of a new town or certain housing associations and trusts. Where a tenancy is entered into after 1 April 1990 it is a tenancy at a low rent if it is £1,000 or less in Greater London or £250 or less elsewhere.

The effect of the 1954 Act is to continue the tenancy automatically where the tenant is in occupation after the date when it would normally expire, on the same terms as before, until either the landlord or the tenant terminates it by one of the notices prescribed by the Act; though the landlord and the tenant can agree on the terms of a new tenancy to take the place of the long tenancy.

A long tenancy can be terminated by the landlord by giving one of two types of notice, each of which must be in prescribed form. If the landlord is content for the tenant to stay in the house, he must serve a landlord's notice proposing a statutory tenancy. If he wishes the tenant to leave he must serve a landlord's notice to resume possession.

Should a tenant wish to terminate a long tenancy, he must give not less than one month's notice in writing.

A landlord's notice proposing a statutory tenancy would set out the proposed terms as to rent and repairs, including "initial repairs". The landlord and tenant can negotiate on these terms and come to an agreement in writing. If they cannot do so, the landlord can apply to the County Court to decide those items which are in dispute.

The 1977 Act procedure already referred to for the determination and registration of fair rents applies to tenancies arising from these provisions, but this applies only until just after the end of 1998. Long tenancies expiring on or after 15 January 1999 become Assured Tenancies with open market rents (see Section 186 Local Government and Housing Act 1989 – Schedule 10).

Where a tenant remains in possession after the end of the long tenancy he is relieved of any outstanding liability in respect of repairs arising under that tenancy. The terms proposed by the landlord may provide for the carrying out of repairs when the new terms come into force. These repairs are known as "initial repairs" and the tenant may have to bear some or all of the cost of them. Where a tenant leaves at the end of the long tenancy, his liability under the tenancy is not affected by the 1954 Act. "Initial repairs" may be carried out either by the landlord or by the tenant or partly by one and partly by the other; neither need do any unless he wishes. If the landlord carries out the repairs he is entitled to recover from the tenant the reasonable cost of the repairs in so far as they are necessary because the tenant did not meet his obligations under the long tenancy. Payment can be made by the tenant either by a lump sum or by instalments, as agreed between the parties or as determined by the County Court.

Where a landlord serves notice to resume possession the tenant, if he wishes to remain in the house, should so inform the landlord. If he does not agree to giving up the house, the landlord can apply to the county court for a possession order. The grounds upon which a landlord can apply for possession include the following:

(i)   that suitable alternative accommodation will be available for the tenant;

(ii)  that the tenant has failed to comply with the terms of his tenancy as to payment of rent or rates or as to insuring or keeping insured the premises;

(iii) that the tenant, or a person residing with him, or any sub-tenant of his has caused nuisance or annoyance to adjoining occupiers;

(iv)  that the premises, or any part of them, which the tenant is occupying are reasonably required by the landlord for occupation as a residence for himself or any son or daughter of his over 18 years of age or his father or mother.

In the last case the Court must not make an order for possession where the landlord purchased the property after 21 November 1950, or where it is satisfied that having regard to all the circumstances of the case, including the availability of other accommodation, greater hardship would be caused by making the order than by refusing to make it. Where the landlord's interest is held by any of certain specified public or charitable bodies (Leasehold Reform Act, 1967, Section 38), there is an additional ground for obtaining possession, namely that the landlord proposes to demolish or reconstruct the premises for purposes of re-development, and will require possession for this purpose at the end of the long tenancy, and has made reasonable preparation for the re-development.

The Leasehold Reform Act 1967 came into force on 27 October 1967, and represented a very radical departure from previous property law. The White Paper[2] which preceded the legislation stated the Government's view that "the basic principle of a reform

---

[2] Cmnd. 2916 of 1966 "Leasehold Reform in England and Wales".

which will do justice between the parties should be that the freeholder owns the land and the occupying leaseholder is morally entitled to the ownership of the building which has been put on and maintained on the land". This principle has, however, only been extended to the limited range of properties to which the 1967 Act applies; and this (unlike the Landlord and Tenant Act 1954, Part I) excludes flats and maisonettes, similar units produced by subdivisions of buildings where the dividing line is not vertical, and higher value properties. These exclusions are now covered by the 1993 Act, which is less advantageous to the tenant.

The 1967 Act enables qualified leasehold owner-occupiers either to purchase the freehold reversion from the ground landlord or to obtain an extension of the term of the lease.

The following requirements must be met before a leaseholder is qualified and entitled to the benefits conferred by the Act:

(a)  The term of the existing lease must be more than 21 years and the rent reserved must be less than two-thirds of the rateable value. Where the lease was for more than 35 years the low rent test will not apply.[3]

(b)  The leaseholder must have occupied the house for three of the last 10 years as his main residence. Use of part of the premises for another purpose, for example as a shop, does not necessarily disqualify.[4]

(c)  The rateable value of the house on 23 March 1965, or when first rated thereafter, must not exceed £400 in Greater London and £200 elsewhere. The Housing Act 1974, Section 118(1), increased these limits, with effect from 1 April 1973, to £1,500 and £750 respectively for tenancies created on or before 18 February 1966, and to £1,000 and £500 respectively for tenancies created after that date. [For tenancies granted after 1 April 1993 the Leasehold Reform, Housing and Urban Development Act 1993 applies (see below, pp. 314–315)].

"Shared ownership leases" granted by various public authorities and housing associations are excluded from the operation of the 1967 Act (Schedule 4A, added by the Housing and Planning Act 1986).

---

[3] Housing Act 1996
[4] *Lake* v *Bennet* (1969) 21 P&CR 93.

*Enfranchisement by way of Purchase of the Freehold*

Where a leaseholder is qualified under the 1967 Act and gives his landlord written notice of his desire to purchase the freehold interest, then, except as provided by the Act, the landlord is bound to make to the leaseholder and the leaseholder is bound to accept (at a price and on the conditions provided) a grant of the house and premises for an estate in fee simple absolute, subject to the tenancy and to the leaseholder's encumbrances but otherwise free from encumbrances.

For higher value properties (i.e. those with an RV of over £1000/£500) the price payable, as defined in Section 9(1A) of the 1967 Act as amended by Section 23 of the Housing and Planning Act 1986, is the amount which the landlord's reversion to the house and premises might be expected to realise on the assumption that it is to be sold in the open market by a willing seller (the tenant and members of his family who reside in the house having no right, for the purpose of this assumption, to buy the freehold or an extended lease but they are assumed to be in the market). It must be assumed also that the freehold interest is subject only to the existing lease, ending on the original date of termination even if extended under the 1967 Act.

Within one month of the ascertainment of the price payable the tenant may give written notice to the landlord that he is unable or unwilling to acquire at that price, in which case the notice of his desire to have the freehold ceases to have effect. In such circumstances the tenant must pay just compensation to the landlord for his costs. The tenant can start the procedure again after 3 years.

*Valuations to Determine Enfranchisement Price – "Original Method"*

The valuation approach to determine the price payable for lower value properties (i.e. below RV £1000/£500) under Section 9 of the Act (as amended) has proved in practice to be a highly contentious area and there have been many cases referred to the Lands Tribunal.

One of the earliest cases turned on whether the valuation approach should reflect the tenant's position as a special purchaser and so allow for marriage value.[5] This led to the amendment which provided that the tenant's bid should be disregarded (but see, in

[5] *Custins* v *Hearts of Oak Benefit Society* (1968) 209 EG 239.

relation to higher value properties, the amendments introduced by the Housing Act 1974 and the "new method" of valuation referred to later).

Subsequent cases have turned chiefly on two aspects, the method of determining the modern ground rent and the choice of the capitalisation rates, within the valuation of the freehold interest.

The typical situation is where the freehold interest is subject to a ground lease granted several years ago at an annual ground rent which is now considered to be a nominal sum. At the end of the lease it must be assumed that the tenant will extend the lease for a further period of 50 years at the current ground rental value (a "modern ground rent") subject to review after 25 years. Thereafter the house and land will revert to the freeholder. This involves a three-stage valuation:

(a)  *Term of Existing Lease*

    This will be the valuation of a ground rent for the outstanding period of the lease. From an investment viewpoint this is an unattractive proposition since the income is fixed and it is small so that management costs are disproportionately high. This has led to disputes as to the appropriate capitalisation rates to be adopted.

    As a general rule, the Lands Tribunal decisions tended to adopt rates between 6 and 8%,[6] but since 1980 most leasehold valuation tribunal decisions have adopted 7%. Higher rates have been employed where the outstanding term is relatively long or where there is market evidence.[7]

(b)  *Reversion to Modern Ground Rent*

    At the end of the existing lease the rent will rise to a "modern ground rent" which will be receivable for 50 years, subject to one review after 25 years.

    The "modern ground rent" is in essence the current ground rental value, being the annual equivalent of the cleared site value. However, in many built-up areas evidence of site values may be non-existent and therefore one approach has emerged which is commonly adopted whereby the value of the freehold interest in the house and land is apportioned between land

---

[6] See for example *Carthew* v *Estates Governors of Alleyn's College of God's Gift* (1974) 231 EG 809; *Nash* v *Castell-y-Mynach Estate* (1974) 234 EG 293.

[7] See for example *Leeds* v *J&L Estates Ltd* (1974) 236 EG 819.

and buildings and the modern ground rent is then derived from the value of the land as apportioned. This is known as the "standing house approach".[8]

Alternative methods may be adopted, involving a valuation direct to site value or to site rental value and are generally to be preferred where evidence exists.[9]

The standing house approach starts therefore with the valuation of the freehold interest in the house and land developed to its full potential. The valuation assumes vacant possession, ignoring for example the fact that part is let to tenants at a controlled or Fair Rent, or any disrepair.[10] The reason for this is that what is sought is the full unencumbered value of the property. For the same reason it will reflect any potential for conversion into separate flats to be sold with vacant possession.[11]

Once this full value has been determined it is necessary to apportion the value between land and buildings. The amount apportioned to the land will depend on the facts of the case, but in general the proportion attributable to the land value of houses in high value areas will be greater than that in low value areas in the same way that values per plot for high value houses are greater than values per plot for cheaper housing. There can be no hard and fast rules but tribunal decisions have tended to adopt around 40% for houses in London and around 30% for houses elsewhere as a reflection of this general proposition.[12]

Having determined the land value, the final stage is to determine the modern ground rent. This is found by decapitalising the value applying the appropriate yield. This again has been an area of dispute, coupled with the choice of yield with which the modern ground rent will be capitalised. Clearly, if both yields are the same then the whole exercise can

---

[8] See, for example, *Hall* v *Davies* (1970) 215 EG 175; *Kemp* v *Josephine Trust* (1971) 217 EG 351; *Nash* v *Castell-y-Mynach Estate, supra.*

[9] *Farr* v *Millerson* (1971) 218 EG 1177; *Miller* v *St John the Baptist's College, Oxford* (1976) 243 EG 535; *Embling* v *Wells and Campden Charity's Trustees* [1978] 2 EGLR 208.

[10] *Official Custodian for Charities* v *Goldridge* (1973) 227 EG 1467.

[11] *Ibid.*

[12] See, for example, *Graingers* v *Gunter Estate Trustees* (1977) 246 EG 55.

be short-circuited by simply deferring the land values at the chosen yield. However, a line of argument was developed which required the decapitalisation rate to be lower than the re-capitalisation rate, an approach known as the adverse differential.

The Lands Tribunal adopted this approach in many early decisions but the correctness of the approach was challenged in the Court of Appeal in one case[13] when the Court held that there was no justification for adopting different rates.

Since then, the tribunal decisions generally have not adopted the adverse differential, although they may do so if they feel the particular circumstances justify it.[14]

When the adverse differential approach is adopted, the land value is commonly decapitalised at around 6% and the resultant modern ground rent then recapitalised at around 8%. In other cases a common rate will be adopted normally between 6% and 8%. These yields do not form absolute precedents, the yield depending on the facts of the case and the valuation implications thereof. Usually all three yields are taken at the same rate.

(c)  *Reversion to House*

Following the expiry of the 50 years deemed extension the whole property reverts to the freeholder. However, if the existing lease does not expire for a considerable period, by the time account is taken of the 50-year extension, the reversion to the standing house will have an insignificant value and can be ignored so that the modern ground rent is valued to perpetuity. However, if the unexpired term of the existing lease is short, the reversion in just over 50 years may be sufficiently significant to be included, as in the case of *Haresign* v *St John the Baptist's College, Oxford*.[15] Such a reversion is termed a "Haresign reversion".

---

[13] *Supra*, n.10.

[14] *Gallagher Estates Ltd* v *Walker* (1974) 230 EG 359; *Leeds* v *J&L Estates Ltd* (1974) 236 EG 819; it is suggested that the Lands Tribunal will normally be reluctant to accept the adverse differential.

[15] (1980) 255 EG 711.

*Market Transactions*

One of the principal causes for the valuation arguments which have emerged is that the statutory approach is not usually found in the open market, thereby leaving the valuer to draw upon valuation principles to carry out the valuation. The only comparables which will generally be found are of sales to tenants who are also enfranchising and where it may be assumed that the statutory approach has been adopted. However, in practice, a tenant may be prepared to pay over the odds to effect a purchase rather than go to the local Leasehold Valuation Tribunal and then possibly on appeal to the Lands Tribunal with the contingent risk of additional costs and uncertainty of outcome. The Lands Tribunal has recognised this factor when considering comparables by discounting the prices paid since the case of *Delaforce* v *Evans;*[16] the pressure on tenants and the consequent over-bidding being termed the "Delaforce effect".[17] [This terminology has now found its way into general valuation terminology and refers to any "over bid" due to the anxiety of the bidder to settle rather than to litigate].

*Example 17–1*

The leaseholder of a late 19th century, brick-built four-bedroom semi-detached house with garage, in a fairly good residential area, wishes to purchase the freehold interest. The lease is for 99 years from 1 January 1909, at an annual ground rent of £5.

On 1 January 1996, the leaseholder, who is qualified under the Act, serves the necessary statutory notice under the Leasehold Reform Act 1967, claiming the right to have the freehold.

The house occupies a site with an area of 300 m². Comparable residential building land in the area has recently changed hands at £825,000 per hectare for development at 38 houses per hectare and with services costing £125,000 per hectare. Similar houses in the area have been sold on the open market with vacant possession at figures ranging from £75,000 to £90,000.

---

[16] *Delaforce* v *Evans* (1970) 215 EG 315.
[17] For an example of the application of the "Delaforce effect" see *Guiver* v *Francine Properties Ltd* (1974) 234 EG 741.

*Valuation*

In order to value the reversion after the existing lease it is necessary
to estimate a modern ground rent:

| (a) Standing House approach | | |
|---|---|---|
| Standing House Value, say | £82,000 | |
| Land Apportionment, say 30% | 0.3 | |
| Site Value | £24,600 | |
| Modern Ground Rent at 7% | 0.07 | £1,722 p.a |
| (b) Capital value of site with | | |
| all services, say | £25,000 | |
| Modern ground rent at 7% | 0.07 | £1,750 p.a. |
| Modern Ground Rent | | |
| Say | | £1,750 p.a. |

In this example both methods have produced a similar modern
ground rent. Method (b) is the preferred method where there is
available evidence because it is less contrived, but method (a) is the
more common because of the comparative ease in obtaining
evidence as to the standing house value. Both methods could
produce a result where the modern ground rent, known as the
Section 15 rent, exceeds the Fair Rent for the actual property. This
was discussed in the Dulwich College case[18] and found to be
acceptable.

| *Term* | | |
|---|---|---|
| Present ground rent | £5 p.a. | |
| YP 12 years at 7% | 7.94 | £40 |

| *Reversion* | | |
|---|---|---|
| After 31st December, 2007, to | | |
| modern ground rent based on | | |
| 1995 site value | £1,750 p.a. | |
| YP in perp at 7% def'd | | |
| 12 years | 6.34 | £11,095 |
| Price payable for enfranchisement | | £11,135 |

---

[18] *Carthew* v *Estates Governors of Alleyn's College of God's Gift* (1974) 231 EG
809.

*Alternatively*

| Term | | £40 |
|---|---|---|
| *Reversion* to Section 15 rent | £1,750 | |
| YP 50 yrs @ 7% × PV of | | |
| £1 in 12 yrs @ 7% | 6.13 | £10,728 |
| *Reversion* to vacant possession value | £82,000 | |
| × PV of £1 in 62 yrs @ 7% | 0.015 | £1,230 |
| Price payable for enfranchisement | | £11,998 |

The enfranchisement price will therefore be £11,998 (say £12,000) because in this case the Haresign effect comes into play.

## Amendments under Housing Act 1974

As was stated above, the Housing Act 1974 extended the rights of enfranchisement to houses with a rateable value up to £1,500 in Greater London (£750 elsewhere) or, for tenancies created after 18 February 1966, £1,000 (£500). The rateable value may be adjusted so as to ignore tenants' improvements and a certificate may be obtained from the Valuation Officer for this purpose.[19] Thus a house in London with a rateable value of say £1,570 may come within the 1974 Act if, for example, the tenant put in central heating and added a garage, all of which has added more than £70 to the rateable value.

## Valuations to Determine Enfranchisement Price – "New Method"

In respect of houses within the extended rateable value limits set by the 1974 Act, the valuation approach for enfranchisement differs in two important respects from "the original method" already considered. The first is that the price may reflect the tenant's bid, so allowing for marriage value.[20] The second is that the valuation should reflect the tenant's right to remain in possession under Part I of the Landlord and Tenant Act 1954 and not his right to extend the lease by 50 years at a modern ground rent. This basis of

---

[19] Section 118 of and Schedule 8 to the Housing Act 1974 as amended by Section 141 of and Schedule 21 to the Housing Act 1980, now Section 1(4A) of the 1967 Act.

[20] *Norfolk v Trinity College, Cambridge* [1976] 1 EGLR 215.

valuation is described as "the new method" and the provisions are in Section 9(1A) of the 1967 Act.

The valuation is therefore significantly different.


*Example 17–2*

The same facts apply as in Example 17–1 save that the RV after adjustment for tenants improvements was £1,150 in London and the vacant possession value ignoring tenants improvements is £350,000 for the freehold and £95,000 for the unexpired lease.

| | | | | | |
|---|---|---|---|---|---|
| Vacant possession Value – Freehold | | | | | £300,000 |
| Leasehold interest | | | £95,000 | | |
| Freehold interest | | | | | |
| Rent Reserved | £5 | | | | |
| YP 12 yrs @ 7% | 7.94 | £40 | | | |
| Reversion to | £300,000 | | | | |
| PV of £1 in 12 yrs | | | | | |
| @ 7% | 0.44 | £132,000 | £132,040 | | £227,040 |
| Marriage Value | | | | | £72,960 |
| Assume divided equally | | | | | 0.5 |
| | | | | | £36,480 |
| Freeholder's interest as above | | | | | £132,040 |
| Enfranchisement price | | | | | £168,520 |

Where the reversion is prior to 15 January 1999 the valuation would reflect the right to remain in possession at a Fair Rent by discounting the VP reversion by at least 10% to reflect the possibility that the lessee would not vacate and would become a Regulated Tenant with a Registered Fair Rent.[21] The discount would not apply if the reversion was later because it would be unlikely for the tenant to wish to remain as an Assured Tenant.[22] However, once past 1999 it will be open to a tenant to agree that he will remain in the premises even at an Assured rent. If the reversion was to be within a year or two this agreement might be successful.

---

[21] *Lloyd-Jones v Church Commissioners for England* [1982] 1 EGLR 209.
[22] *Eyre Estate Trustees v Shack* [1995] 1 EGLR 213.

## Extension of the Existing Lease (lower value properties)

Where a leaseholder is qualified under the Act and gives the landlord written notice of his desire to extend his lease then, except as provided by the Act, the landlord must grant and the leaseholder must accept a new tenancy for a term expiring 50 years after the term date of the existing tenancy, with a rent review after 25 years.

With the exception of the rent, the terms of the new tenancy will be significantly the same as the terms of the existing tenancy although some updating will be allowed. Rules for the ascertainment of the new rent are laid down in Section 15 of the 1967 Act but the main point to be noted is that the rent must be a ground rent in the sense that it represents the letting value of the site excluding anything for the value of the buildings on the site. This modern ground rent which is payable as from the original term date can be revised if the landlord so requires after the expiration of 25 years. The rent will be calculated in exactly the same way as for calculating the enfranchisement price and the valuation date is the original term date of the lease (or 25 years later) with the calculation done at that time.

## Landlord's Overriding Rights

In certain circumstances the leaseholder's right to purchase the freehold or to extend the lease may be defeated. If the landlord can satisfy the Court that he requires possession of the house for redevelopment or for his own occupation or occupation by an adult member of his family, the Court may grant an order for possession. The tenant is entitled to receive compensation for the loss of the buildings. The compensation will be the value of the 50-year lease at a modern ground rent which will normally be calculated by deducting from the vacant possession value of the freehold what would have been the cost of enfranchising on the last day of the lease.

Where there is development potential this is reflected in the enfranchisement price by assuming that at the end of the current lease there would be no renewal but compensation will be payable.

## Example 17–3

The same facts as in example 17–1 but assuming that the house stood on 4,000 m$^2$ of ground worth £300,000 for development and the extra garden only added £10,000 to the value of the freehold house if there was no possibility of development.

| | | | |
|---|---|---|---|
| Rent Reserved | | £5 | |
| YP 12 yrs @ 7% | | 7.94 | |
| | | | £40 |
| Reversion to: | | | |
| Site value | | £300,000 | |
| *Less* | | | |
| Compensation | | | |
| Value of F/H houses | £92,000 | | |
| Enfranchisement Price | | | |
| if no landlords rights | | | |
| 30% × £92,000 | £27,600 | £64,400 | |
| | | £235,600 | |
| × PV of £1 @ 12 yrs @ 7% | | 0.44 | £103,664 |
| Enfranchisement Price | | | £103,704 |

*Retention of Management Powers*

In circumstances specified in Section 19 of the 1967 Act, landlords could retain certain powers of management on enfranchisement in order to maintain adequate standards of appearance and regular redevelopment in an area.[23]

*The Leasehold Reform, Housing & Urban Development Act 1993*

This Act has extended the right to enfranchisement to all houses regardless of their rateable value and it also allows leaseholders of flats to collectively purchase the freehold of their block or individually extend their leases by 90 years. Only qualified lessees can exercise these rights in the same way as under the earlier enfranchisement legislation, but these include investors with three or fewer flats and to be qualified the lessee must have owned the lease for not less than 12 months. The qualifying leases must be

---

[23] For a more detailed consideration of the Leasehold Reform Act 1967, see Chapter 3 of *Valuation: Principles into Practice* (4th Edition) Ed. W.H. Rees (Estates Gazette), and *The Handbook of Leasehold Reform* by C. Hubbard and D. Williams (Sweet & Maxwell).

longer than 21 years at a low rent or longer than 35 years regardless of the rent payable (Housing Act 1996).

In the case of houses the purchase price is calculated in the same way as for the higher rateable value houses save that the Act provides that not less than 50% of the marriage value will be payable to the freeholder. In all the cases determined by the Leasehold Valuation Tribunals (LVT) bar one, and in the Lands Tribunal, this has been interpreted as being 50%, as in Example 17–2. Logic dictates that in the standard case this should continue, but further Lands Tribunal decisions may change the position.

The provision for flats is that the existing lease is surrendered and a new lease is obtained at a peppercorn rent for the original term plus 90 years. There are no limits on the number of 90-year extensions which can be obtained. The premium payable is calculated in the same way as the enfranchisement price for the higher value houses with again not less than 50% of the marriage value going to the freeholder. Where there are a series of leaseholders the enfranchisement price is decided in accordance with a formula laid down in the Act.

Instead of acquiring lease extensions individually, the lessees can collectively purchase the freehold of their block. The price is effectively the sum of all of the leasehold extension prices for all of the flats in the block, plus any payments for additional land which the landlord requires to be purchased plus any further payment required to compensate the landlord for any other losses he may sustain as a result of the acquisition. This may be in relation to losses suffered in respect of service charges on other blocks where there was an estate-wide service charge scheme or for loss of development value. In order to purchase, not less than two-thirds of qualifying lessees must sign the notice of whom not less than 50% must occupy their flats as their main or principal residence (not just in the UK) and these must account for not less than 50% of all the flats in the block. There are special rules where part of the property is commercial, and where not all qualifying lessees are involved in the purchase, the remaining flats are purchased at their investment value with no marriage value. In the latter case an unsuccessful argument has been made to an LVT that for those flats being purchased by qualifying lessees more than 50% of marriage value should be payable. At the time of publication no decision has yet been given on this proposition on appeal to the Lands Tribunal.

## 5. Smaller Residential Properties and Tenements

*(a) General Method of Valuation*

The first question to be asked in advance of any mathematical approach to a valuation is whether the property is saleable, with or without improvement or repair, with vacant possession to owner occupiers. If the answer is in the affirmative then the value will be more a function of the potential to vacant possession than a capitalisation of the net income. Thus a country cottage let to a widow of 85 living alone will be significantly more valuable than if let to a couple in their 40s at the same rent. However, in the case of a tenement, a sale of an individual flat for owner occupation is unlikely and the rental income is much more important to a determination of value.

The gross rent will be determined by either the Rent Officer or by market forces and will in all cases reflect the age, character, locality and state of repair and will usually be paid either weekly or monthly.

In free market conditions slight differences in style and accommodation – for instance, a few feet of front garden with a dwarf brick wall and railings, or the fact that the front living room opens out off a small hall-passage, instead of direct on to the street – may cause a considerable difference in rental between houses within a short distance of each other.

Where vacant possession is the determining factor in the valuation no allowance need be made for bad debts because this could lead to a windfall early possession. In other cases the differential between a Fair Rent and a market rent will also ensure payment of the former or give rise to a windfall increase in income, and so alleviate the need for bad debt provision. In other cases a bad debt allowance of say 10% might be appropriate because of the low quality of covenant provided by many tenants in this part of the market.

If, owing to overcrowding or tenant ignorance, a house is let at what may be regarded as an excessive rent, the excess should be disregarded. This is because it will certainly be lost should the local authority exercise its powers under the Housing Act 1985[24] to pay housing benefit at an assessed level and not necessarily at the same

---

[24] Part X; see also the provisions in Part XVII relating to compensation for compulsory purchase.

level as the rent payable, bearing in mind that some or all of the rent for this low-quality accommodation is often paid by either local or national government.

Repairs, insurance and management will be the main outgoings and, in some cases the landlord may also pay rates and water and drainage charges.

Yields vary widely according to the circumstances of the particular case and the state of the local market. In the case of freeholds they may range from 5% or less where vacant possession is likely in the comparatively short term and where the security is good to figures of 15% or more where property is old, in poor repair and occupied by an unsatisfactory type of tenant. Where the rate of interest is low this is because of the large increase in capital value which occurs if vacant possession is obtained. In this connection it may be noted that the market for what may be described as "tenements", i.e. large old buildings let out to a number of tenants, is usually poor and uncertain compared with that for small, fairly modern houses let to single families. This is particularly so at the present time, owing to the high cost of repairs in the case of the "tenement" type and to the large increase in capital value if vacant possession is obtained of a house let to a single family.

Indeed there has been a sharp decline in the number of investors who are interested in holding residential properties as long-term investments, unless they reserve the right to obtain vacant possession through the grant of assured shorthold tenancies. The tendency is either for owners to sell to "sitting tenants", or not to re-let properties when they become vacant but to sell them off with vacant possession. A sale to a sitting tenant would realise a figure between investment value and vacant possession value whereas a sale with vacant possession would realise the full vacant possession value. In most cases where a tenanted property is being valued, it is suggested that a normal investment valuation should be performed, i.e. the net income should be multiplied by the appropriate YP, and that the resultant value should be checked as a percentage of vacant possession value. In the case of houses saleable with vacant possession and let at Fair Rents, the value will be found to fall within a range of 40% to 60% depending on how the market assesses the possibility of vacant possession being obtained. Where the property is let on an Assured Tenancy (i.e. at market rent but with the tenant having security of tenure) the percentage of vacant possession value will be higher because of the higher yield available; the value could be as high as 80% dependent on yield. In other cases the property may be regarded as a poor investment with unlimited outgoings and a high yield will be required. Where there is certainty of obtaining possession, as in the case of a shorthold tenancy, the vacant possession value should be deferred for the period of waiting and thus the value would not normally exceed 90–95% of vacant possession value.

## (b) Outgoings

The principal outgoings are repairs, insurance and management. In some cases the landlord may also pay Council Tax and water and drainage charges. Under the Rent Act 1977, any increase in rates due either to changes in assessment or poundage can be passed on to the tenant.

Apart from any contractual liability to repair, in the case of leases granted after 24 November 1961 for terms of less than seven years of dwelling-houses which the statutory protection of business or agricultural tenancies does not cover, the Landlord and Tenant Act 1985, Sections 11–16, implies a covenant by the lessor to keep in repair the structure, exterior and installations for water, gas and electricity supplies, for space and water heating and for sanitation (contracting out is prohibited except with the approval of the County Court).

It is generally preferable to make the allowance for repairs by deducting a lump sum per property which should be checked by reference to past records and to the age, extent and construction of the premises.[25] For example, the allowance in the case of a house stuccoed externally and requiring periodic painting will naturally be higher than in the case of a similar house with brick facings in good condition. It has been customary in the past to estimate the amount to be allowed for repairs as a percentage of the rent but this method is unreliable and illogical.

Where in the past the work of repair has been left undone it may be necessary to include in the valuation a capital deduction for immediate expenditure to allow for the cost of putting the property into a reasonable (but not necessarily a first-class) state of repair. This is called an end allowance.

---

[25] See Chapter 6.

The annual cost of insurance can usually be determined with sufficient accuracy, either by using the actual premium paid or by assessing it on the basis described in Chapter 20. Fire insurance premiums are becoming significant sums so that, where the landlord is responsible, care must be taken to ensure that the appropriate deduction is made.

Management may cost from 10% upwards plus VAT according to the circumstances and the services rendered.

Reference so far has been made to the estimation of net income by calculating each outgoing separately. It is useful as a check to determine the percentage which they represent of the gross rents.

### (c) State of Repair

Careful consideration must be given to the actual state of repair as affecting the annual cost of repairs, the life of the building and the possibility of heavy expenditure in the future. Particular attention should be paid to the structural condition, the presence or absence of dampness in walls or ceilings, the proper provision of sanitary accommodation and of cooking and washing facilities.

Regard should be had to the possibility of the service of a dangerous structure notice in respect of such defects as a bulged wall.

Where premises are in such a condition as to be a nuisance or injurious to health, the local authority may require necessary works to be done.[26] They may also serve a notice to repair a house, which is in substantial disrepair or unfit for human habitation.[27] The standard for deciding whether a house is "unfit for human habitation" is that laid down by Section 604 of the Housing Act 1985.

Where a house cannot be made fit for human habitation at a reasonable cost, the local authority may serve a demolition order. In respect of any part of a building unfit for habitation, including any underground room, or any unfit house from which adjoining property obtains support, or which can be used for a non-residential purpose, or is of special historic or architectural interest, they may serve a closing order[28] instead of a demolition order.

---

[26] Building Act 1984, Section 76; Environmental Protection Act 1990, Part III.

[27] Housing Act 1985, Part VI.

[28] *Ibid.*, Part IX.

In lieu of making a demolition order, a local authority may purchase a house which is unfit if they consider that it can be rendered capable of providing accommodation of a standard which is adequate for the time being.[29]

Special provision is made in the Housing Act 1985 as regards houses in multiple occupation (Part XI), and as regards houses which are overcrowded (Part X).

Where, on inspection, a property is considered not to comply with statutory provisions currently in force, allowance should be made in the valuation for the cost of compliance.

The allowance for annual repairs included in the outgoings is usually based on the assumption that the property is in a reasonable state of repair, at least sufficient to justify the rent at which it is let. If it is not, a deduction should be made from the valuation for the cost of putting it into reasonable repair, less any grant payable.

### (d) Clearance Areas

There are many houses, let either as a whole or in parts, which are of considerable age, indifferent construction, or so situated as likely to be included in a Clearance Area in consequence of which the local authority may acquire houses and buildings for the purpose of demolishing them and re-planning the area.[30]

Particular regard must be had to the possible application of this procedure, and to the statutory provisions applying to individual houses in a clearance area, when valuing old or defective property.

It must be remembered that a house, although not unfit in itself, may be included in consequence of its being in or adjoining a congested or overcrowded area.

---

[29] *Ibid.*, Part VI, Section 192.

[30] Housing Act 1985, Part IX. They are empowered to preserve unfit houses from demolition which can provide accommodation that is of standard "adequate for the time being" (Sections 300-302); but this does not confer on them an immunity from proceedings before the magistrates if any such house is either "prejudicial to health" or "a nuisance" under what is now Part III of the Environmental Protection Act 1990 (*Salford City Council* v *McNally* [1976] AC 379).

## (e) Duration of Income

In addition to the factors already considered, regard must also be had to the length of time during which it is expected that the net income will continue.

In the case of houses of some age, it is sometimes suggested that an estimate should be made of the length of life of the property, and the net income valued for that period with a reversion to site value. However, as explained in Chapter 9 it is more usual for the factor of uncertainty of continuance of income to be reflected in the rate per cent adopted.

In such cases a net yield of up to 20% may well be expected, whereas, in the case of more modern properties, the yield may be as low as 5%.

Where, however, there are strong reasons for assuming that the life of the property will be limited, allowance should be made for that factor in the valuation.

If demolition of existing buildings is likely in the future under the provisions of the Housing Act 1985 (Part IX), the income may properly be valued as receivable for a limited term with reversion to site value.

## Example 17–4

A terrace house in a suburban area built about 70 years ago comprises ground, first and second floors, with two large and one small room on each floor. There is a WC on each floor and each of the small rooms has been adapted for use as a kitchen with a sink and ventilated food storage provided. The house is situated in a neighbourhood where the development is open in character. The general structural condition is good, with the exception of the flank wall of the back addition, which is badly bulged, and the front wall where there are extensive signs of rising damp.

There are three tenants: the ground floor producing £10.00 per week, the first floor £12.00, and the second floor £10.00. These rents are the maximum rents permitted under the Rent Act 1977 and exclude Council Tax.

Drainage and water charges total £400 p.a.

What is the present market value?

*Valuation*

| Gross Income: | | per week | per annum |
|---|---|---|---|
| Ground Floor | £10.00 | | |
| First Floor | £12.00 | | |
| Second Floor | £10.00 | £32.00 | £1,664 |
| *Less* Outgoings: | | | |
| Drainage and water | £400 | | |
| Repairs and Insurance | | | |
| (incl VAT) | £600 | | |
| Management at 10% + VAT | £192 | | £1,192 |
| Net Income | | | £472 |
| YP in perp at say 10% | | | 10 |
| | | | £4,720 |
| Estimated cost of repair of | | | |
| flank wall and damp in front | | | |
| wall, net of grant and | | | |
| inclusive of VAT, say | | | £3,000 |
| Value, say | | | £1,700 |

*Note*: This is a very poor investment. There is no early prospect of vacant possession and the only potential is for a Regulated Tenant to be replaced by an Assured Shorthold tenant thus increasing the income, which is reflected in the comparatively low yield of 10%. However, the value must still be "tested" against its VP value as there may be some logic in keeping the property vacant on a piecemeal basis until full VP is obtained. If there is a large potential gain to be made then the value would be much higher than £1,700. If the VP value was £40,000 then the market value would still be in excess of £10,000 equal to say 25% × VP.

## 6. Larger Residential Properties

*(a) Generally*

When valuing houses with vacant possession, as already pointed out in Section 1 of this chapter, the capital value must be fixed by direct comparison with prices actually realised on sale.

In the case of larger houses which are let, the method recommended earlier in this chapter in relation to smaller houses, of performing a normal investment valuation and comparing the resultant figure with the vacant possession value, is again appropriate.

*(b) Factors Affecting Value*

The factors in question may be briefly summarised as follows:

(i)   *Size and number of rooms.* A prospective occupier is primarily in search of a certain amount of accommodation. He is likely, in the first instance, to restrict his enquiries to properties having the number of rooms of the size he requires.

(ii)  *Position.* The price he is prepared to pay for this accommodation will be influenced by all the factors associated with position.

Proximity to shops, travelling facilities, places of worship, open spaces, golf courses and schools: the character of surrounding property, building development in the neighbourhood, the presence and cost of public services and the level of Council Tax.

(iii) *Planning, etc.* In viewing property itself the prospective occupier will also consider such points as the arrangement, aspect and lighting of the rooms, the adequacy of the domestic and sanitary offices, the methods of heating, the presence or absence of a garage, the size and condition of the garden and the prospects for improving and enlarging.

The value of the accommodation provided by a house will be increased or diminished according to the way in which, in the details enumerated, it compares with other properties of a similar size.

(iv)  *Age and state of repair.* Changes in taste and fashion and the greater amenities provided in modern houses tend to reduce the value of older houses, but this is not always the case as "period houses" sometimes have an additional value.

The state of repair must be considered both as regards the cost of putting the premises into a satisfactory state of repair now and the cost of maintenance in the future.

A number of the points already mentioned in connection with smaller property must be considered.

The principal points include the condition of the main structure and roof; the penetration of damp, either by reason of a defective damp-proof course or insufficient insulation against wet of the external walls; the presence of conditions favourable to dry rot; the presence or absence of such rot or of wet rot or attack by woodworm or beetle; the arrangement and condition of the drainage system; the condition of external paintwork; internal decorations.

Houses with a large expanse of external paintwork or complicated roofs with turrets and domes will cost more in annual upkeep than houses of plainer design. The annual cost of repairs will also be influenced by the age of the premises.

When houses are let, the incidence of the liability for repair as between landlord and tenant is an important factor in their valuation. In the case of the larger houses let on long lease the tenant usually undertakes to do all repairs; but in the case of short-term lettings of small or moderate sized houses, the bulk of the burden has usually fallen on the landlord. The implied repairing obligations placed on the landlord by the Landlord and Tenant Act 1985, Sections 11–16 (already referred to under small houses), must also be borne in mind.

### (c) Methods of Valuation

At the present day, formal estimates of open market value for house property are seldom made except for mortgage, divorce or probate purposes. If a house is vacant, it will almost certainly be put on the market with vacant possession, and its capital value arrived at by direct comparison with recent sales, without reference to rental value. Comparisons are sometimes made between flats to find fair rents on the basis of so much per unit of net floor space but this is very restricted in terms of its more general application. Purchasers in the United Kingdom do not base their purchase price by a comparison with other houses or flats on a rate per square metre basis. It is therefore very dangerous to value on this basis as it strays from reality. Non-usable areas of accommodation such as a large hall may add to the overall value but the area of such accommodation is not included in the net floor space and thus an analysis will produce a high rate per square metre which will not be applicable to a house or flat without this feature. Similarly, the effect on value of a luxurious standard of finish will cause difficulties. Houses and flats should be valued by direct comparables of like with like; small differences in room size will have no effect on value. Where direct comparables are not available then the best available comparables should be used and suitably adjusted for location, size, etc., but this cannot be done on any mathematical formula basis.

New house builders often apply a rate per square metre to their proposed houses taken from their experience of houses elsewhere on other estates they have developed. This approach works because of the factual similarities of their product in terms of room sizes, standard of finish, etc. and this is a short-hand way of using the comparables. However, the end result must then be scrutinised to ensure that the resultant value compares with other local properties.

The Lands Tribunal have rejected valuations using this mathematical approach and have preferred the entirety approach of comparing the subject property to other specific properties.[31]

The exception lies with luxury new developments in Central London which are marketed abroad where the practice is to bid for the property on a price per square metre.

*Example 17–5*

A small semi-detached, brick-built modern house in good repair, on a plot with a 30-ft frontage, contains two reception rooms and WC on ground floor, two large and one small bedroom, bathroom and WC on the first floor.

What is its market value?

*Valuation*

Direct comparison of capital value. Similar houses in the neighbourhood sell for prices ranging from £55,000 to £60,000; by direct comparison of position, size of rooms, condition of repair and the amenities provided, the value is estimated to be £57,500.

In the case of tenant-occupied houses where the tenant is protected by the Rent Acts or where he holds the house under a lease or agreement with a reasonable term to run at a low rent, the tenant himself may be anxious to purchase the property. In these circumstances the price he will pay is likely to be the result of bargaining between the parties. The tenant is obviously not going to pay full vacant possession value; the landlord is not likely to be content with investment value only; the result will usually be a purchase price between these two extremes.

---

[31] *Toye v Kensington and Chelsea Royal London Borough Council* [1994] 1 EGLR 204.

*(d) Sales Records*

The systematic recording of the results of sales is essential to the valuer; the form such records can take varies widely.

## 7. Blocks of Flats

*(a) Generally*

The valuation of a block of flats does not differ in principle from the valuation of properties already considered.

The problem is only complicated by the greater difficulty in forming a correct estimate of gross income and of outgoings.

*(b) Gross Income*

In a few cases flats are let at inclusive rents, with the tenants not being liable for any outgoings, although extra charges for special services are usually encountered.

In estimating gross income, it should be borne in mind that fair rents can be reviewed every two years or, if tenants are not protected, existing agreements may permit rents to be reviewed more frequently.

Among the many factors affecting the present rental value are the presence of lifts, central heating, constant hot water, the adequacy of natural lighting, the degree of sound insulation between flats, convenience of proximity to main traffic routes offset by possible nuisance from noise.

*(c) Outgoings*

In estimating the allowance to be made for outgoings on an existing block of flats it is necessary to study the tenancy agreements carefully in order to determine the extent of the landlord's liabilities. In recent years there has been a marked tendency to make tenants responsible for all possible repairs, but the provisions of the Landlord and Tenant Act 1985[32] must be kept in mind.

---

[32] Sections 11–16 (referred to earlier in this chapter).

## Taxes and Charges

The present cost of Council Tax and water and drainage charges can easily be ascertained. Generally flats are let at rents exclusive of Council Tax. VAT payable on items such as repairs is not recoverable and these items should be included gross of VAT.

## Repairs

It is difficult to give any general guide to the allowance to be made for repairs, since the cost will depend upon many factors.

An exterior of stucco work, needing to be painted approximately every five years, will cost much more than one of plain brickwork, which is only likely to involve expenditure on repointing every 25 or 30 years. Regard must also be had to the entrance hall, main staircase and corridors, those having marble or other permanent wall coverings will have a low maintenance cost; other types of decoration requiring considerable expenditure to keep in a satisfactory condition will have a much higher maintenance cost.

## Services

Until the Rent Act 1957, rents were frequently inclusive of services. Now these are frequently covered by a separate service charge payable in addition to the rent in the case of Regulated tenancies. Such service charges are usually variable according to cost. It is necessary for the valuer to check that the charge is sufficient to cover costs and depreciation. Where services are included in rents paid, the cost must be deducted from the gross rents together with other outgoings to find the net income. Where Sections 11-16 of the Housing Act 1985 apply, the variable service charge element can only cover non-repair items.

The Housing Acts and the Landlord and Tenant Acts 1987 and 1996 give tenants paying service charges a large measure of protection. They must be consulted in advance of expenditure currently above £50 per unit or £1,000 (but subject to variation from time to time according to tabled amendments) in total, independent estimates are required for intended work and they are usually entitled to receive a written summary certified by a qualified accountant justifying the amount charged. Service charges must be such as a court considers reasonable both as to standard of services and amounts charged. Before ordering any "qualifying works", the landlord or his agent must obtain not less than two estimates, one of which must be from a genuinely independent source, and notify the tenants (including any association representing the tenants). The Housing Act 1996 provides for a Local Valuation Tribunal to adjudicate on a tenant's liability for a service charge before a landlord can commence forfeiture proceedings for non-payment.

Similar protection is extended to "secure tenants" (i.e. tenants in the public sector) by the Housing and Building Control Act 1984, Section 18 and Schedule 4, amplified by the Housing and Planning Act 1986, Sections 4–5. The latter Act (Section 4) inserts an elaborate code into the Housing Act 1985, Part V, concerning information which tenants are entitled to receive from landlords when exercising the "right to buy". This code, contained in Sections 125A, B and C inserted into the 1985 Act, deals with "improvement contributions" as well as service charges.

### (d) Net Income and Yield

A reasonable deduction from gross income having been made in respect of the estimated outgoings, a figure will be arrived at representing prospective net income.

However carefully this figure may have been estimated, it is likely to vary from time to time, particularly in relation to expenditure on repairs. The valuer will have regard to this fact in selecting the rate per cent of Years' Purchase on which his valuation will be based.

The rate per cent yield on the purchase prices of blocks of flats varies considerably. At the time of writing first-class flats are showing about 8%. A considerably higher return is likely to be expected from "converted" houses, probably 10% to 12% (or even more) but once again the prospect of vacant possession must be considered.

### Example 17–6

You are instructed to value a large freehold property comprising 76 flats in eight blocks, some of three storeys and some of two, erected about six years ago. The Vendor's agents inform you that the flats are all let, at recently reviewed Fair Rents of between £2,000 and £2,400. The standard form of tenancy agreement provides that the tenants are liable for internal decorative repairs.

The following particulars are also supplied. Total rent roll, £171,200 (including 18 garages producing £3,600). Outgoings last year: repairs, £10,000; lighting, £1,000; gardening, £1,500; insurances, £4,000. (All costs are inclusive of VAT where payable.)

As a result of your inspection the following additional facts are established:

The tenants pay the Council Tax on both flats and garages, with the rents for the latter averaging £4 per week. The agreements are for three years.

The gardens are extensive. The general condition of repair is satisfactory.

The flats are brick-built with tiled roofs, concrete floor, and a modern hot water system; there is no lift.

There is good demand for flats with vacant possession at an estimated price of £45,000 per flat excluding a garage.

The landlord provides no services other than lighting of common parts and upkeep of the gardens.

The outgoings seem to be reasonable except for the amount for repairs and maintenance which is obviously too low.

| | | | |
|---|---|---:|---:|
| Total Rent Roll | | | £171,200 |
| *Less* Garages | | | £3,600 |
| | | | £167,600 |
| *Less* | | | |
| Outgoings: | | | |
| Lighting | | £1,000 | |
| Garden | | £1,500 | |
| Insurances | | £4,000 | |
| Repairs (incl. of VAT) say | | £20,000 | |
| Management, say | | £8,000 | £34,500 |
| Net Income | | | £133,100 |
| YP in perp at 8% | | | 12.5 |
| | | | £1,663,750 |
| 18 garages, gross | | £3,600 | |
| *Less* | | | |
| Outgoings | | | |
| Repairs | £400 | | |
| Management, say | £360 | £760 | |
| Net Income | | £2,840 | |
| YP in perp at 12.5% | | 8 | £22,720 |
| | | | £1,686,470 |
| Value, say | | | £1,650,000 |

The value analyses at £21,891 per flat which equates to close to 50% of the vacant possession value per flat which is within the expected band.

## (e) Flats Held as Investments

The provisions of the Housing Acts in relation to statutory controls on management of blocks of flats have already been mentioned. The next part of this chapter outlines the provisions of the Landlord and Tenant Act 1987 which tightens these management restrictions and also provides for tenants to be given first refusal when a landlord wishes to dispose of his interest. Although the Government's intention to protect tenants from bad landlords (and bad managers) is both understandable and laudable, it is difficult to see that there will be any incentive in the future for holding flats as investments. At the same time, it will be difficult for existing investors to withdraw (except by selling to the tenants) at what is submitted must be a basic investment value. This is not to say there is no future for private blocks of flats but it seems likely that each flat will be individually owned, most probably on a long lease, with the freehold held and management arranged on a communal basis.

## Landlord and Tenant Act 1987

This statute deals with privately owned blocks of flats, and implements a number of recommendations contained in a report on the management of such flats, produced by a committee chaired by Mr E.G. Nugee, QC.

Part I of the Act (Sections 1–20), which was brought into effect on 1 February 1988, applies to flats occupied by "qualifying tenants", i.e. those whose tenancies are not shorthold tenancies (under the Housing Act 1980, Section 52), or business tenancies (protected under the Landlord and Tenant Act 1954, Part II), or tenancies which go with the tenant's employment, or tenancies comprising more than one flat; but a "qualifying tenant" can hold two or more tenancies of up to 50% of the flats in the block. In regard to such tenants, the immediate landlord must serve on them a notice giving them first refusal before making a "relevant disposal". This means any disposal by way of a legal estate or equitable interest (which may extend to "common parts" of the block), other than certain prescribed exceptions such as disposals to the Crown, to a compulsorily purchasing authority or to an associated company, and non-commercial disposals within the landlord's family or for the purposes of trusts, charities, bankruptcy, etc. "Qualifying tenants" must occupy more than half the flats in the block. Flats of which more than 50% of the internal floor area is used for non-residential purposes are excluded. If the landlord also holds under a tenancy which will be or can be terminated within seven years, the superior landlord also counts as "the landlord", and so on up the scale. Flats owned by public sector landlords are not within the scope of these provisions, nor are those owned by resident landlords.

The "notice conferring rights of first refusal" which must be served on the "qualifying tenants" must set out the terms of the proposed disposal and serves as an offer which may be accepted by a majority vote of the tenants, i.e. more than 50% of the votes on the basis of one flat one vote. A period of at least two months must be allowed for acceptance, and a further period of at least two months must be added if the tenants (by a majority) nominate a person or persons for the purposes of acquiring the interest to be disposed of. If the offer is not accepted within the specified time, the landlord may within 12 months thereafter dispose of the "protected interest" to be disposed of (i.e. the interest which is subject to the tenant's right of first refusal) to anyone he chooses, so long as he does not undercut the terms of his offer to the tenants. If the tenants do not propose to accept the offer they can, within the time limit specified, make a counter-offer. The landlord may accept or reject this; but his rejection may also contain a counter-offer of his own, provided that it sets out the terms of such counter-offer and specifies a further period of time for acceptance. If the tenants do not accept within that time, the landlord may within 12 months dispose of the protected interest in the manner stated above. Short of the conclusion of a binding contract, either the landlord or the person(s) nominated by the tenants may withdraw, and the nominated person(s) must do so if the number of tenants wishing to proceed falls below a majority.

If the landlord wrongfully disposes of the protected interest in disregard of the tenants' right to have "first refusal" – it is a criminal offence under the Housing Act 1996 – the new landlord, i.e. the transferee, can be required to give them details of the offending transaction within one month. The tenants by a majority (as above) may then within three months serve a "purchase notice"

on the new landlord requiring him to sell the interest he has
acquired to a person or persons nominated by them on the same
terms as those on which it was acquired by him. If any dispute
arises over the terms of such a purchase notice, it is to be referred
to the local rent assessment committee (as constituted under the
Rent Act 1977). The "new landlord" may be the previous landlord's
own landlord if the wrongful disposal took the form of a surrender.
If a prospective purchaser from the landlord is unsure whether
rights of first refusal exist or not, he may serve notices on the
tenants concerned so that the question can be resolved in advance.
It should be noted that the Secretary of State for the Environment is
empowered to modify the operative provisions concerning the
"notice conferring rights of first refusal" and consequential
procedures, by making regulations to that effect. There are special
rules if a landlord wishes to sell by auction.

There are special rules provided for in the Housing Act 1996 for
selling by auction. A landlord can sell in this way without going
through the above detailed procedure, but the tenants have the
right to purchase at the achieved price and the successful bidder
therefore has a conditional contract. The procedure laid down by
the Act must be followed.

Part II (Sections 21–24) concerns the appointment by order of the
court of managers for flats in buildings or parts of buildings
containing at least two flats not held on business tenancies
protected under the Landlord and Tenant Act 1954, Part II. The
High Court or County Court must be satisfied that the landlord is
in breach of some obligation owed by him to the tenant which
relates to the management of the premises, and that these
circumstances are likely to continue, and that in all the
circumstances it is just and convenient to make the order. These
provisions do not apply to flats which are owned by public sector
landlords or resident landlords or "included within the functional
land of any charity".

Proceedings start with a preliminary notice served by the tenants
on the landlord stating that an application to the court is intended
on grounds referred to above, as appropriate. Only if the landlord
does not take steps to rectify the situation within a reasonable
period of time, which the tenant must specify (if applicable), can
the tenant apply to the court, which can then make an order
appointing a manager on such terms as it thinks fit, in a similar
manner to the appointment of a receiver. The order may be

provisional or final, may be suspended, and may extend to more or less than the premises in respect of which it has been sought.

Part III (Sections 25–34) confers on tenants a power of compulsory purchase of the interest held by their landlords. As with Parts I and II, premises owned by public sector landlords, and also resident landlords, are excluded. Included are buildings or parts of buildings containing two or more flats occupied by "qualifying tenants", i.e. tenants holding under "long leases" (21 years or more or leases granted under the "right to buy" in the Housing Act 1985, Part V) but not business tenancies protected under the Landlord and Tenant Act 1954, Part II. At least 90% of the flats in such a block must be held by "qualifying tenants" (all but one, if there are five to nine flats; all, if there are four flats or less). If more than 50% of the internal floor area of a block is in non-residential use (excluding common parts) then it too is excluded. A tenant under a lease covering more than a single flat is not a "qualifying tenant", though qualifying tenants are not restricted to one lease of one flat.

The power given to "qualifying tenants" is to nominate a person to acquire the landlord's interest under an "acquisition order" to be made by the court. At least 50% of the "qualifying tenants" (one tenant one vote) must first serve a "preliminary notice" on the landlord, stating that they will apply to the court for such an order; but will not do so if the landlord remedies specified defaults which are capable of remedy within a reasonable time where this is the case. The pre-requisite for the procedure is that the landlord is in breach of his obligations to the tenants under their leases in respect of repair, maintenance, insurance or management, in circumstances which are likely to continue, and the appointment of a manager under Part II, "would not be an adequate remedy". If the landlord fails as stated above to remedy specified defaults (if that is the case), or if his defaults are not remediable, the "requisite majority of such tenants" must then apply to the court for the "acquisition order", the application being registrable as a "pending land action" (i.e. a land charge), and inhibitions are obtainable if it is registered land. The court must not make the "acquisition order" unless satisfied that the above requirements are broadly met and that it would be appropriate to do so in all the circumstances. If a manager has already been appointed under Part II, it is necessary to wait until that appointment has lasted three years. The landlord's interest will, under the "acquisition order", be vested in the nominated person referred to above, on terms settled (in default of agreement) by the local rent assessment committee; but if the landlord is a leaseholder who needs the consent of a third party to the transfer of his own lease and this consent is genuinely not forthcoming, the order will not take effect. The market price to be paid must assume that the landlord's interest is subject to all the existing leases and that none of the tenants are buying it or seeking to buy it; this therefore excludes any marriage value. Subject to any agreement or court decision otherwise, mortgages, liens and charges – other than rent charges – are to be paid off in accordance with a procedure laid down in Schedule 1. If the landlord is not traceable, the acquisition terms are settled by the court, and the price is paid into court after being assessed by a surveyor chosen by the President of the Lands Tribunal. Conversely, if acquisition proceedings are abortive, the order can be discharged and the landlord, in a proper case, can recover reasonable costs.

Part IV (Sections 35–40) deals with variation of leases of flats. In theory, its scope is wider than the foregoing provisions in that only tenancies protected as business tenancies under the Landlord and Tenant Act 1954, Part II (this does not here include assured tenancies) are excluded; but since the provisions apply only to long leases, i.e. for 21 years or more, public sector leases will only rarely be affected in practice. As with the foregoing provisions, leases covering more than individual flats are excluded, but there is nothing to exclude tenants who hold more than one lease in any building. Anyone who is a party to a long lease within these provisions may apply to the court for an order varying, or directing parties to vary, that lease if it is defective because it does not (for whatever reason) make satisfactory provision in regard to the premises for any of the following matters:

(a) Repair or maintenance of the premises or of ancillary features such as garages, access, etc.;
(b) Insurance;
(c) Repair or maintenance of installations which are reasonably necessary for proper enjoyment of the premises;
(d) Provision or maintenance of services which are reasonably necessary for proper enjoyment of the premises;
(e) Reimbursement of any party to the lease for expenditure reasonably incurred or to be incurred by him for the benefit of any other party.

Parties to any lease, other than the applicant, may thereupon apply to the court for other orders making corresponding variations of other leases, particularly in cases where any variation would not be satisfactory in isolation. If the court is satisfied, it may by order make the variations applied for, or any other variations which it thinks should be made; but no variation is to be made if any person would be substantially prejudiced thereby in a way that cannot fairly be met by financial compensation, or if on any other grounds the variation would not be reasonable. Third parties (including predecessors in title) are bound by any variations which are made. No variation in regard to insurance must interfere with provisions in a lease for nominated or specified insurers; on the other hand, the right to apply to vary in respect of insurance is extended to dwellings *other* than flats.

Part V (Sections 41–45) relates to management of leasehold property more generally in the residential private sector. Section 45, which took effect on 1 February 1988, amends the Housing Associations Act 1985, Section 4 (but not in Scotland), by extending the "permissible objects" of housing associations registered with the Housing Corporation so as to include management of any house or any block of flats (i.e. a building with two or more flats "held on leases or other lettings" which is "occupied or intended to be occupied wholly or mainly for residential purposes").

Sections 41–44 amend the Landlord and Tenant Act 1985 in respect of "service charges" payable by residential tenants. Detailed procedural amendments are made to the 1985 Act to put tenants who are represented in dealings with their landlord by a recognised tenants' association on the same footing as those who are not (long leases are no longer to be treated differently from other leases). Also worthy of note is a provision that the retrospective effect of service charges is to be limited to 18 months (Section 41 and Schedule 2). The basic point about service charges is that they relate to variable costs towards which two or more tenants are required by their tenancies to contribute periodically because of the benefits which accrue to them (thus "fair rents" registered under the Rent Act 1977 are to be excluded except where there are elements producing variable rents). The 1987 Act (Section 42) now provides that without prejudice to any pre-existing trust, the recipient of service charges – the landlord or management company, or other person – is to hold the money (with interest) on trust to defray all the relevant costs, and then to hold the balance for the tenants in shares in proportion to their liabilities (presumably to set against subsequent costs, because persons ceasing to be tenants will not recover their shares). Further safeguards are provided for tenants (Section 43 and Schedule 3) in respect of their insurance cover as paid for by service charges. Tenants' associations are given rights of consultation in relation to the appointment of managing agents for premises where service charges are payable (Section 44 inserting a new Section 30B into the 1985 Act).

Part VI (Section 46–51), which took effect on 1 February 1988, requires the giving of information to tenants of dwellings other than premises covered under the protection of business tenancies in the Landlord and Tenant Act 1954, Part II. Written demands for rent or other payments must contain the landlord's name and address (and also if necessary "address for service" in England and Wales) failing which no service charge can be recovered (Sections 47–49, which also modify Section 196 of the Law of Property Act 1925). If the landlord's reversion is assigned, the assignor continues to be liable to any tenant until either the assignor or the assignee gives notice, to that tenant, of the assignee's name and address; and the old and new landlords are jointly and severally liable for breaches occurring between the date of assignment and the date when notice is given (Section 50). In regard to registered land, the tenant of any dwelling is given the right (on payment of a fee) to search the proprietorship register in the Land Registry for details of his landlord's name and address (Section 51, adding a new Section 112C to the Land Registration Act 1925).

Part VII (Sections 52–62) contains various general provisions. It should be noted that, except in regards to Sections 41, 43–45, and 49–51, "the court" means the county court, for the purposes of this Act; and in the absence of some specific justification anyone bringing High Court proceedings will be penalised in costs (Section 52). The Act does not bind the Crown, but applies to Crown Land in the sense that a Crown interest may subsist in the premises provided that it is both separate from and superior to the private landlord's (leasehold) reversion as well as the interests of the various tenants of that landlord (Section 56).

## 8. Grants for Improvement of Private Sector Housing

Part I (Sections 1–103) of the Housing Grants, Construction and Regeneration Act 1996 (replacing earlier legislation at dates prescribed by the Secretary of State) makes provision for a system of local authority grants for improving or repairing residential accommodation. Public sector housing bodies cannot apply. The four grants (not available to anyone under 18) are:

(i) Renovation grants;
(ii) Disabled facilities grants;
(iii) Common-parts grants;
(iv) HMO (houses in multiple occupation) grants.

Of these, only disabled facilities grants, and also HMO grants provided by conversion as distinct from construction, are available for dwellings less than 10 years old.

Applications for renovation grants must be accompanied by a certificate, which will be one of the following as appropriate:

(i) an *owner occupation certificate* that the dwelling is intended to be the main residence of the applicant or a member of his family for five years; or
(ii) a *certificate of intended letting* that the applicant intends to be landlord for five years of a tenant who is neither a member of his family nor a "long" lessee for a term exceeding 21 years; or
(iii) a *tenant's certificate* that the dwelling is intended to be the main residence of the tenant or a member of his family.

The owner-occupation certificate or the tenant's certificate also requires that the applicant has occupied the dwelling as his main residence for the past three years. Normally a certificate of intended letting and a tenant's certificate should be submitted together. Licences are included, but not holiday lettings.

Applications for disabled facilities grants must be accompanied by either an *owner's certificate* that the dwelling is to be occupied for five years (health permitting) by a disabled occupant, or a *tenant's certificate* to the same effect (whether or not the tenant is the disabled occupant).

Applications for common parts grants must be accompanied by a *landlord's application* (from the owner of the building containing the dwellings and the "common parts") or a *tenant's application* from at least three-quarters of its occupying tenants who are under an obligation to carry out some or all of the proposed works and are statutorily protected.

Applications for HMO grants must be accompanied by a *certificate of future occupation* from the present or prospective owner that the building (or a specified part of it) will be residentially occupied or available for such occupation "by persons who are not connected with the owner for the time being of the house" under tenancies (except "long" leases exceeding 21 years). Licences are included, but not holiday lettings.

The proposed works must, for disabled facilities grants, be for any of various specified purposes to facilitate use by disabled occupants, and for the other grants be appropriate for various specified purposes which include thermal insulation, fire escapes, space heating, internal arrangement, other requirements as specified by the Secretary of State, or compliance with repair notices under sections 189 and 190 of the Housing Act 1985 (also Section 352(1A) for HMO grants).

Prior approval by the local authority for the proposed works is essential except where special reasons exist (e.g. urgency) or the works are needed to comply with the Housing Act 1985. Subject to any maxima prescribed by the Secretary of State the local authority decides what amount to pay in relation to the works which are approved. These must be completed to the authority's satisfaction within a year of being approved and before the grant is paid (as a whole or by instalments). Conditions, such as availability for letting, are imposed where appropriate; and grants must be repaid wholly or partly if the recipient (or successor) disposes of the property within five years (except where this is justified in certain specified cases) in respect of all except disabled facilities grants.

There are also "group repair schemes" under which a local authority renovates the exterior of groups of dwellings or makes them structurally stable, or both, provided that the various owners contribute to the cost. Local authorities will also be able to give "home repairs assistance" to persons over 18 whose income depends on social security, in the form of money or materials subject to limits to be prescribed by the Secretary of State.

### 9. The Borderline of Private and Public Sector Housing: Housing Associations

In addition to the grants and loans for assistance to private housing mentioned earlier in this chapter which are furnished by local authorities, public financial help is also made available by central government for the provision of dwellings by "housing associations". These are non profit-making bodies composed of private persons, corporations and charitable organisations; and the financial assistance is chiefly available in the form of grants from the Secretary of State for the Environment and loans from the Housing Corporation. The system of grants is similar in many respects to the housing subsidies paid by the Secretary of State to local authorities and new town corporations under the Housing Act 1985 (Part XIII), and earlier legislation (these subsidies are not within the scope of this book); but it is legitimate in this matter to regard housing associations as private rather than public bodies operating in the housing field. The law as to housing associations is now consolidated in the Housing Associations Act 1985, as amended by Part II of the Housing Act 1988 and Part I of the Housing Act 1996 (dealing with "social landlords", i.e. housing associations which are registered as charities or other non-profit-making bodies).

The Housing Corporation is a national body which has the function of registering housing associations and making (or guaranteeing) loans to them or to persons buying or leasing dwellings from them or from itself (it can provide dwellings and hostels directly). Such loans may be made to the housing associations that it registers, and also to those that it does not register provided that they are "self-build societies", that is to say associations which provide dwellings for purchase or occupation by their members using chiefly labour of such members in the work of construction or improvement. This system of registration applies only to housing associations which are already registered either as charities or as industrial and provident societies. Further financial provision for the assistance of registered housing associations takes the form of grants payable by the Secretary of State for the Environment. These are: "housing association grants" towards the cost of approved general expenditure on the provision of dwellings; "revenue deficit grants" to make up for deficits in associations' annual revenue accounts; and "hostel deficit grants" in respect of revenue deficits in the management of hostels. This system of grants replaces the subsidies payable to housing associations under previous legislation.

Apart from loans and grants there is the question of control of rents and security of tenure. Section 5 of the Rent Act 1968 excluded from the protection afforded to tenants by that Act tenancies held from housing associations where the dwellings had been provided with the assistance of grants of public money (under legislation pre-dating the Housing Act 1974) or where the associations were registered as charities or industrial and provident societies. Accordingly, there was a basic distinction between housing associations in the public sector, which were exempt from the Rent Act, and those in the private sector, to which that Act applied. Thus "private sector" lettings by housing associations enjoyed rent control under the "fair rent" system and security of tenure; in other words, they were included within the category of "regulated tenancies".

Housing association tenants are within the private sector for statutory protection purposes if their association is *not* registered with the Housing Corporation, and such tenants enjoy protection under the Rent Act 1977 or the Housing Act 1988, as described earlier in this chapter. Registered housing associations are within the public sector, and in consequence their tenants are "secure tenants" under the Housing Act 1985 like other public sector tenants, unless the association is in addition a "registered society" under the Industrial and Provident Societies Act 1965.

### 10. Public Sector Residential Tenancies

Parts II to V of the Housing Act 1985, subsequently amended by Leasehold Reform, Housing and Urban Development Act 1993, re-enacted previous legislation concerning provision of housing accommodation by local housing authorities – "council housing". Government policy is currently to reduce this area of housing, but it will presumably continue to exist in a reduced form. Generally speaking, council housing is not within the scope of this book; but one or two aspects of it may be touched on very briefly.

Part IV of the 1985 Act provides that, in general, residential tenancies granted by local housing authorities, housing action trusts, new town corporations, urban development corporations and the Development Board for Rural Wales enjoy security of tenure. This broadly resembles security of tenure in the private sector, so that recovery of possession by the landlord authority necessitates a court order. Rent increases can be made but in accordance with the terms of the existing tenancy (which, if it is a fixed term, will continue after its termination as a periodic tenancy in accordance with the rental periods of the expired tenancy). Public sector tenants protected by these provisions are "secure tenants".

Part V of the 1985 Act confers a "right to buy" upon secure tenants who, if they exercise it, then cease to be "secure tenants". Tenants of houses can acquire the freehold, if the landlord authority owns it; if not, they can acquire a lease for 125 years (expiring 5 days before the landlord authority's lease, if shorter than 125 years). Tenants of flats can acquire a lease of 125 years or less according to the length of the landlord authority's freehold or leasehold. The purchase price, in accordance with a determination of value by the District Valuer, is discounted by 32% plus 1% for each year of occupation over two years, to a maximum of 60% (for houses), or 44% plus 2% for each year of occupation over two years, to a maximum of 70% (for flats). Alternatively, the tenant can buy or "rent to mortgage terms", i.e. continue to pay rent but have the rental payments treated as repayment by instalments of a capital sum borrowed and paid to the landlord as if that sum were borrowed on mortgage.

Part III of the Housing Act 1988 (Sections 60–92 and Schedules 7–11) created a new entity, Housing Action Trusts, as from 15 November 1988, for the purpose of taking over public sector housing in a "designated area" for each Trust. The main aims are management, repair and improvement of the housing stock, the improvement of social and living conditions, and diversification of ownership. There must first be a ballot of secure tenants who will be affected – but only if there is a majority of all eligible tenants against such designation, will it be prevented. Each trust (either a new or a pre-existing body) will be a body corporate established by the Secretary of State and not democratically elected. Each Trust has been under a duty to draw up proposals, which will be for the exercise of various local housing authority powers under the Housing Act 1985 and related statutes, parallel with or instead of their exercise by the relevant local authority. Local authorities and their public health and planning controls, etc. may be by-passed. The Secretary of State is empowered to transfer local housing stock to any Trust, which may also be given compulsory purchase powers. Tenants will be "secure tenants" with a "right to buy"; but a Trust is expected to sell its landlord's interest to other landlords approved by the Housing Corporation (or Housing for Wales), and eventually to dispose of its housing stock and assets (with the Secretary of State's consent) and be dissolved.

Part IV of the Housing Act 1988 (Sections 93–114 and Schedule 12) introduced a system for transferring the landlord's freehold interest in public sector housing from local housing authorities, new town corporations, and even the new housing action trusts, commencing at dates prescribed by the Secretary of State. The freehold of such property will be transferred to "approved persons" in the private sector, with approval being given by the Housing Corporation or Housing for Wales after prescribed consultations. Disputes over the transfer price are to go to the District Valuer; and public housing authorities may have to pay the "approved persons" to take the property instead of being paid by them. It should be noted that, if the Secretary of State makes regulations for the purpose, individual tenants may be given the right to "opt-out", thus affecting the price of transfer. Subsequent tenancies, and tenancies of premises transferred to the new owner because the premises "cannot otherwise be reasonably managed or maintained", are deprived of all statutory protection (other than the minimum 4 weeks' notice to quit). Tenants, including those with "long" leases (over 21 years), must with some exceptions be consulted. The new private sector landlords, however, will with some exceptions *not* be able to dispose of the property rights they acquire without the Secretary of State's consent.

## 11. Places of Worship

Schedule 6 to the Leasehold Reform Act 1967 amends the Places of Worship (Enfranchisement) Act 1920 which gives rights to persons holding leasehold interests in places of worship or ministers' houses to acquire their freeholds.

Where premises are held under a long lease to which this Act applies and are held upon trust to be used for the purposes of a place of worship or, in connection with a place of worship, for the purpose of a minister's house, whether in connection with other purposes or not, and the premises are being used in accordance with the terms of the trust, the trustees notwithstanding any agreement to the contrary (not being an agreement against the enlargement of the leasehold interest into a freehold contain in the lease granted or made before the passing of this Act), shall have the right as regards to their freehold interest to enlarge that interest into fee simple, and for that purpose to acquire the freehold and all intermediate reversions; provided that:

(a)  if the premises exceed two acres in extent, the trustees shall not be entitled to exercise the right in respect of more than two acres thereof; and

(b)  if the person entitled to the freehold, or an intermediate reversion, requires that underlying minerals are to be excepted, the trustee shall not be entitled to acquire his interest in the minerals if proper provision is made for the support of the premises as they have been enjoyed during the lease, and in accordance with the terms of the lease and of the trust; and

(c) this Act shall not apply where premises are used or are proposed to be used for the purposes of a place of worship in contravention of any covenant contained within the lease and which the premises are held, or in any lease superior thereto; and

(d) this Act shall not apply where the premises form part of the land which has been acquired by or invested in any municipal, local or rating authority or in the owners thereof for the purposes of a railway, dock, canal or navigation under any Act of Parliament or Provisional Order having the force of an Act of Parliament and the freehold reversion in the premises is held or retained by such owners for those purposes.

The leases to which this Act applies are leases (including underleases and agreements for leases or underleases), whether granted or made before or after the passing of this Act, for lives or a life or for a term of years where the term as originally created was for a term of not less than 21 years, whether determinable on a life or lives or not.

The procedure for acquisition of the reversionary interest is set out in Section 2 of the Act and the basis of compensation is the same as if the trustees acquiring were an authority authorised to acquire the premises by virtue of a compulsory purchase order, made under what is now the Acquisition of Land Act 1981, subject to a number of amendments.

*Chapter 18*

# Commercial Properties (1)
# Landlord and Tenant Acts and Rent Reviews

## 1. Generally

There has been considerable intervention by governments in the operation of the letting of commercial properties. This legislation provides that leases may not end when they should end under the terms of the lease, rents that landlords might wish to charge cannot be charged, tenants may be compensated because they are forced to leave at the end of their leases, and may be compensated for improvements when they leave.

A valuation of commercial property cannot be prepared unless the possible impact of the legislation has been considered. It is found mainly in the Landlord and Tenant Acts of 1927 and 1954 (as amended by the Law of Property Act 1969 and the Landlord and Tenant (Licensed Premises) Act 1990).

## 2. Landlord and Tenant Act 1927

*(a) Landlord and Tenant Act 1927 (Part I)*

This Act gives tenants of premises let for trade, business or professional purposes the right to compensation for improvements made during the tenancy. The provisions as to compensation for improvements are subject to certain amendments made by Part III of the 1954 Act which are comparatively unimportant in practice and in their effect on values.

The tribunal for the settlement of questions of compensation for improvements under the Act is the County Court, but claims are referred by the Court to a referee, who is selected from a special panel.

The notes which follow do not attempt to give details of procedure under the Act, but merely to indicate the possible effect on value of the tenant's right to claim.

345

*A tenant's claim for compensation for improvements*[1] is limited to (a) the net addition to the value of the holding as a whole which is the direct result of the improvement, or (b) the reasonable cost of carrying out the improvement at the termination of the tenancy, allowing for the cost of putting the existing improvement into a reasonable state of repair, whichever is the smaller. In determining the compensation, regard must be had to the purpose to which the premises will be put at the end of the tenancy and to the effect that any proposed alterations or demolition or change of user may have on the value of the improvement to the holding.

It may be said, therefore, that in principle the basis for determining the compensation payable to the tenant is the benefit which the landlord will derive from the improvement.

In order that an improvement may carry a right to compensation under the Act, the tenant must first have served notice on the landlord of his intention to make the improvement, together with full particulars of the work and a plan. If the landlord does not object, or, in the event of objection, if the tribunal certifies that the improvement is a "proper" one, the tenant may carry out the works. The tenant may not carry out the work in advance of obtaining the Certificate or his right to compensation would be lost. Alternatively, the landlord may offer to carry out the improvement himself in consideration of a reasonable increase in rent.

Compensation is only payable if the tenant quits the holding at the end of the term during which the improvements were made. This is the main reason why, in practice, it is found that, despite the tenant's initial right to compensation for improvements, very few claims under the Act are served on landlords; in addition very few notices are actually served.

Nonetheless, in valuing any given property, it should be established whether or not improvements have been made by the tenant within the terms of the Act which may give rise to a claim for compensation at the end of the tenancy. The way in which a potential claim for compensation should be taken into account when preparing a valuation must depend on the circumstances. An example is given later in this chapter.

---

[1] Landlord and Tenant Act 1927, Sections 1–3.

*(b) Landlord and Tenant Act, 1927 (Part II)*

Section 18 of the Act in Part II makes provision to limit the amount payable by a tenant to his landlord as damages for dilapidations on the conclusion of the lease. Most leases provide for a tenant to leave the premises in repair at the conclusion of the lease. Failure to comply results in the tenant paying the landlord compensation for the accrued dilapidations. A claim by a landlord for dilapidations is broadly the cost of making good the disrepair. However, Section 18(1) provides that the landlord cannot receive more than the amount of the diminution in the value of his reversion caused by the dilapidations. Thus, a dilapidation claim needs to compare the cost test with the value test, with the landlord receiving the lesser of the two figures. In extreme cases, such as where the landlord proposes to pull down the building and re-develop, the disrepair has no effect on the value of the reversion. In such cases there is no liability for dilapidations on the part of the tenant since the diminution is nil, so the compensation is nil.

Section 18(1) only applies to damages arising out of a breach of the covenant to repair. It does not apply to damages for any other breach such as failure to re-decorate in the last year of the term or to reinstate at the end of the term.

## 3. Landlord and Tenant Act 1954

Subject to certain exceptions and qualifications, Part II of the Act (as amended by subsequent legislation, in particular the Law of Property Act 1969, Part I), ensures security of tenure where any part of a property is occupied by a tenant for the purposes of a business carried on by him – provided it is not carried on in breach of a general prohibition in the lease.

The term "business" in the 1954 Act means any trade, profession or employment and includes any activity carried on by a body of persons corporate or unincorporate.

The Act therefore covers not only ordinary shops, factories and commercial and professional offices, but also premises occupied by voluntary societies, doctors' and dentists' surgeries, clubs, institutions, etc.[2] Security of tenure under the Act was extended to

---

[2] A tennis club registered under the Industrial and Provident Societies Act 1893 was held by the High Court (Queen's Bench) to be a business letting

licensed premises by the Landlord and Tenant (Licensed Premises) Act 1990. The scope of premises to which the 1954 Act applies is wider than that of the 1927 Act which applies to premises let for trade, business or professional purposes.

The Act extends not only to lettings for fixed terms, e.g. five-, 10- 21- or 99-year leases, but also to periodic tenancies, e.g. quarterly, monthly or weekly tenancies. But the following types of tenancy are excluded:

(i)   tenancies of agricultural holdings;
(ii)  mining leases;
(iii) tenancies within the Rent Act 1977, or which would be but for the tenancy being a tenancy at a low rent;
(iv)  "service" tenancies, i.e. tenancies granted by an employer only so long as his employee holds a certain office, appointment or employment;
(v)   tenancies for a fixed term of six months or less, with no right to extend or renew, unless the tenant and his predecessor (if any) have been in occupation for more than 12 months.

The general principle upon which this Part of the Act is based is that a tenancy to which it applies continues until it is terminated in

---

[2] *continued*

within the 1954 Act in *Addiscombe Garden Estates Ltd* v *Crabbe* [1958] 1 QB 513; but sub-letting part of premises as unfurnished flats with a view to making a profit out of the rentals was held by the Court of Appeal not to be a business letting within the Act in *Bagettes Ltd* v *GP Estates Co Ltd* [1956] Ch 290. In the former case, the use of the premises was "an activity carried on by a body of persons", but in the latter case it was ordinary residential use. In some circumstances a residential letting could be a business, where the landlord preserves a major degree of control over sub-tenants and provides services (see *Lee-Verhurst (Investments)* v *Harwood Trust* [1973] 1 QB 204). Yet occupation of premises by a medical school for the purpose of student residences was held by the Court of Appeal to be a protected business use in *Groveside Properties* v *Westminster Medical School* [1983] 2 EGLR 68. A government department can have a protected business tenancy even if the premises are occupied on its behalf by another body, rent free: see *Linden* v *DHSS* [1986] 1 EGLR 108. A lessee of an enclosed market was held not to have a protected tenancy where all of the "stallholders" themselves had protection even though the lessee had control of the walkways and common parts; *Graysim Holdings Ltd* v *P&O Property Holdings Ltd* [1996] 1 EGLR 109. This case also cast doubt on the *Lee-Verhurst* decision.

one of the ways prescribed by the Act. Thus, if it is a periodic tenancy, the landlord cannot terminate it by the usual notice to quit. If it is for a fixed term, it will continue automatically after that term has expired, on the same terms as before, unless and until steps are taken under the Act to put an end to it.

One way in which a tenancy can be determined is by the parties completing on agreement on the terms of a new tenancy to take effect from a specified date. A tenancy may also be terminated by normal notice to quit given by the tenant, or by surrender or forfeiture. In the case of a tenancy for a fixed term, the tenant may terminate it by three months' notice, either on the date of its normal expiration or on any subsequent quarter day (a Section 27 notice) or by vacating before the term date.[3]

Apart from the above cases, the methods available to landlord or tenant to terminate a tenancy to which Part II of the Act applies are as follows.

The landlord may give notice, in the form prescribed by the Act, to terminate the tenancy at a specified date not earlier than that at which it would expire by affluxion of time or could have been terminated by notice to quit. Not less than six months' nor more than 12 months' notice must be given. It is to be noted that a landlord must give six months' notice to terminate a quarterly, monthly or weekly tenancy. The notice must require the tenant to specify by counter-notice whether or not he is willing to give up possession and must state whether, and, if so, on what grounds the landlord would oppose an application for a new tenancy.

After notice has been served the parties may negotiate on the terms of a new tenancy or may continue negotiations begun before the notice was served. If they cannot agree on the terms, or if the tenant wishes to remain in the premises but the landlord is unwilling to grant him a new tenancy, the tenant can apply to the County Court for a new tenancy and the Court is bound to grant it unless the landlord can establish a case for possession on certain grounds specified in the Act.

A tenant holding for a fixed term of more than one year can initiate proceedings by serving a notice on his landlord (a Section 26 notice) in the prescribed form requesting a new tenancy to take effect not more than 12 months, nor less than six months, after the

---

[3] See *Esselte AB* v *Pearl Assurance plc* [1997] 1 EGLR 73.

service of the notice, but not earlier than the date on which the current tenancy would come to an end by effluxion of time, or could be terminated by notice to quit. This notice must state the terms which the tenant has in mind. If the landlord opposes the proposed new tenancy, the tenant can apply to the Court for the grant of a new tenancy, the terms of which can be agreed between the parties or, in default of agreement, will be determined by the Court. The valuation date is the date of the Court Order and the lease will commence three months and four weeks after the Court hearing (the four weeks is to allow for any appeal).

The Court must have regard to the following points in fixing the terms of a new tenancy:[4]

(i)   The tenancy itself may be either a periodic tenancy or for a fixed term of years not exceeding 14;

(ii)  The rent is to be fixed in relation to current market value and the following are to be disregarded in determining it:

   (a) the fact that the tenant is a sitting tenant;
   (b) any goodwill attached to the premises by reason of the carrying on of the tenant's business on the premises;
   (c) any improvement carried out by a person who was the tenant at the time of improvement but only if the improvement was made otherwise than in pursuance of an obligation to the immediate landlord and, if the improvement was not carried out during the current tenancy, that it was completed not more than 21 years before the application for the new tenancy;
   (d) in the case of licensed premises, any additional value attributable to the licence where the benefit of the licence belongs to the tenant.

(iii) The terms may include provision for varying the rent even where there was no review clause in the old lease [Section 34(3)].

(iv)  Other terms of tenancy must be determined having regard to the terms of the current tenancy and to all relevant circumstances [Section 35].

---

[4] Landlord and Tenant Act 1954, Sections 34 and 35, as amended by Law of Property Act 1969, Sections 1 and 2.

In practice, most terms will be agreed between the parties so that the Court is required only to settle those not agreed. For example, they may agree that the new lease shall be for 25 years rather than the maximum period of 14 years which the Court can fix. Commonly all terms are agreed other than the rent payable. The Courts, in "having regard to the terms of the current tenancy" tend to retain the same covenants as in the existing lease unless sound reasons can be advanced for a change (*O'May* v *City of London Real Property Co Ltd*.[5] This does not mean that the terms of the lease are fossilised; new terms can be introduced such as the requirement to pay interest on late payment of rent or VAT, or to take into account the provisions of the Landlord and Tenant (Covenants) Act 1995.

Subject to certain safeguards, the Court will revoke an order for the new tenancy if the tenant applies within 14 days of the making of the order, so that a tenant is not bound to accept the terms awarded by the Court.

During the period while an existing tenancy is continuing only by virtue of the Act, and provided the landlord has given notice to determine the tenancy or the tenant has requested a new tenancy, the landlord may apply to the Court for the determination of a "reasonable" rent (an Interim Rent).[6] The interim rent is determined in accordance with the same rules which apply to a rent for a new tenancy but is on a year-to-year basis having regard to the rent payable under the expiring lease. The interpretation of these rules has led to a general approach whereby the rental value established for the new fixed term lease is amended by a given percentage to convert it to a year-to-year basis, to allow for the valuation date (i.e. the commencement day of the interim period), and then reduced by a further percentage so as to have regard to the existing rent, known as cushioning (examples are *Janes (Gowns) Ltd* v *Harlow Development Corporation*[7] and *Ratners (Jewellers) Ltd* v *Lemnoll Ltd*.[8] However, the percentage additions or subtractions are not fixed and the Courts may vary the approach depending on the facts of the case. For example, in *Charles Follett Ltd* v *Cabtell Investments Ltd*,[9]

---

[5] [1982] 1 EGLR 76.
[6] Law of Property Act 1969, Section 3, inserting Section 24(A) into the 1954 Act.
[7] [1980] 1 EGLR 52.
[8] [1980] 2 EGLR 65.
[9] [1987] 2 EGLR 88.

the new lease rent was fixed at £106,000. This was reduced to £80,000 as being the rent on a year-to-year basis. The further reduction so as to have regard to the existing rent was fixed at 50%, producing an Interim Rent of £40,000. This was because the rent is not a reasonable rent but "the rent which it would be reasonable for the tenant to pay" and this, coupled with the need to "have regard to the rent payable under the (old) tenancy" (in this case £13,500 p.a.) justified such a large percentage deduction. This level of reduction is wholly unusual and the traditional deduction was approximately 12.5%.

A landlord can successfully oppose an application for a new tenancy on the following grounds:

(a) That the tenant has not complied with covenants to repair;
(b) That the tenant has persistently delayed paying rent;
(c) That there are substantial breaches of other covenants by the tenant;
(d) That the landlord can secure or provide suitable alternative accommodation for the tenant, having regard to the nature and class of business and to the situation and extent of, and facilities afforded by, his present premises;
(e) That, in the case where the tenancy was created by a sub-lease, the rental value of the whole is greater than the rental value of the parts (the effect on capital value is irrelevant);
(f) That the landlord intends to substantially demolish or reconstruct the premises;[10]
(g) That the landlord requires the premises for his own occupation – but a landlord is debarred from using this ground if he acquired the premises less than five years before the end of the tenancy. This five-year limitation does not apply when purchasers wish to reconstruct premises for their own occupation,[11] but Section 31A would most definitely come into play.

---

[10] There are many decided cases relating to this ground. Note that the Law of Property Act 1969, Section 7, inserts an additional section (31A) into the 1954 Act which entitles a tenant to a new tenancy despite reconstruction works by the landlord if terms reasonably facilitating the performance of those works can be included, or if "an economically separable part of the holding" can be substituted for the entirety of the holding.

[11] See *Atkinson* v *Bettison* [1955] 1 WLR 1127 and *Fisher* v *Taylor's Furnishing Stores Ltd* [1956] 2 QB 78 (both Court of Appeal decisions).

In some cases the landlord is not ready to take possession for redevelopment because his plans are not yet complete. In such cases the landlord can ask the Court to impose a redevelopment clause in the new lease which would enable him to determine the lease when he is ready to develop. The Court of Appeal has considered this in a number of cases and has held that, provided the landlord can show a *prima facie* case, such a clause should be included.[12]

In cases where the landlord wishes to occupy for himself but the period of ownership is too short, i.e. less than five years, the Court may grant only a short term.[13]

A tenant who is refused a new tenancy is entitled to compensation from his landlord if either the landlord or the Court refused on grounds (e), (f) or (g) above, and there is not an agreement which effectively excludes compensation. Such compensation is in addition to any the tenant may be entitled to under the Landlord and Tenant Act 1927, in respect of improvements.

If the business has been carried on in the premises for less than 14 years, the compensation is equal to a multiplier fixed by Statutory Instrument. If it has been carried on for 14 years or more, the compensation is twice the multiplier. The multiplier is applied to the rateable value of the premises; the current multiplier is 1.[14] A tenant has the choice of adopting the pre-1990 rateable value and applying a mutliplier of 8 where the Section 25 Notice or Section 26(b) counter notice is given before 1 April 2000 and the tenancy was entered into before 1 April 1990.

In general it is not possible to contract out of Part II of the Act, but Section 5 of the Law of Property Act 1969 introduced an important exception whereby the Court may, on the joint application of landlord and tenant prior to the grant of the lease, authorise an agreement to exclude the provisions of the 1954 Act regarding security of tenure (and indeed other provisions). Section 38 of the 1954 Act allows the parties to exclude the compensation provisions where the tenant or his predecessors in business have occupied the premises for less than five years. Some leases contain

---

[12] See *Adams* v *Green* [1978] 2 EGLR 46 and *National Car Parks Ltd* v *The Paternoster Consortium Ltd* [1990] 1 EGLR 99.

[13] *Upsons Ltd* v *E Robins Ltd* [1956] 1 QB 131.

[14] Landlord and Tenant Act 1954 (Appropriate Multiplier) Order 1990 (SI 1990 No 363).

a covenant which excludes payment of any compensation but such a covenant is void once occupation reaches five years.

The effect of all these contracting out provisions on the value of business premises which do fall within Part II of this Act has, generally speaking, been slightly to lower rents obtainable from sitting tenants when leases are renewed. There is no doubt at all that the rights under the Landlord and Tenant Act considerably strengthen the hand of a tenant in negotiations for a new lease or in negotiations for the surrender of an existing lease in return for a new lease. Nevertheless, lessees generally prefer to have a lease for a definite term of years rather than to rely solely on their rights to a new lease under the Act. Where the goodwill of a business relates to particular premises (e.g. a retail business) the tenants who have only a few years of their lease unexpired and who wish to sell the goodwill of their businesses or to invest in the property or the business nearly always have to surrender their existing short leases to secure a sufficiently long term to realise the full potential of their business. This is particularly so where costs of removal are expensive or when goodwill can be lost by a move. In other cases, tenants prefer the flexibility of manoeuvre which a short lease coupled with security of tenure gives them. Purchasers frequently require the security of a term of years rather than the rights under this Act.

With few exceptions, it would seem that reversions after an existing lease should be valued on a capitalised rental value based on the terms – referred to above – to which a Court must have regard when settling disputes between the parties for new tenancies. If a landlord's interest is valued on the assumption that a new tenancy will not be granted, allowance should be made for compensation which may have to be paid to the tenant.

There is little doubt that the security of tenure afforded by Part II of the Act increases the value of a tenant's interest. If his interest is valued upon the basis that he will have to give up possession at the end of the lease for one of the reasons set out above which attract the payment of compensation, an appropriate addition should be made. Conversely, the effect is to decrease the value of a landlord's interest. Only in a comparatively few cases can reversions be valued on the basis of vacant possession.

In general, the 1954 Act has had a much greater effect in practice than the 1927 Act.

The following example illustrates the possible effect of both the 1927 and the 1954 Acts in various circumstances.

*Example 18–1*

Shop premises in a secondary location are let by the freeholder on a lease granted 15 years ago and with three years unexpired at a rent of £4,000 p.a. Improvements were carried out seven years ago at a cost of £30,000 by the tenant, who served notice of his intention on the landlord under the Landlord and Tenant Act 1927, and received the landlord's consent thereto.

It is considered that without these improvements the rental value is £5,000 p.a., but that in consequence of the work done the annual rental has been increased to £7,500 p.a. The present Rateable Value is £6,000. Prepare a valuation of the freehold interest.

*Valuation*

As the tenant has a right to a new lease which excludes the value of the improvements made by him, the rental value of £7,500 p.a. should be ignored for the probable period of the new lease, a maximum of 14 years if fixed by the Court. The landlord might not wish to grant a lease for longer than this so as to obtain the rental value of the improvements at the earliest opportunity (21 years from improvement being made). He would weigh this consideration against possible disadvantages of a lease for a "non-standard" term.

| | | | |
|---|---|---|---|
| Term – rent payable under lease | | £4,000 p.a. | |
| YP for 3 years at 9% | | 2.53 | £10,120 |
| Reversion – to new lease (say 14 years) at rental value ignoring improvements | | £5,000 p.a. | |
| YP 14 years at 9% | 7.79 | | |
| PV £1 in 3 years at 9% | 0.77 | 6.00 | £30,000 |
| Reversion: to rental value reflecting improvements | | £7,500 p.a. | |
| YP in perp at 9% deferred 17 years | | 2.57 | £19,275 |
| | | | £59,395 |
| Value, say | | | £59,000 |

If it is known that the landlord wishes to occupy the premises himself when the present lease terminates, the value would be:

| Term – rent reserved under lease | £4,000 p.a. | |
| YP for 3 yrs at 9% | 2.53 | £10,120 |
| Reversion – to | £7,500 | |
| YP in perp at 9% deferred | | |
| 3 years | 8.58 | £64,350 |
| | | £74,350 |

*Deduct:*

| Compensation under the Landlord and Tenant Act, 1954 – say 2 × RV £6,000 | £12,000 | |

Compensation under the Landlord and Tenant Act 1927

| (a) Increase in rental value of landlord's reversion | £2,500 p.a. | |
| YP in perp at 9% | 11.11 | |
| say | £ 27,775 | |

| (b) Cost of carrying out improvements at end of lease, say | £45,000 | |
| *Less* depreciation, say 30% | £13,500 | |
| | £31,500 | |

| Take lesser of (a) or (b) | £27,775 | |
| | £39,775 | |
| PV £1 in 3 years at 9% | 0.77 | £30,627 |
| | | £43,723 |
| Value, say | | £44,000 |

It is assumed for the purposes of the above that the landlord is not debarred by the five-year restriction. It is also assumed that the tenant has been in occupation for over 14 years and that the amount of compensation under the Landlord and Tenant Act 1954 is twice the RV. The cost of meeting the compensation has been deferred at the remunerative rate as it is the landlord's choice to take possession and the compensation is the payment required to obtain the advantages sought.

The value of £44,000 is the value to the landlord who wishes to occupy. This represents the worth to the landlord as only he can obtain possession for occupation. The open market value remains £59,000 since this is the value for investor purchasers of the freehold interest.

If the landlord intended to demolish the premises, the valuation would be:

| | | |
|---|---|---|
| Term – rent reserved under lease | £4,000 p.a. | |
| YP for 3 years at 9% | 2.53 | £10,120 |
| Reversion to site value, say | £200,000 | |
| Deduct:- | | |
| Compensation under the Landlord | | |
| and Tenant Act 1954 | £12,000 | |
| Net value of reversion | £188,000 | |
| PV £1 in 3 years at 9% | 0.77 | £144,760 |
| | | £154,880 |
| Value, say | | £155,000 |

In this case, no compensation would be payable under the 1927 Act as the demolition would eliminate the effect which the improvements would have on the value of the holding.

## 4. Health and Safety at Work etc. Act 1974

This Act "to make further provision for saving the health, safety and welfare of persons at work, for protecting others against risks to health and safety in connection with the activities of persons at work" provided[15] for the progressive replacement by a system of Health and Safety Regulations and Approved Codes of Practice of such legislation as the Factories Act 1961, and the Offices, Shops and Railway Premises Act 1963 which remain in force.

Health and Safety Regulations may be made for any of the general purposes of Part I of the Act[16] and Schedule 3 sets out some of these matters in detail, including such things as structural condition and stability of the premises, means of access and egress, cleanliness, temperature and ventilation, fire precautions and welfare facilities.

These requirements must be borne in mind when leases are being granted and freehold and leasehold interests are being valued. In appropriate circumstances, deductions may have to be made to allow for the cost of complying with the regulations.

---

[15] Section 1.
[16] Section 15.

## 5. Rent Reviews

The introduction of rent reviews into leases in the post-war years was brought about by the steady increase in rental values at that time. It was previously common practice for landlords to grant leases for as long as possible at a fixed rent. In the early 1960s landlords began to give close attention to the fact that, when a lease is granted for several years, if the rent is fixed it becomes less valuable in real terms as inflation takes its toll, and further that the steady increase in rental value accrues wholly to the lessee who is not, in essence, a property investor.

Two solutions are obvious. One is to shorten the length of the lease so that the landlord can adjust the rent frequently on the grant of a new lease. This is unattractive to investors and also frequently to occupiers because, for the investor it increases the possibility of voids since lessees have regular opportunities to vacate, whilst the occupier has no long-term certainty of occupation. Not only that, the frequent grant of new leases is costly. The other solution is to retain the long term of years, but provide for rent increases during that term. Initially, such increases were often effected by fixing the rent for the first few years at the agreed rental value, and then fixing a higher rent for the subsequent periods. For example a shop might be let for 21 years at £1,200 p.a. for the first seven years, rising to £1,500 p.a. for the following seven years and £1,800 p.a. for the final seven years.

This approach went some way to overcoming the problem, but it soon became apparent that it was far from ideal. At the time when the first increase came into operation, it was seen to be quite fortuitous if the new rent represented rental value at that time. Quite often it would be less, so that the landlord was still losing ground. Where it was more, the tenant would be unhappy at the injustice, as he saw it.

This led to the introduction within leases of a machinery for the rent, at intervals, to be re-determined at the then prevailing rental value. The machinery was contained in rent review clauses. In the absence of precedents, rent review clauses took many different forms reflecting the differing views of lawyers, some of whom initially raised doubts as to whether such clauses were valid.

In time, however, the practice of providing for rent reviews grew and became accepted. In occupation leases the reviews tended to be at seven-year intervals but this came down to five years, and in the early 1970s pressure grew for three-yearly reviews due to the high incidence of inflation. The market settled in the main for five-year reviews, with the consequential effect that leases tend to be granted for multiples of five years. Ground leases have seen a similar development, initially the review period being 33 years or perhaps 25 years, but now five-year reviews tend to be commonly found.

The introduction and acceptance of rent reviews took several years, but by 1969 they were the rule rather than the exception. This was recognised by the introduction of Section 34 (3) into the Landlord and Tenant Act 1954 by the Law of Property Act 1969. The period 1970 to 1973 saw a sharp rise in property values, both capital and rental, until an economic crisis brought on by the miners' strike and the "three day week" brought it to an end. This led to the introduction of a rent freeze on business properties in 1972, which lasted until 1974. In that period, rent review clauses became ineffective, so that new rentals were agreed but could not be claimed by the landlords. This meant that, following the lifting of the freeze, rents rose in 1974 and into 1975 when capital values were declining rapidly. It is not possible to prove, but experience does suggest that, in that period, tenants began to be far more sensitive to the fact of increasing rents and since then rent reviews have attracted considerable attention. This has been magnified by the fact that the first effects of the practice of five-year reviews begun in the early 1970s coincided with the falling in of many leases granted in the 1950s at fixed rents, as well as joining with the period of business depression. Tenants, and particularly companies such as multiple retailers, who are tenants of many properties, faced with sharp increases in their rent bills in such circumstances naturally seek to restrict increases to the minimum.

The debate over rent reviews was intensified in the years after 1988 when rental values generally stopped rising, and in many instances fell, in some cases very sharply to half or less of the 1988 levels. If the purpose of rent review clauses is accepted as ensuring that the rent payable follows changes in rental value then there should be no greater problem when rental values are falling than when they are rising. However, the usual practice, right from the introduction of rent review clauses, was that the rent payable from the time of the rent review should not be less than the rent payable in the period before the review. These "upward only" review clauses cause no friction when values are rising but, after 1988, they acted significantly against tenants. A tenant with a rent review in, say, 1994, when rental values may have halved from the rent payable under the lease, was forced by the upward only provisions to continue to pay twice the then prevailing rental value. This type of situation resulted in a large number of over-rented properties and leases with sharply negative values (see Chapter 9 for valuations of such interests).

The debate has also switched to the Courts in relation to rent review provisions in renewals under the Landlord and Tenant Act. The sentiments that "what is sauce for the goose is sauce for the gander" are often expressed and upwards/downwards reviews have been determined in a number of cases starting with *Janes (Gowns) Ltd v Harlow Development Corporation* in 1979. There is no fixed rule and the eventual result depends upon the evidence in the local market at the date of the court hearing. The final result is often a trade-off between the landlord wanting the interest to have a long lease and the tenant not wanting to be locked in to what might be an uneconomic rent, i.e. an upward/downward review.

Inevitably, with closer attention being paid to the actual wording of the rent review clause in conjunction with the other terms of the lease, valuers developed valuation theories and ideas which has led to a body of valuers specialising in rent reviews. Disputes between the parties have become more common, leading to a marked increase in the use of independent determination and also to a considerable number of cases heard by the Courts. There have thus emerged certain principles applicable to rent reviews.

A typical rent review clause will cover the following areas:

(a)  The machinery for putting into effect the review;
(b)  The factors to be taken into account and assumptions to be made in arriving at rental value;
(c)  The machinery for determining disputes in the absence of agreement;

Most of these aspects can be considered in turn.

### (a) Machinery for review

In general, the review is triggered by the service of a notice by the landlord on the tenant, following which the parties' valuers have a period of time in which to reach agreement. Failing agreement, the matter is referred to independent determination which can be either by Arbitration or reference to an Independent Expert. Although a time scale is laid down for all these matters, the time scale is not usually so restrictive as to compel absolute compliance therewith,[17] unless it is clearly stated or indicated that time is to be of the essence,[18] or some other terms of the lease (such as a tenant's option to break the lease following review) make it essential for time to be of the essence.[19] Sometimes there is no mechanism for the review but merely a provision for change which comes in automatically.

### (b) Rent payable

The rent review clause sets down the factors which are to be taken into account in arriving at the rent. These take many forms and each clause must be read carefully to see which apply to any particular case. In general they have tended to vary between two extremes. At one extreme, the various factors affecting rent are set out in great detail, including a definition of the rent payable – "rack rent", "open market rent", "open market rent as between willing lessor and willing lessee", etc.; assumptions as to use; whether to disregard tenant's improvements; whether to disregard tenant's goodwill; whether to assume the lease has its full term to run or merely the actual unexpired term; assumptions as to general lease covenants; assumption that the rent free period which would be given on a new letting has expired. At the other extreme, the clause is limited to saying that "market rent" or some similar phrase shall be adopted.

The assumptions may have a considerable effect on rent. In interpreting rent review clauses the landlord and tenant are treated as hypothetical parties[20] so that the status of the actual parties is ignored unless the wording so requires.[21] For example:

---

[17] *United Scientific Holdings Ltd* v *Burnley Borough Council* [1977] 2 EGLR 61.
[18] *Drebbond Ltd* v *Horsham District Council* [1978] 1 EGLR 96 and *Mammoth Greeting Cards Ltd* v *Agra Ltd* [1990] 2 EGLR 124.
[19] See, for example, *Al Saloom* v *Shirley James Travel Service Ltd* [1981] 2 EGLR 96 and *Stephenson & Son* v *Orca Properties Ltd* [1989] 2 EGLR 129. For a full discussion of time of the essence, see 284 EG 28.
[20] *FR Evans (Leeds) Ltd* v *English Electric Co Ltd* [1978] 1 EGLR 93. In this case the Court held that, as each party was willing, the actual market conditions which showed an absence of any demand were to be ignored so as to reflect the existence of a tenant willing to take the premises. A more realistic view of market conditions can be taken following *Dennis & Robinson Ltd* v *Kiossos Establishment* [1987] 1 EGLR 133.
[21] *Thomas Bates & Son Ltd* v *Wyndham's (Lingerie) Ltd* [1981] 1 EGLR 91.

## (i)  Rent description

Generally where the phrase used is one commonly adopted and widely understood no problems arise. However, where less familiar terms are used then they may have a considerable impact on the rent finally determined. For example, "a reasonable rent for the premises" led to the Court fixing a rent reflecting the improvements carried out by the tenant, even though they would be ignored on renewal of the lease.[22]

## (ii)  User

Normally the use assumption follows the user covenant of the lease under the general lease covenant assumptions, but sometimes an assumption is imposed which is different. In either case, where a lease severely limits the use to which a property may be put, this will tend to reduce the rent below the level acceptable where the tenant has flexibility of approach. Thus, the rental value of a shop whose use is strictly limited to the retail business of cutlers is clearly worth less than one where the tenant may adopt other activities with the landlord's consent which is not to be unreasonably withheld.[23] This contrasts the difference between restricted user clauses and open user clauses. The class of person who may use a building may be restricted. Thus, offices held under a lease which limits occupation to civil engineers are likely to be worth less than the same offices if they can be occupied by anyone in the market, since the competition therefore is so restricted. Such was the situation in the Plinth Property case[24] when it was agreed that the "open" use rental value was £130,455 p.a. whereas in the market limited to civil engineers it became £89,200 p.a. The possibility that landlords might well relax the provisions was ignored.

Where the use is even more restricted to that of a named tenant only, then it is assumed that there is a "blank lease" with the name to be inserted, and the generally prevailing rental values for that type of property will be adopted.[25] If there is an assumed use for

---

[22] *Ponsford* v *HMS Aerosols Ltd* [1978] 2 EGLR 81.
[23] *Charles Clements (London) Ltd* v *Rank City Wall Ltd* [1978] 1 EGLR 47.
[24] *Plinth Property Investments Ltd* v *Mott, Hay & Anderson* [1979] 1 EGLR 17.
[25] *Law Land Company Ltd* v *Consumers' Association Ltd* [1980] 2 EGLR 109.

rent review purposes which displaces the actual use covenant in the lease it is assumed that planning permission exists for that use[26] although the need to carry out works to achieve the assumed use would need to be reflected in the rent.[27]

## (iii) Improvements

It is widely accepted that tenants should receive the benefit of any improvements carried out by them not as a condition of the lease, a view reflected in successive Landlord and Tenant Acts, including the 1954 Act which requires their effect on rents to be disregarded. The wording of the 1954 Act is commonly adopted in rent review clauses.

A problem arises as to how the valuer should disregard the effect of improvements. Approaches commonly adopted include assessing the rental value as if the improvements had never been carried out, or to assume that the improvements are to be carried out at the time of review at current building costs, amortising the cost over the hypothetical term of years, allowing for the tenant's right to a new lease or compensation, and deducting the amortised cost from the rental value which reflects the improvements. A further approach is to adopt a "gearing" approach whereby if, at the time of the improvements being carried out, the rental value was increased by 50%, then the rent at review, ignoring the effect of the improvements, is taken to be two-thirds of the rental value reflecting the improvements. The Courts have rejected these approaches and have suggested that the parties put themselves in the same factual matrix at the date of review as they were when the lease was granted,[28] but, notwithstanding such criticisms, the methods are still commonly applied as providing a practical valuation solution even though not perhaps a true legal interpretation.

---

[26] *Bovis Group Pension Fund Ltd* v *GC Flooring & Furnishing Ltd* [1984] 1 EGLR 123.
[27] *Trusthouse Forte Albany Hotels Ltd* v *Daejan Investments Ltd* [1980] 2 EGLR 123.
[28] *GREA Real Property Investments Ltd* v *Williams* [1979] 1 EGLR 121 and *Estates Projects Ltd* v *Greenwich London Borough* [1979] 2 EGLR 85.

## *(iv)  Goodwill*

Although valuers are commonly required to ignore goodwill attributable to the carrying on of the tenant's business, this is not normally a matter which raises specific valuation issues. It would generally apply only to special properties, such as hotels and restaurants, where the history of trading by the tenant generates rental valuation evidence by analysis of the trading record of the tenant.

## *(v)  Length of lease*

A rent review clause commonly provides that, on review, the actual unexpired term is to be adopted in determining the rent. Thus, on a 20-year lease with five-yearly reviews, at the first review it is to be assumed that a new lease of 15 years is being offered (with five-yearly reviews), at the second one of 10 years, and at the third a lease of five years. This is unlikely to have any significant effect apart from the final review when such a short term might be argued to be very unattractive to tenants. Where the tenant has carried out major improvements which are to be ignored, then the effect might be significant at later reviews as the time left to recoup the expenditure by adjustment of rent could require a considerable reduction. However, the possibility of the tenant obtaining a new lease under the 1954 Act should be reflected in the approach and this will tend to negative the tenant's argument for a lower rent.[29]

On the other hand, some leases do provide that it shall be assumed that on each review a lease for the original term is being offered. However, it seems that this is to be interpreted as assuming that the lease offered is for the actual term of years but commencing when the actual lease started,[30] unless the facts of the case require otherwise.[31] This of course coincides with the reality of the case.

---

[29] *Pivot Properties Ltd* v *Secretary of State for the Environment*  [1980] 2 EGLR 126.

[30] See, for example, *Ritz Hotel (London) Ltd* v *Ritz Casino Ltd* [1989] 2 EGLR 135; *British Gas plc* v *Dollar Land Holdings plc* [1992] 1 EGLR 135.

[31] *Prudential Assurance Co Ltd* v *Salisburys Handbags Ltd* [1992] 1 EGLR 153.

## (vi) General lease covenants

The rent review clauses will generally provide that the other terms of the lease are to be assumed to be included in the hypothetical new lease. Thus covenants as to repair, insurance and alienation are to be the same as in the lease subject to review. The more restrictive or onerous they are to one party, so the rent will be adjusted upwards or downwards to reflect this quality, as is the case on the grant of a lease.

## (vii) Expiry of rent-free period

In the years after 1988, landlords were forced to offer inducements to tenants to take leases. These often took the form of a long rent-free period, way beyond the traditional few months at the start of a lease when tenants are fitting out premises. The rent payable at the end of the rent-free period became known as the "headline rent". So it might be said that "offices have been let at £10 per sq ft (£108 per m$^2$) with a rent-free period".

Two developments emerged from this introduction of long rent-free periods. One was the problem of analysing the terms of the letting to seek to obtain a rent on review which was equivalent to the headline rent obtained on the letting of comparable properties. The approach is considered below.

The second was the interpretation of rent review clauses where landlords sought to obtain the headline rent on a rent review, but which rent was recognised as being above the true rental value if rent is payable without a rent-free period, as is the case with rents arising from a rent review.

As an example, the rent review clause of a lease provided that the rent on review should be determined ". . . upon the assumption that . . . no reduction or allowance is to be made on account of any rent free period or other rent concession which in a new letting might be granted to an incoming tenant". This clause, and other clauses with similar wording, was considered by the Court of Appeal when it heard four cases together. The Court adopted a "presumption of reality" in interpreting the clauses which seek to avoid business commonsense. As a result, in three cases[32] it was

---

[32] *Prudential Nominees Ltd* v *Greenham Trading Ltd* [1995] 1 EGLR 97; *Co-operative Wholesale Society Ltd* v *National Westminster Bank plc* [1995] 1 EGLR 97; *Scottish Amicable Life Assurance Society* v *Middleton Potts & Co* [1995] 1 EGLR 97.

held that the rent payable on review was not the headline rent, but a rent derived from analysis of the comparable transactions (see below). In one case, however,[33] it was held that a headline rent was payable on the particular wording of the rent review clause.

The general conclusion is that a headline rent will not be payable unless no other interpretation of the rent review clause is possible. If the short fitting-out period is to be disregarded then suitable wording to this effect will be successful.

### (viii)  Whether rent review in hypothetical lease

One problem area concerns a phrase commonly adopted, whereby the hypothetical lease is assumed to contain the same covenants as in the existing lease "other than as to rent" or some such exclusion. It seems probable that the purpose of this exclusion is to indicate that the rent itself will be different. However, the Courts have considered the meaning of this exclusion clause in the various forms it takes, and in particular whether it meant that the actual clause requiring the rent to be reviewed should be assumed to exist or not.

Initially, the Courts adopted a strict literal interpretation and decided that ignoring the provisions as to rent meant that the rent review clause itself should be ignored.[34] Later cases adopted a more commercial view so that, unless the words in the clause clearly required the provisions as to rent review to be disregarded, then a rent review clause should be assumed.[35] In a yet later decision on appeal to the Court of Appeal, the more literal approach was favoured but this was for rather special reasons.[36] Finally, two cases in 1991[37] appear to have settled the matter, requiring the commercial view to be adopted, unless the wording of the clause makes it beyond question that no rent reviews shall be adopted.

---

[33] *Broadgate Square plc* v *Lehman Brothers Ltd (No 2)* [1995] 2 EGLR 5.

[34] *National Westminster Bank plc* v *Arthur Young McClelland Moores & Co* [1985] 2 EGLR 13.

[35] *British Gas Corporation* v *Universities Superannuation Scheme Ltd* [1986] 1 EGLR 120.

[36] *Equity & Law Assurance Society plc* v *Bodfield Ltd* [1987] 1 EGLR 124.

[37] *Arnold* v *National Westminster Bank plc* [1993] 1 EGLR 23 and *Prudential Assurance Co Ltd* v *99 Bishopsgate Ltd* [1992] 1 EGLR 119

Indeed a judge in the Arnold case stated that the Arthur Young decision in 1984 was clearly wrong.[38]

These examples illustrate that the valuer needs to weigh up the effects of the assumptions in determining the rent, as the rental valuation is of the hypothetical world which the assumptions create rather than the real world if the property were being put on the market with full flexibility of action. Even where the clause requires the rent to be the market rent as between willing lessor and willing lessee, if the parties are not willing then a hypothetical willing lessor and lessee must be assumed to exist.[39]

## 6. Valuation Consequences

The considerable attention given to aspects of rent reviews has led to some standardisation of the contents of rent review clauses. Nonetheless, they still vary considerably and the valuer needs to pay close attention to their actual wording in each case as the rent review clause represents the instructions to the valuer as to how he shall prepare his valuation. As a general rule, the more onerous or restrictive the covenants are on the freedom of action of the tenant, the lower the rent, and vice versa. The consequences on the rent payable can be considerable. For example:

### (a) Restricted User

Although there are two cases where the effect on rent of a restricted user has been determined each case depends on the fact of that particular case. For example, a lease of a shop where the user is restricted to greengrocer might have minimal effect where the shop is one in a small neighbourhood centre. On the other hand, such a restriction where the shop is close to a supermarket selling a full range of food products including fruit and vegetables could justify a huge reduction from the generally prevailing level of shop rents.

Normally there will be no direct evidence to support any rent adjustment so that it becomes a matter of valuation judgement.

---

[38] See n.33 *supra*. However, even this may not turn out as badly as would have been thought. In *Prudential Assurance Co Ltd* v *Salisburys Handbags Ltd* (n.30 *supra*) the absence of an assumed term of years allowed the court to hypothesise a term equal to what was normally offered in the market.
[39] See n.19 *supra*.

There are cases where a restrictive user can justify an increase in rent. This will be so where the retail units in the area or centre are under the control of a single landlord who maintains a pattern of use through an estate use plan so that the tenant is guaranteed a monopoly or a limited amount of competition.

### (b)  Period Between Reviews

Another problem which may arise in the welter of different lease covenant assumptions is the actual period between reviews. Clearly, if the lease in question has five-year rent reviews and the rental evidence is of similar leases with five-year rent reviews, then a direct comparison can be made. But what if the review period is shorter, say three years, or the period is actually longer, say seven or 21 years, or assumed to be longer by the interpretation of the phrase "other than as to rent" referred to above, when no rent review is to be assumed? As a general rule it can be said that the longer the period between the reviews, the higher should be the rent, and vice versa. The mathematical analysis by using discounted cash flow techniques is considered in Chapter 10.

Landlords tend to resort to a mathematical solution which puts them in the same position on a discounted rent basis whatever the review period. Where the period is relatively long this can produce a rent fixed well above the rents paid with the common five-year review pattern. In the Arthur Young case referred to above[40] the Court recognised that the uplift was 20.5%. In practice, there is strong tenant resistance to significant increases as rents reach levels which tenants argue they cannot afford, particularly in the early years after the review. They may accept an uplift up to 10%, but argue that tenants would not afford more than this. A "rule of thumb" approach has emerged whereby 1% for each additional year beyond five years is added to the rent on a five-year basis. These arguments can only be tested by marketing properties on different rent review bases but this is rarely attempted.

### (c)  Analysis of Incentives

When the market supply outweighs demand, as happened particularly after 1973 and 1988, landlords offer incentives to

---

[40] See n.33 *supra*.

applicants to persuade them to take their space. These incentives take various forms but typically include long rent-free periods and capital payments to the new tenant – "reverse premiums". In such market conditions these lettings are the only evidence usually available to the valuer since rent reviews, being upward only, are not activated. The problem for the valuer is how to analyse the transactions.

*Example 18–2*

A landlord has recently let 10,000 sq ft (930 sq m) of offices on a lease for 10 years with a review to full rental value after five years. No rent is payable for the first three years, the rents payable in years four and five being £100,000 p.a. The offices are similar to offices for which a rent review is now due.

The valuer needs to determine what rent the tenant would have paid if there had been no rent-free period which is the assumption of the rent review of the similar offices. It is assumed that the headline rent of £10 per sq ft (£108 per m²) is not payable under the rent review clause (if it is a substantially different answer would be achieved).

Various approaches are available to the valuer. One is to find the annual equivalent of £100,000 p.a. payable in years four and five which would be:

$$\frac{\text{£100,000 p.a.} \times \text{YP 2 years} \times \text{PV 3 years}}{\text{YP 5 years}}$$

(At a yield of 8% this produces a rent of £35,463.)

A DCF comparison which establishes a common NPV between an assumed day one rent and the actual letting would produce the same result.

However, such analyses are derived from an assumption that the present value of either approach should be the same. This is a landlord's point of view. A tenant might not agree with such an analysis since the initial rent-free period is a form of business loan which might be of particular value, especially for a newly established business.

In practice, an arithmetical average approach has become commonly adopted. Thus, taking the period up to the rent review, the analysis would be:

Rent payable (Years 4 and 5)     $= 2 \times £100,000$   $=$   200,000
Total years to review                                                      5

$\therefore$ Annual Average = £40,000 p.a.

Similar approaches can be adopted for the analysis of a reverse premium. Suppose, for example, the landlord in the above example also offered an initial sum of £200,000. Here, even more than with the rent-free period, the attraction to a tenant of such a capital sum clearly produces a different approach between the parties and thus a different analysis.

There are yet further aspects to consider. Although the landlord is receiving no rent for three years, the fact that the property is occupied means that he avoids outgoings such as rates, insurance, and repairs so that, by the letting, he converts the negative cash flow of an empty property into nil costs. The tenant takes on this burden so that the comparison between the two lease approaches should probably be on the basis of the cost of occupation rather than rent only, particularly where a discounted cost / value analysis is adopted.

No generally accepted form of analysis has yet emerged although the problem has been well detailed in articles.[41]

It may well be that the simple arithmetical average, though divorced from valuation principles of discounting, produces as reliable an answer as other methods given that the many factors involved make an approximate day one rent the best that can be achieved. It is up to the valuer to apply his judgment to the result of the analysis to form a view as to its reasonableness.

### (d) Comparable Evidence

In a normal market the valuer will have evidence of actual market lettings. These form the most reliable guide as to the level of values. Clearly, the valuer needs to know the details of the whole agreement.

In the absence of market lettings, the valuer may have details of rent review agreements negotiated between the parties which can be analysed to produce comparable evidence.

---

[41] For example, R. Goodchild [1992] 08 EG 85; J. Lyon [1992] 23 EG 82; J. Rich [1992] 43 EG 104; T. Asson [1994] 36 EG 135; D. Epstein [1993] 14 EG 90, 15 EG 120, 16 EG 83, 17 EG 76; and Inaugural Professorial Lecture by Neil Crosby at Oxford Brookes University on 6 October 1993.

The only other sources of evidence are the rents awarded by experts or arbitrators in the case of disputed rent reviews, and rents determined by the Courts for new leases granted under the Landlord and Tenant Act 1954. Such transactions provide the least reliable evidence and arbitrators' determinations cannot be relied on as evidence when presenting a case in support of a rental value to such parties.[42]

## (e) Machinery for Settling Disputes

Where the parties cannot agree then the clause provides for the matter to be referred to an independent third party, commonly a valuer agreed between the parties or one appointed by the President of the Royal Institution of Chartered Surveyors. He may be required to act either as an arbitrator or as an expert.

---

[42] *Land Securities plc v Westminster City Council* [1992] 2 EGLR 15.

# Chapter 19

# Commercial Properties (2)
# Types of Property

## 1. Generally

There are several categories of property which come under the general heading of commercial properties, and the categories can in turn be sub-divided into types within each category. As a start the main categories can be identified as:

(a)  Retail
(b)  Industrial
(c)  Warehouses
(d)  Offices.

Not only can each category be sub-divided but there is also considerable overlap between categories. Take, as an example, Retail and Warehouses. Retail properties come in various forms. The obvious example is a shop in a street, but even there they vary from kiosks to standard shop units to departmental stores. Over recent years there has been extensive development of shopping centres where the shopper enters off the shopping street into a development of various retail units under one roof, frequently with its own multi-storey car park. A further development has been the creation of edge-of-town centres where the shopping centre is surrounded by extensive car parking. A yet further trend has been the development of out-of-town stores of large size, generally food based, again with extensive car parking, described as superstores. In the non-food sector there has been the creation of one-off or groups of retail warehouses selling furniture, DIY or carpets, each of a size larger than the traditional shop units, with extensive car parking as a key element. These are often referred to as retail warehouses, to distinguish them from the traditional distribution warehouse where goods (wares) are housed in bulk prior to distribution to other smaller traders or retail outlets.

Thus "retail" covers several types of properties where goods are sold to the public, with some retail properties being similar to warehouses save that the public at large can visit them and purchase goods; whilst a non-retail warehouse is a property where goods are centrally assembled prior to distribution to trade purchasers.

Another example is the comparison between offices and industrial. Offices themselves range from small shop-type units in shopping streets used by building societies or estate agents, to floors above shops occupied by solicitors, accountants or employment agencies providing a local service, to free-standing buildings given over to office activities which themselves can range up to the very large and high office blocks found in major city centres. Recent years have seen the development of offices whose location is related to motorways, away from the traditional urban centres. There are often several buildings contained within a business park and where office activity runs alongside research and manufacturing activity but where the whole building is of a more or less uniform standard of fitting and finish, commonly termed "high-tech" or B1 buildings. This manufacturing activity, generally in electronics or similar activities, can properly be described as industrial, but differs substantially from other, more traditional industrial activities, in small to large factories, and even more from the large "heavy" industrial plants given over to engineering and other large scale manufacturing activities.

The traditional dividing line between these various activities of retail, industrial, warehouse and office is, in the interests of business efficiency, becoming less rigid – a fact acknowledged by the introduction of new Use Classes for planning purposes by the Town and Country Planning (Use Classes) Order 1987 (SI 1987 No 764). Nevertheless, most properties tend to relate principally to one of these activities.

The feature common to all the above four classes of property is that, generally speaking, they are occupied for the purpose of carrying on an industry, trade or profession in the expectation of profit: and it is the profit which can reasonably be expected to be made in the premises which in the long run will determine the rent a tenant can afford to pay for them. In economic terms rent = f (profit [P]) or = f (turnover [T]).

The principal factors affecting each class are considered in detail later and they fall under three broad categories:

(1)  the quality;
(2)  the quantity of the accommodation; and

(3)  the location of the premises.

There are some types of property which, although capable of inclusion in one or other of the above classes, are seldom let to a tenant for trading purposes and in respect of which there may be little or no evidence of rental transactions to guide the valuer in arriving at a proper valuation. Typical examples are premises given over to a special purpose, such as chemical, gas or electricity plants or works, and premises occupied for charitable purposes.

In many cases such properties are only likely to come onto the market if the use for which they were built has ceased. In these circumstances it will be necessary to consider the alternative use to which they can be put – if planning permission is likely to be forthcoming to enable such a change of use to be made – the cost of converting them to or redeveloping them for such use and the value they will have when converted or redeveloped. Where any have to be valued for the particular purpose for which they are used, resort may be had to the Depreciated Replacement Cost approach described in Chapter 20.

## 2. Retail Premises

### (a) Location

These may vary from small shops, with upper floors, in secondary positions likely to let at rents of a few thousand pounds or so per annum, to out-of-town superstores with floor areas of thousands of square metres and town centre enclosed air-conditioned centres with shops on a number of levels and the rent of a small standard unit measured in terms of tens of thousands of pounds per annum.

As already indicated, the chief factor affecting values is that of position and its effect on trade. The prospective purchaser or tenant is likely, in nearly every case, to attempt some estimate of the trade in those premises in that position. The degree of accuracy that he will attempt to achieve will depend largely on circumstances. It is well known that many of the bigger concerns, with a large number of branches, have arrived from experience at certain methods of assessing, with a fair degree of accuracy, the turnover they are likely to be able to achieve in a given shopping position. From such an estimated turnover it is necessary to deduct the cost of goods to be sold; from the gross profit remaining will fall to be deducted wages, overhead charges and a margin for profit and interest on capital. There remains a balance or residue which the tenant can afford to pay in rent, rates and repairs.

The use of shop premises for one particular retail trade can, with some exceptions, be changed to its use for another retail trade without the necessity of obtaining planning permission. If, therefore, premises are capable of occupation for a number of different purposes, and are likely to appeal to a number of different tenants, there will be competition between the latter, and the prospective occupier whose estimate of the margin available for rent, etc, is largest, is likely to secure the premises by tenancy or purchase.

It is not suggested that in the majority of cases a valuation of shop premises should be based on an analysis of the probable profits of any particular trade in order to arrive at the rent a tenant can afford to pay for them, largely because the probable profits are based on amounts which in practice are very difficult to assess with any reliable degree of accuracy. But the general factors likely to influence prospective occupiers in their estimate of turnover and margin available for rent will certainly have to be taken into account by the valuer.

It is common sense that shops in a prime position such as the main thoroughfare of an important town with large numbers of passers-by will command a much higher rent than those in secondary positions. Again, the potential purchasing power of the passers-by must be considered; for instance, the amount of money they have available to spend in the shops is likely to be greater in Bond Street, London, than in the main shopping street of a small provincial town.

In some instances, large variations in value may be found within a comparatively short distance. A location at the corner of a main thoroughfare and a side street may be of considerable value, whereas a short distance down the side street the shops will be of comparatively small value. Location is therefore a factor to which most careful consideration must be given. It has been said that the three main qualities which determine value are "location, location and location", and this is particularly so in the case of shops.

Some of the most important points in regard to location are: in what class of district is the shop situated? What is the type of street and what sort of persons use it? What is the position of the shop in the street? Is it a focal point such as the corner of an important road junction? Are there any multiple shop branches nearby or any "magnet" such as a large store? Is the property close to premises which break the continuity of shopping, such as a town hall, bank or cinema? How does it compare with the best position in the town? How close is it to car parks or public transport?

## (b) Type of Premises

The next most important factor is the type of premises and their suitability for the display and sale of goods. The available frontage for the display of goods needs first consideration, together with the condition and character of the shop front. The suitability of the interior must then be considered; the adequacy of its lighting; the presence or absence of access at the rear for receipt or delivery of goods (this is becoming increasingly important as parking, waiting and loading restrictions become imposed over more and more roads in the centres of towns); the character of the upper floors, whether separately occupied or only usable with the shop; the general condition of repair and the ancillary staff facilities.

In the case of newly erected premises, shops are often offered to let on the basis that the tenant fits out the shop and installs his own shop front at his own expense; what is offered is a "shell" often with the walls and ceiling to be plastered and the floor in rough concrete. In more established districts, premises will be offered to let complete with shop front. If the latter is modern and likely to be suitable for a number of different trades, the value of the premises may be enhanced thereby. If, however, the shop front is old-fashioned, or only suitable for a limited number of types of business, it will be necessary to have regard to the fact that any prospective occupier will need to renew the shop front. Most multiple retailers have their own style of shop front so that they will be indifferent to the existence and quality of any existing shop front. This will apply also to the internal fittings since extensive shop fitting work is carried out before trading begins.

Sometimes a factor of importance is the right of a tenant to display goods on a forecourt. There are many shops where the enforcement of a building line or the like has created a forecourt of considerable extent and of value to a number of trades, such as DIY, greengrocers and others, for the display of goods.

In considering particular premises it may be necessary to have regard to the possibility of structural alterations in the future for the purpose of improving the premises. In the case of large premises the division of the premises to provide a few "standard" units or, for very large premises, the creation of a shopping precinct should be considered.

In valuing premises of some age, regard may have to be had not only to their probable useful life but also to the possibility of their being redeveloped in whole or in part or of additions being made, for example to extend the selling space to a greater depth, subject to the necessary planning consent.

As regards accommodation on upper floors, it is necessary to consider the best purpose to which it can be put. In the more important positions it is frequently found that such floor space can be used for the display and sale of goods. In less prominent positions it may be more profitable for the upper part to be let off separately for use as offices or dwelling purposes, but in such cases it must be borne in mind that planning permission for a change in use may not be forthcoming.

The extent to which large basement areas are of value will again depend largely upon position. In valuable situations where rents are high, any prospective tenant is likely to wish to use the whole of the ground floor area for the display of goods and an extensive and dry place for storage will enhance the rental value of the premises considerably. In less important positions the use of the rear portion of the ground floor for storage purposes may be more practicable, and the enhancement of value due to the presence of a basement may be quite small.

The comments so far have concentrated on the traditional shop premises found in shopping streets. Nonetheless the same considerations apply to out-of-town superstores or retail warehouse parks or to shopping centres, but will have a different emphasis.

For example, the quality of location for out-of-town premises will be tested more against accessibility for customers who will be mainly motorists, although the availability of public transport may be important. Thus the location should be well placed for the road network, with sites near to motorway junctions being preferred. Extensive parking will also be required to accommodate the motorist customer on arrival, ideally at ground level only. In the case of in-town shopping centres a good road access is less easy to achieve but extensive parking is also required, commonly on several levels within the shopping centre.

Similarly, the type of premises is important. Retail warehouses require large open lofty spaces, although a high standard of internal finish is not generally sought. On the other hand, food superstores require a high standard of design both internally and externally. Indeed the cost of fitting out such stores can match or exceed the cost of the building itself. Similar requirements as to design and finish apply to shopping centres, not only to the shop units themselves but increasingly to the common parts. The early shopping centres are more and more being extensively refurbished with the roofing over of public malls, the upgrading of floor surfaces, and improvements to lighting and general design being areas of special attention. The effect on rental values of such works can be significant, with tenants accepting a considerable rental increase in recognition of the improvement to trading potential which the works generate.

### (c) Type of Tenant

This is of considerable importance to the value of property from an investment point of view.

A letting to a large multiple concern is considered to give exceptional security to the income, since the mere covenant to pay rent by such an undertaking can be relied on for the duration of the lease even if the profits from the particular branch are less than was expected or even if the lease is assigned to another trader (subject to the provisions of the Landlord and Tenant (Covenants) Act 1995).

This contrasts with premises let to individuals or small companies with limited resources where the chance of business failure is far higher.

Since multiple concerns tend to occupy prime property and the tenants of poor covenant the poorer property, this contributes to the sharp difference in yield between prime property at low yields and secondary or tertiary property at high yields.

### (d) Terms of the Lease

Shop property may be let on various repairing terms; on the one hand, the tenant may be responsible for all repairs and insurance, and at the other extreme, the landlord may be responsible for all repairs. In between these extremes a variety of different responsibilities is found. The tendency is for the responsibility to be put on to the tenant. Where the landlord has any responsibility for outgoings, an appropriate allowance must be made in valuations.

The length of the lease is important. An investor will prefer a long lease with provision for regular rent reviews.

The terms of letting of the upper portion (if separate from the shop) must also be noted and the presence of any protected tenancies under the Rent Acts.

*(e)  Rental Value*

When making comparisons of rental value between shops in very similar positions, regard must be had to size. Lock-up and small or medium size shops may provide sufficient floor space to enable a large variety of businesses to be carried on profitably and may be in great demand whereas, in the case of larger shops for which there is less demand, it may be found that the rent is not increased in proportion to the additional floor space available.

For example, in a given position a lock-up shop with a frontage of 7 m and a floor space of 140 $m^2$ may command a rent of £14,000 p.a., equivalent to £100 per $m^2$; whereas, nearby, in a very similar position, a larger shop with a frontage of 10 m and a depth of 20 m may only command a rent of £16,000 p.a., equivalent to a rent of £80 per $m^2$.

In recent years the demand has grown for large units as patterns of retailing change so that it could be that, in prime positions, the rent for larger units will be proportionately larger than that for small units. The valuer will need to assess the relative demand for different sized units within the shopping area before preparing a valuation.

Although various "units of comparison" can be adopted by a valuer in comparing rents paid for other units with the unit to be valued, such as frontage or standard unit comparisons, the method most commonly adopted is that of "zoning" as described in detail in Chapter 21.

In valuing extensive premises comprising basement, ground, and a number of upper floors, there is some difficulty in deciding upon the comparable rental values of the different floors. Generally speaking, the zoning method allows for this by adopting a proportion of the Zone A rent. Thus a first floor retail area may be valued at one-sixth of the ground floor Zone A rental value, whereas first floor storage might be one-tenth. The actual rent per $m^2$ would need to be considered to assess whether it is unrealistic for such a use. For example, if the ground floor Zone A rent is

£1,500 per m², then one-tenth, or £150 per m², might be regarded as excessive for storage space on the first floor if warehouse or storage buildings in the area have a rental value of only £40 per m².

The method of approach to problems of this type is illustrated by the following example.

*Example 19–1*

You are asked to advise on the rental value of shop premises comprising basement, ground, first and second floors, in a large town. The premises are old but in a fair state of repair; the basement is dry, but has no natural light; the upper part is only capable of occupation with the shop. The net floor space is as follows:

Basement, 110 m²; Ground Floor, 98 m²; First Floor, 80 m²; Second Floor, 30 m². The shop has a frontage of 7 m and a depth of 14 m. The first floor is retail and the second floor staff rooms.

The following particulars are available of recent lettings of shops in comparable positions, but all of them are more modern premises:

(1)  Premises similar in size let at £36,000 p.a.: areas – Basement, 100 m²; Ground Floor, 100 m²; First Floor, 70 m² retail space; Second Floor, 60 m² storage. The ground floor has a frontage of 5 m and a depth of 20 m.
(2)  A lock-up shop, 4 m frontage and 5 m depth, let at £9,000 p.a.
(3)  A shop and basement, frontage 15 m and depth 28 m, with a basement of 100 m² let at £86,500 p.a. The upper part is used as offices and is separately let in floors as follows:

First Floor, 140 m² at £7,000 p.a.; Second Floor, 130 m² at £5,200 p.a.

(4)  The basement under lock-up shop No. 2 and three more shops, is dry and well lit with direct street access; floor space is 120 m² and it is let for storage purposes at £3,600 p.a.

*Note*:   In practice these lettings might be on different terms and adjustments would be required to the rents to reduce them to net figures.

*Analysis:*

Analysis of the first letting is best deferred until other lettings of various parts of premises have been dealt with:

| Letting | Floor | Area in m$^2$ | Rent £'s | Rent £'s per m$^2$ |
|---------|----------|------|-------|-----|
| 2 | Ground | 20 | 9,000 | 450 |
| 4 | Basement | 120 | 3,600 | 30 |
| 3 | First | 140 | 7,000 | 50 |
| 3 | Second | 130 | 5,200 | 40 |

The evidence from these lettings can be used to analyse the letting of shop No. 3 as follows:

| | |
|---|---|
| Rent for shop and basement | £86,500 |
| *Less* | |
| Basement | |
| 100 m$^2$ at say £25 per m$^2$ | £2,500 |
| Rent for shop | £84,000 |

Adopting 6 m zones-

| | | |
|---|---|---|
| Zone A | 15 m × 6 m = 90 m$^2$ – in terms of Zone A = | 90 m$^2$ |
| Zone B | 15 m × 6 m = 90 m$^2$ – in terms of Zone A/2 = | 45 m$^2$ |
| Remainder | 15 m × 16 m= 240 m$^2$ – in terms of Zone A/4 = | 60 m$^2$ |
| Total area in terms of Zone A ('ITZA') | | 195 m$^2$ |

195 m$^2$ at £84,000 gives Zone A value of          £431  per m$^2$

This evidence can then be used to check the letting of shop No. 1, as follows:

| | | |
|---|---|---|
| Basement: | 100 m$^2$ at £25 | £2,500 |
| Ground Floor: | 55 m$^2$ ITZA at £431 | £23,705 |
| First Floor: | 70 m$^2$ at £71.83 | £5,028 |
| Second Floor: | 60 m$^2$ at say £25 | £1,500 |
| | | £32,733 |

| | | |
|---|---|---|
| Actual Rent | £36,000 | equating to £480 per m$^2$ |

*Note:* The Ground Floor is Zone A (5 × 6) + Zone B (5 × 6)/2 + Remainder (5 × 8)/4 = 66.5 m$^2$ in terms of Zone A. The First Floor being retail is valued at A/6. The Zone A rent is higher than Shop No. 3 which is larger than usual.

*Valuation*

Bearing in mind age, position and all other relevant factors, the valuation might be as follows:

| | | |
|---|---|---|
| Basement: | 110 m² at £20 | £2,200 |
| Ground Floor: | 66.5 m² ITZA at £450 | £29,925 |
| First Floor: | 80 m² at £75 | £6,000 |
| Second Floor: | 30 m² at £20 | £600 |
| | | £38,725 |
| Rental Value Say | | £39,000 |

The preceding comments on the rental valuation approach apply to shops in town centres and shopping parades. However, the zoning method has limitations and is not generally applied to the larger types of shop found particularly in town centres such as supermarkets, chain stores and sometimes departmental stores. In these cases ground floor retail areas will be valued at an overall rate per m² and upper floor retail areas again at an overall rate per m² which might be related to the ground floor rate, or on the basis of a percentage of turnover.[1]

In the case of retail warehouses or out-of-town food stores, the rental value will generally be determined by comparison with similar units in the area or even in comparable locations in relation to other urban centres. The rental value analysis will produce rental values per m² applied to the total retail floor space, perhaps with a differential rental value for storage space. Unlike town centre shopping centres, the factors in determining rental value will depend more on accessibility and customer parking and visibility.

A further form of rental approach may be found where the rent payable is related to the level of business done by the occupier. Commonly this is a "turnover rent" since the gross takings will be the determining factor. Generally, the rent payable has two elements, a base rent payable whatever the turnover, and a further rent payable being a proportion of the turnover in excess of a specified amount.[2]

---

[1] See Chapter 7 of *Valuation: Principles into Practice* (4th Edition) Ed. W.H. Rees.

[2] For a detailed examination of turnover rents see Paper Five, "Turnover Rents for Retail Property" by R.N. Goodchild in the *Property Valuation Method Research Report* published by the RICS and South Bank Polytechnic in July 1986.

*(f)  Capital Value*

Similar methods of analysis and comparison can be used in determining the appropriate rate per cent to use in capitalising rental value.

Shops let to good tenants can generally be regarded as a sound security since rent ranks before many other liabilities, including debenture interest and preference dividends in the case of limited liability companies.

Present-day rates per cent may vary widely from around 4% to 11%. The lower rates reflect the above average and above inflation rate of growth anticipated for shops in prime positions which has been seen over recent decades, with occasional periods of low growth or even falling values as experienced, for example, in some centres in the early 1990s.

Where the upper part of freehold shop premises is let separately, e.g. for residential purposes, it is probably desirable to value it separately at a different rate from that used for the income from the shop. In the case of leaseholds, however, this would involve an apportionment of the ground rent, and a more practical method may be to deduct the ground rent capitalised at a fair average rate per cent for the whole premises.

In any particular case, evidence of actual transactions is usually the best basis for capitalising rental values.

## 3.  Industrial Premises

*(a)  Generally*

Reference was made in Chapter 5, Section 4, to the method of ascertaining the rental values of this type of property on a floor space basis in terms of $m^2$ of gross internal area.

The range of properties to be considered is extensive and varies from shop or residential property or railway arch, converted for use as storage or for factory purposes, to well-constructed, well-lighted, up-to-date premises with many amenities.

Situation is one of the chief factors affecting value, including such points as access to motorways or link roads and railways, also proximity to markets and, in particular, to the supply of suitable labour.

In making comparison between premises otherwise similar, the influence of these matters must be carefully considered.

## (b) Construction

Good natural lighting is essential in many trades; and premises in which the use of artificial light during day-light hours can be kept to a minimum might well command a higher rent than those where a good deal of work has to be done under artificial light.

Adequate working heights are essential and in new single storey factories 6 to 7 m to eaves is regarded as a minimum.

The strength of floors is an important factor as many trades use machinery which requires a high loading factor.

The rates of insurance for many trades are comparatively high. The risk of fire will be minimised by fire-proof construction and by the presence of sprinklers.

The means of escape from fire will need to be considered, and also the provision of adequate sanitary accommodation and adequate ventilation. It must be ascertained that the premises under consideration comply with the requirements of the Factories Acts and other statutory regulations.

The mode of construction may also be important in relation to the annual cost of repair. The rental value on lease will be diminished, for example, where the roof is old and by reason of age or construction is likely to leak.

Consideration must also be given to the services available, the type of central heating plant, electricity supply and equipment and the availability of a gas supply.

The layout of the premises should be such as to give clear working space, to facilitate dealing with the delivery and despatch of goods, and the handling of goods within the premises without undue labour. Ideally the space should be uninterrupted by columns or other roof supports and modern building techniques allow the construction of large clear areas.

For many trades the floor space on the ground floor is of considerably greater value than that on upper floors, although this may be offset in the case of valuable sites, particularly in central districts, by adequate provision for handling goods by lifts or conveyors between floors. Particularly in the case of ground floors, it is a disadvantage if the whole of each floor of each building is not at the same level, or not at the same level as the exterior, i.e. if there are steps or breaks in the floor because of the extensive use of fork lift trucks.

The artificial lighting system should be considered, to see that there is an adequate standard of illumination to enable work to be done satisfactorily.

In large factories the provision of adequate canteen, welfare and car parking facilities is an essential point.

Loading and unloading facilities should allow easy access for fork lift trucks and large vehicles, particularly for larger premises. In some urban areas access to premises is often poor, and, at worst, off-street loading and unloading may be impossible. This has a significant adverse impact on values.

## (c) *Rental and Capital Values*

In some cases it may be found that land is sold and factories erected to the requirements of an occupier who purchases the freehold, and there may be little evidence of rental values. In such cases, regard must be had to the selling price rather than the rental value of the accommodation when making a valuation. Wherever possible, values arrived at on this basis should be checked by consideration of rental value.

In the centres of towns a large building is often let off in floors, with the tenants paying exclusive rentals and the landlord being responsible for the maintenance of common staircases, lift, and sometimes the provision of heating or the supply of power. The method of valuation in such cases will be along the lines indicated in this chapter in respect of office premises; an estimate of the net income is made by deducting the outgoings from the gross rentals and, where necessary, the actual rentals are checked against comparable values for other similar properties. The cost of services may, however, be recovered by a service charge. Service charges are also found on industrial estates where, although individual factories will be let on full repairing and insuring terms, the landlord provides other services the cost of which is recovered through a service charge.

Currently the yields for modern single storey factories are around 8%–10% but are significantly higher for poor quality premises. The range of rents for similar properties shows a similar wide disparity – from up to £100 per $m^2$ for the best down to £20 per $m^2$ or even less for poor quality premises.

In making comparisons between one property and another, careful note must be made of the factors already mentioned. It may be found that one factory in a similar position and of a similar area and construction has a modern central heating system, sprinklers, and good road access, whereas the other lacks these amenities.

Such differences are usually taken into account by modifying the price per unit of floor area. But it must not be overlooked that the provision of additional facilities may be possible at a comparatively small cost and it may be more realistic to value the property as if the improvement had been carried out and to deduct the cost.

It may sometimes be found that extensive factory premises comprise new and old buildings, some of which may be of little value in the open market. They may, for instance, have been erected for the purposes of a particular trade and, although the covered floor space area may be extensive, they may have a comparatively low market value.

The following example is typical of fairly modern factory premises.

*Example 19—2*

You are instructed to prepare a valuation for purposes of sale of a large factory on the outskirts of an important town.

The premises are in the occupation of the owners, by whom they were acquired in 1968; the premises are of one storey only for the greater part, with offices, canteen and kitchens in a 2-storey block fronting the main road. The factory is brick built with steel span roof trusses covered externally with corrugated asbestolux and lined internally. Height to eaves is 7m. There is good top light. The floor is of concrete finished with granolithic paving. There is an efficient central heating system. Gas, electricity and main water supplies are connected. The site has a frontage of 100 m to a main road and has a private drive-in with two loading docks. There is adequate lavatory and cloakroom accommodation. The site area is 1.2 ha.

Temporary buildings for storage purposes with breeze-slab walls and steel-truss roofs carried on steel stanchions have been erected since the property was purchased.

The available areas are as follows:

Main factory, 3,000 m$^2$; boiler house, 70 m$^2$; loading docks, 120 m$^2$; outside temporary stores, 200 m$^2$; ground floor offices, 200 m$^2$; first floor offices, 120 m$^2$.

There are a number of similar factories in the vicinity, of areas from 1,000 m$^2$ to 2,500 m$^2$, let within the last five years at rents of from £25 to £40 per m$^2$. The rents show a tendency to increase.

*Valuation*

It is usual for valuations of industrial premises to be based on the gross internal area (inside external walls).

As a result of an inspection, it will be possible to decide a fair basis of valuation in comparison with those factories where there is evidence of rental value.

A reasonable basis for a valuation might be as follows:

|  |  | m² |  | £ |
|---|---|---|---|---|
| Ground Floor: | Main factory floor | 3,000 | at £35 | 105,000 |
|  | Loading docks | 120 | at £35 | 4,200 |
|  | Outside stores | 200 | at £11 | 2,200 |
|  | Offices | 200 | at £50 | 10,000 |
| First Floor: | Offices | 120 | at £40 | 4,800 |
| Full rental value – net |  |  |  | 126,200 |
| YP in perp, say |  |  |  | 10 |
|  |  |  |  | £1,262,000 |
| Value, say |  |  |  | £1,300,000 |

Alternatively, the gross internal area can be valued at an overall rate of £36 per m² plus the outdoor stores taken at £11 per m². This can be done provided that the ancillary space is of average size for an industrial building of this size.

*Notes:*

(1) The increased value placed on floor space used as offices reflects the higher standard of finish and amenity compared with the main factory. The office content, just under 10% of the total floor space, accords with the norm.
(2) It will be observed that the total area covered by buildings, allowing for outside walls, etc. is probably about 4,000 m² so that on a site of 1.25 ha and working on a site coverage of 40%, there is room for extension. Depending on the particular circumstances, this might justify some specific addition to the value but has been accommodated in this case in the rounding up.

## 4. "High-Tech" Industrials

Reference was made earlier in this chapter to the types of building which are provided to house high-technology based industry.

Although many of the locational requirements which apply to light industry apply also to this type of industry, there are some differences to be noted. The first is general environment, where higher standards are required. The second is proximity to residential areas and other facilities appropriate to the level of staff employed, which again must be of a higher standard. The third, and this affects the building form more than immediate location, is a much higher standard of internal working environment. "High-tech" buildings tend to be two or perhaps three storeys, and the internal finish is of office rather than industrial standard. The essential feature, however, is flexibility so that any part of the building can be used for any of the various activities which form part of the total process. Finally, a higher standard of car parking is required.

Rents for these types of premises range at the time of writing from £40 to £100 per m$^2$ overall and yields tend to be slightly lower than for the best industrial, from about 7%.

A matter which has led to some difficulty has been the terms of leases. Landlords seek the traditional pattern of 20 or 25 years with five-year reviews, but the type of occupier concerned often desires a shorter term commitment, particularly foreign owned companies.

There is no consensus of opinion as to whether the rental value should be applied to the gross internal area or net usable floor space. The more the premises appear to be offices the more likely it will be that net usable floor space is appropriate and vice versa.

## 5. Business and Science Parks

Reference has already been made to the industrial estate which has been a feature of the industrial scene since the early part of the 20th century.

The business park is a more recent development which exploits the greater flexibility between business uses permitted by the 1987 Use Classes Order. Access to a motorway tends to be the chief locational requirement but, in so far as "high-tech" and office uses will be accommodated, their specific locational requirements must also be met.

The science park caters primarily for research and development activities and the essential locational requirement is proximity to a university or other academic institution with which the occupiers co-operate.

## 6. Warehouses

A warehouse is, by definition, a place where people house their wares. At one time warehouses were found in the main at docks. Goods were brought to the docks where they awaited being loaded on to ships, or they lay there having been unloaded from ships and prior to being distributed around the country. Indeed, in many of the older docks there may still be found former spice warehouses, sugar warehouses, and others designed for specific commodities.

After a decline, warehouses have now taken on a greater significance in the development of the economy, so that today they represent an important element in the economic structure. For example, the growth of major retailing groups has led to demands by them for large warehouses to which the various goods they offer for sale can be delivered by the manufacturers and there assembled with other items for delivery to their stores.

Similarly, shippers, having moved towards containerisation, require warehouses where they can have goods delivered and where they can join with others to fill the containers for various destinations. Indeed, the many developments in the manufacturing, retailing and distribution industries have led, in recent years, to considerable demand for warehousing facilities which were not contemplated previously and which led to the development of warehouses and whole warehouse estates, to meet these demands. In consequence, the warehouse, or shed as it is often termed, has become a major attraction to investors in contrast to former times when it was generally of secondary importance and interest.

These comments on warehouses refer to what are commonly known as wholesale warehouses to distinguish them from retail warehouses which were considered earlier in this chapter.

The investor in warehouses is looking for various qualities when judging any particular warehouse; qualities which will also attract the occupier. The principal qualities are:

(i)  *Location*
     Ideally a warehouse should be well located within the general transport network. In general this means it should be close to the motorway network.

(ii) *Site Layout*
     A warehouse is a building to which products are brought and subsequently removed. This involves lorries bringing in goods and then removing them. Thus there will be a flow of vehicles

and the best planned warehouses have facilities to accommodate lorries waiting to unload, good unloading facilities, and easy means of access. Contrast this with old warehouses where lorries have to park in the streets when waiting to deliver; a narrow access requires several manoeuvres before the lorry can back into the narrow access to the yard space, and all set in areas of narrow streets.

(iii) *Design*

A modern warehouse will have easy access to vehicles, and loading bays, or docks, so that lorries can readily be brought up to the premises to load or unload. It will be single-storey to avoid raising goods by lift or crane or gantry. It will have clear floor space to allow fork lift trucks to transfer goods around the property unhindered by columns or walls. The headroom will allow bulky goods to be moved around without hindrance and will allow smaller items to be stocked to the height limit of the fork lift trucks or other mechanical devices.

If these qualities are brought together it is seen that a modern warehouse is normally a single-storey building, of clear space, with minimum height to eaves of around 7 m, and up to 12 m or more with loading/unloading facilities to accommodate the largest lorry, high floor loading capacity, and located close to a motorway and urban centres. Such a building is totally different from the "traditional" warehouse, usually multi-storey, close to docks or a railhead, in an area with poor street access and no off-street loading/unloading facilities. It is important to keep this sharply contrasting picture in mind when warehouses are mentioned.

The valuation of a warehouse reflects these various qualities. In determining the rental value, the actual quality judged against the desirable qualities must be considered. The common approach is to apply a rental value based on the gross internal floor area. Thus, in a locality, warehouses which exhibit the most desirable qualities might command a rent of up to £90 per m², whereas old multi-storey warehouses command rent of £20 per m² on the ground floor with reducing rental values on the upper floors.

This appears to be the general approach, so that cubic content tends not to be considered in assessing rental value even though it would appear that the capacity for storage should be a critical factor. It is true that warehouses with low headroom tend to attract lower rental values than warehouses with high overall clearance, but nonetheless the cubic capacity is rarely determined and noted. An exception to this is specialist warehouses. For example, cold stores, which are essentially warehouses of a special type, are generally described in relation to their cubic content. In the case of occupiers using containers or pallets, the required height is a multiple of the height of the containers or pallets – the difference between a height of 4.5 containers to 5 containers is 11% in cubic capacity, but effectively 25% in storage capacity. Apart from such special cases, a ware-house is typically described and considered in relation to its floor area.

As to the yields required by investors in warehouses, these tend to be at or around the yields required for industrial premises of a similar age and quality. There are, however, two qualities which might lead to the acceptance of a lower yield as against a comparable industrial unit. First, the amount of wear and damage which a warehouse suffers by use will tend to be less than an industrial unit. Secondly, a warehouse is an adaptable building. It is usually a simple building, referred to familiarly as "a large shed", which enables it to be switched to other uses speedily and for a low cost. Hence, if the demand for warehousing falls away, a modern warehouse can readily be adapted to industrial use, or to retail use. Indeed, there has been a strong demand for the adaptation and use of warehouses as retail stores for discount stores, superstores or "retail warehouses", which were considered earlier in this chapter. In the case of older multi-storey warehouses, many of these can be converted to residential use in the form of flats or lofts, giving them an extended life and a higher, residential value.

Thus, the warehouse is again an important building in the economy both to users who require well planned storage space for their operations, to investors who feel that they offer a sound investment because of this demand by users, frequently large companies with a sound covenant, and to developers who are happy to meet the demand since the actual development of warehouses generally presents fewer problems than most other forms of development.

## 7. Office Properties

### (a) Generally

The range of properties to be considered is extensive. At the one extreme there are converted dwelling-houses or floors above shops, perhaps in a small provincial town or the suburb of a larger town where the demand for office accommodation is limited; at the other, there are modern buildings in large towns or on landscaped edge-of-town sites constructed specifically for offices, containing considerable floor space and providing within one building many amenities, which may include restaurants, club rooms, shops, post office, etc., and equipped with central heating, air conditioning and lifts. In the latter case the services provided by the landlord may be considerable, and may include not only lighting and cleaning of the common parts of the building but also the cleaning of offices occupied by tenants. The cost of such services is usually covered by a service charge payable in addition to the rent. Recent years have also seen the development of out-of-town offices in spacious, well landscaped sites, sometimes referred to as campus offices.

The method of valuation usually employed will be to arrive at a fair estimate of the gross income, deduct therefrom the outgoings borne by the landlord, and apply to the net income so arrived at an appropriate figure of Years' Purchase. Apart from retail premises, offices tend to have a significantly higher rental value then other uses, with rents for high quality offices ranging from around £100 per m$^2$ up to £500 per m$^2$ or more in the best locations.

In practice, some offices are distinguished from others because they are open to and are visited by members of the public or clients. Typically these are banks, post offices, solicitors and estate agents and located in or close to retail centres, sometimes in retail units converted for their use. The valuation approach to such offices follows that for retail property considered earlier in this chapter.

### (b) Income

The annual value is usually considered in relation to area, and in advertisements of floor space to let in office buildings the rent is often quoted at so much per unit of floor area based on the net internal area. Before buildings are erected, this practice is difficult to avoid but it gives rise to disputes as to the actual area involved, notwithstanding that a code of measuring practice has been produced by the RICS. Once buildings are erected, rentals should be quoted in terms of pounds per annum; this will avoid disputes as to the actual floor area (although a tenant would no doubt check the rent against his own calculation of floor space).

Actual lettings will require careful consideration and analysis and may often appear to be inconsistent. This may be accounted for by the fact that tenancies were entered into at various dates and that the landlord will have sought to make the best bargain he could with each tenant. In recent years, incentives have commonly been offered to new tenants. Such incentives include long rent-free periods and reverse premiums, particularly for office lettings. The need to have the full details of a comparable letting is obvious, and the problems of analysis to provide a rental on a unit basis were examined in Chapter 5.

A further factor is connected with the question of varying areas of floor space. It may be, for example, that the whole of the fifth floor of a building is let to one tenant, whereas the sixth floor is let to seven or eight different tenants, so that the total net income derived from the higher floor may be greater than that from the one below.

The valuer's estimate of a fair and proper rental will depend upon the circumstances of each case and upon his judgment of the way in which the building can best be let. He will be guided primarily by the level of rental value obtaining in the district, found by analysing recent lettings of similar accommodation in comparable buildings in the area.

When dealing with office premises which are empty, or when estimating the rents at which a proposed new building will let, it is usual to make comparison with other properties on the basis of the terms on which such accommodation is usually let. In general, offices are let on full repairing and insuring leases. In the case of buildings in multiple occupation, the tenant will be limited to internal repairs but a service charge will be levied in addition to cover external repairs and maintenance of common parts and services.

It is important that the same basis of area is used in analysis as is employed in the subsequent valuation. This may be gross internal area, or net usable area excluding lobbies, toilets and the like. The RICS/ISVA Code of Measuring Practice recommends net internal areas. Agreement of the basis of measurement and the area which results is one of the most contentious factors in negotiations between valuers.

## (c) Terms of Tenancy

As has been stated, terms of tenancy will vary considerably. Lettings are today almost always exclusive of rates. The cost of external repairs, maintenance of staircases, lifts and other parts in common use, in the case of buildings let in suites, will be covered by a separate service charge. The tenants may be liable for all repairs to the interior of the offices. Where whole buildings are let to single tenants, the lease is normally a full repairing and insuring one.

Where accommodation is let in suites it is quite likely that individual variations will be found in the tenancies within the same building, and careful examination of the terms of tenancy for each letting is necessary.

## (d) Outgoings

The use of a percentage deduction from gross rentals to arrive at a net income, or even to give any accurate indication of costs in regard to the principal outgoings likely to be found, is in most cases very unreliable. Where available, it will usually be found desirable to examine the actual costs of the outgoings for the property under consideration, and to consider them critically in the light of experience and by comparison with other similar properties.

### Repairs

The average cost of repairs will depend upon the type of construction, the planning and age of the building, and the extent of the tenants' liabilities under their covenants. Where tenants are only liable for internal decorative repairs (fair wear and tear excepted), the landlord must expect to bear practically the whole of the cost and a comparatively large allowance must be made.

Probably the soundest method is to make an estimate of the periodic cost of external and internal repairs by reference to the actual building and reduce this to a yearly allowance, checking the result by comparison with a similar building.

### Rates

In the case of existing buildings in assessment the present rates can be easily ascertained. In the comparatively few cases found today where the landlord pays the rates, a most important point is whether the terms of the tenancy permit any increase being passed on to the tenant.

### Services

There is no hard and fast rule as to the items covered by this term. The general tendency is to increase the scope and to recover from tenants not only the actual cost but also an amount for supervision of services and repairs to buildings let in suites. As stated above, the items are usually covered by a separate service charge payable as additional rent. Where this is the case, the valuer concerned should satisfy himself as to the conditions of the charges. In particular, he should ascertain if increases in costs can be passed on to tenants. He should also ascertain if the amount of the charge is sufficient to cover the cost of the services, allowing for depreciation to installations.

The following items usually fall under the heading of services. The remarks given against each item give the considerations to be kept in mind when examining the cost of individual services to ascertain if the service charge covers the cost of provision adequately.

### Heating and Domestic Hot Water

The annual cost of central heating will depend largely on the type and age of the system and the type and amount of fuel used.

### Air Conditioning Plants

Many offices now have complete air conditioning plants. As a generalisation such plants are expensive both to operate and maintain.

### Lighting

Tenants are usually responsible for the cost of lighting the offices, but the service charge will cover lighting of staircases and other parts of the building in common use. The cost will depend upon the extent of the lighting provided, including the replacement bulbs.

### Lifts

The annual cost of the maintenance of lifts will vary considerably. Much will depend upon the age and type of lift, and the amount of traffic carried.

*Cleaning*

When provided, this will depend on the number of staff employed.

Where tenants are responsible for providing their own services, e.g. heating, no account need be taken of the cost except in so far as it will affect the rental value of accommodation.

*Staff*

The cost of staff again must be ascertained in respect of each particular case, and consideration should be given as to whether or not the actual number of staff employed is appropriate for the building. In large buildings, staff may include receptionists, security personnel, and maintenance engineers.

*Insurance*

There will be a number of insurances included in the service charge. Fire insurance is the main item, but in addition there are insurances for public liability, lifts (if any), boilers and heating.

*Management*

An allowance for management is usually considered to be necessary, and this is likely to vary between 3% and 5% on the gross rents.[3]

In the case of existing buildings, it will usually be possible to ascertain the total service charge payable in terms of a unit cost related to area of occupation.

*(e) Net Income and Capital Value*

The estimates of net income made on the lines indicated above will not necessarily accord with the actual net income of any recent year. It should, however, represent what can be taken to be a fair average expectation over a considerable period.

The basis upon which it is to be capitalised will depend on the type of property, its situation and neighbourhood, the competition of other properties in the vicinity, and the valuer's estimate of the

---

[3] See *Service Charges – Law and Practice* (1997) by P. Freedman, E. Shapiro and B. Slater (Jordans) for a detailed consideration of services and service charges.

general trend of values for the type of investment under consideration. The rate of return for a high class modern building let to first-class tenants, will at the time of writing range from 5.5% in the best locations to around 7%. Similar buildings in multi-occupation will produce slightly higher yields. Older, poorer buildings and conversions will produce upwards of 10%.

Care should be taken, in analysing sale prices to arrive at yields, to ensure that the methods used are similar to those it is intended to apply to the valuation. For example, reliable evidence of years' purchase and yield can only be deduced from the result of a sale where an accurate estimate of net income is first made to compare with the sale price. In addition, special factors may affect the sale price and the yield – for example, there may be an allowance for a void period, or the existing lease may expire shortly with a probable void on a re-letting brought into account, or the property may be over-rented. These problems are addressed in Chapter 9.

In some cases an investment that has to be valued may comprise different types of property, which in themselves might be valued at differing rates per cent. For example, a property might comprise shops on the ground floor, basement restaurant, and offices in the upper parts. The rates per cent applicable separately might be 7.5% for the shops, 9% for the restaurant and 8% for the offices. Where the property is held on lease at a ground rent considerable difficulty may arise in determining how this should be apportioned so as to arrive at a fair net income in respect of each of the different types of letting. It is probably best in practice for the valuer, having capitalised the occupation rents individually at appropriate yields, to deduct the ground rent capitalised at an average of those yields.

*Example 19–3*

You are required to make a valuation for sale in the open market as an investment of a block of offices on ground and five floors over. You are furnished on receiving your instructions with the following details.

The interest in the block to be sold is leasehold, having 56 years to run at a fixed ground rent of £600 p.a.

The landlord supplies services consisting of automatic lifts, central heating to the offices (but not hot water to lavatory basins, as the tenants are responsible for this), and lighting and cleaning those parts of the building not let to tenants. The costs of these services, of repairs for which the tenants are not directly liable and of all insurances are recoverable from the tenants by a service charge which you investigate and find is reasonable. The provisions in the leases as regards the service charge have the normal escalator clause which covers rises in costs. The tenants arrange for their own office cleaning and pay for their own lighting.

All leases make tenants responsible for internal repairs and incorporate five-year upward only rent review clauses.

Ground Floor
Let to an insurance company for offices at £35,000 p.a. exclusive for 20 years from this year.

First Floor
Let at £20,000 p.a. exclusive, for 10 years – now having two years to run.

Second Floor
Let at £19,000 p.a. exclusive, for 10 years – now having two years to run.

Third Floor
Let at £21,500 p.a. exclusive, for 15 years – now having five years to run.

Fourth Floor
Let at £17,000 p.a. exclusive, for 10 years – now having seven years to run.

Top Floor
Let at £4,000 p.a. exclusive, for five years – now having two years to run.

On internal inspection the premises are found to be in good condition and to provide an acceptable modern standard of accommodation. The premises were originally constructed in the 1930s of brick and stone with slated roof and were completely refurbished internally about 10 years ago. You have been given a plan of the building from which you ascertain the following to be the net floor areas:

| Floor | Area |
|-------|------|
| Ground | 350 m$^2$ |
| First | 310 m$^2$ |
| Second | 300 m$^2$ |

SCHEDULE 1

| 1 Floor | 2 Present Rents £'s | 3 Area m² | 4 Present Rent £ per m² | 5 Tenancy Details (a) Term years | 6 (b) Time to Review (or expiry) years | 7 Estimated Rental Value per m² | 8 Estimated Rental Value £'s | Remarks |
|---|---|---|---|---|---|---|---|---|
| Ground* | 35,000 | 350 | 100.00 | 20 | 5 | 100 | 35,000 | Recent Letting |
| First | 20,000 | 310 | 64.52 | 10 | (2) | 75 | 23,250 | |
| Second | 19,000 | 300 | 63.33 | 10 | (2) | 75 | 22,500 | |
| Third* | 21,500 | 285 | 75.44 | 15 | 5 | 75 | 21,500 | Recent Review |
| Fourth | 17,000 | 285 | 59.65 | 10 | 2 | 75 | 21,375 | |
| Top | 4,000 | 100 | 40.0 | 5 | (2) | 50 | 5,000 | |
| Total | £116,500 | 1,630 m² | | | | | £128,625 | |

| Third  | 285 m$^2$ |
|--------|-----------|
| Fourth | 285 m$^2$ |
| Top    | 100 m$^2$ |

## Valuation

The first stage in the valuation is to draw up a schedule collating the information to hand. This is best done in columnar form and, in this instance, would be in accordance with Schedule 1 (opposite).

Columns 1, 2, 3, 5, and 6 present no difficulty. Column 4 is completed from Columns 2 and 3. Column 7 is completed by a careful scrutiny of the whole schedule: those floors marked * are the most recently let and form the basis for the other figures. The valuer would rely, not only on the rents obtained for this building, but also on his local knowledge of similar recent lettings of comparable accommodation. Column 8 is obtained from 3 and 7.

The actual and estimated gross income is now apparent as:

| For the next two years | £116,500  p.a. |
|------------------------|----------------|
| Then onwards           | £128,625  p.a. |

## Outgoings

All outgoings are included in the service charge except management and the ground rent.

| Gross Rents for next 2 years |        | £116,500 |          |
|------------------------------|--------|----------|----------|
| *less*                       |        |          |          |
| Management, say 5%           | £5,825 |          |          |
| Ground Rent                  | 600    | 6,425    |          |
| Net Income                   |        | £110,075 |          |
| YP 2 years at 8 & 3%         |        |          |          |
| adj. for tax @ 35%           |        | 1.194    | £131,430 |
|                              |        |          |          |
| Gross Rents after 2 years    |        | £128,625 |          |
| *Less*                       |        |          |          |
| Management at 5%             | £6,431 |          |          |
| Ground Rent                  | 600    | 7,031    |          |
| Net Income                   |        | £121,594 |          |

| | | | |
|---|---|---|---|
| YP 54 years at 8 & 3% | | | |
| adj. for tax @ 35% | 10.90 | | |
| PV £1 in 2 years at 8% | 0.857 | 9.74 | £1,135,688 |
| | | | £1,267,118 |
| Say | | | £1,300,000 |

*Note*: The above calculation has been adjusted for income tax on the sinking fund element. For discussion of methods of adjustment see Chapter 12.

# Chapter 20

# Valuation of Company Assets; Depreciated Replacement Cost; Mortgages and Loans; and Fire Insurance

## Valuation of Company Assets

### 1. The Red Book

Valuers have been required to provide valuations for company purposes for many years. These purposes include values to be incorporated in the Balance Sheet or other accounts; values of a company's property assets to be incorporated in a prospectus when the company is going public; values of a company's property assets when it is the subject of a takeover bid; and values of property bonds and the like.

The demand for such valuations has grown over recent years. Until 1974 there were few guidelines for valuers as to how to approach such valuations.

In 1974 the Royal Institution of Chartered Surveyors set up an Assets Valuation Standards Committee ("AVSC") to examine the matter in depth and in 1976 the first Guidance Notes were produced.

Until 1976, the approach tended to be either open market value or going concern value. (The going concern method of valuation was rejected by the AVSC but readers who would like to consider such an approach should see *Modern Methods of Valuation* – Sixth Edition.)

Valuers now had guidance on their approach to valuation which was subject to an ongoing review. Even so, the sharp decline in values after 1990 revealed the inadequacy of the guidance when faced with unusual and volatile market conditions.

As a result, the RICS set up a committee chaired by Michael Mallinson which produced the Mallinson Report in 1994. This made many recommendations as to the future of valuers and valuation.

One major consequence was the publication of the *Appraisal and Valuation Manual* in 1995, known universally as the Red Book. The

Book is divided into two principal parts, Part I containing Statements of Practice, and Part 2 containing Guidance Notes. A commentary follows on the contents of the Book but the reader is advised to read it – and, indeed, if a member of the RICS or ISVA or IRRV the reader is required to have full knowledge of its contents under the bye-laws of the respective bodies, since a Statement of Practice must be followed unless exceptional circumstances justify a departure therefrom.

It is likely that, in time, the European standards will be adopted, replacing the Red Book with a Blue Book. This in turn will probably be superseded by internationally approved standards operated on a worldwide basis.

The purpose of the Red Book is to set out what might be described as 'best practice' to be followed by valuers in the preparation and production of a valuation. It does not pretend to provide guidance on methods of valuation to be adopted in any particular circumstance (which is the role of this text book) although it does comment on various situations where the valuer is led towards the only obvious method to apply. Detailed commentary on valuation methods is contained in separate Information Papers published by the RICS, such as one on Commercial Investment Property referred to in Chapter 9.

The Statements of Practice in Part I cover the whole process, from receiving instructions, terms of employment, identifying the purpose of the valuation, assembling data required to prepare the valuation on to the final production of the valuation report and what it should contain. It also covers specific situations such as valuations for mortgage purposes, both residential and commercial, for Property Funds, for Insurance Companies and other cases.

As to valuations themselves, it does identify several types of value that can be determined by valuers and also which type or types are most appropriate for specified purposes. The principal types of value identified by the Red Book are Open Market Value, Market Value, and Existing Use Value.

### (i)  Open Market Value (OMV)

Open Market Value means the best price at which the sale of an interest in a property might reasonably be expected to have been completed unconditionally for cash consideration on the date of valuation assuming:

(a) a willing seller;
(b) that, prior to the date of valuation, there had been a reasonable period (having regard to the nature of the property and the state of the market) for the proper marketing of the interest, for the agreement of the price and terms and for the completion of the sale;
(c) that the state of the market, level of values and other circumstances were, on any earlier assumed date of exchange of contracts, the same as on the date of valuation;
(d) that no account is taken of any additional bid by a prospective purchaser with a special interest; and
(e) that both parties to the transaction had acted knowledgeably, prudently and without compulsion.

### (ii) Market Value (MV)

This is a definition derived from the International Valuation Standards Committee. Notwithstanding the different wording, it is regarded as having the same meaning as Open Market Value. The definition is:

> The estimated amount for which an asset should exchange on the date of valuation between a willing buyer and a willing seller in an arm's-length transaction after proper marketing wherein the parties had each acted knowledgeably, prudently and without compulsion.

Open market value (market value) is the most commonly adopted type of value, as is apparent throughout this book. There are other definitions of market value in various statutes, particularly tax and compensation statutes, which must be applied. However, they are generally interpreted to produce the same value as that derived from the application of 'open market value'.

### (iii) Existing Use Value (EUV)

Where a valuation is required for incorporation in a Balance Sheet or other accounts, the valuation must assume the continuation of the business. Hence, alternative uses or development which would necessarily interfere with the business activities must be ignored. The determination of existing use value will achieve this goal. The definition is the same as for open market but subject to further assumptions which preclude consideration of non-business activities. These are:

(a)  the property can be used for the foreseeable future only for the existing use; and
(b)  that vacant possession is provided on completion of the sale of all parts of the property occupied by the business.

By comparison, open market value which does reflect the possibility of other activities is sometimes termed alternative use value.

A valuation on an existing use value basis does not ignore any potential for development so long as it is compatible with the existing use. So if a valuation is required of an industrial unit occupied by a manufacturing company and there is potential to extend the unit, existing use value would be the value of the industrial unit applying general industrial values plus the development value of the site for the extension.

Where there is a significant difference between OMV and EUV it would be prudent to provide a note to this effect because it might be relevant to consider realising this difference by moving the operation of the company to a different location.

Where the property is unusual it may be appropriate to prepare a valuation on the basis of its depreciated replacement cost. This is considered in detail in Section 2.

The Red Book also identifies other types of value including estimated realisation price which is considered in Section 3.

## Depreciated Replacement Cost[1]

### 1.  Cost-Based Valuation Methods Generally

Reference is made to a number of the cases where cost of construction is the principal or a main factor in the valuation e.g. residual valuations and development appraisals (Chapter 11), and the Contractor's Basis in rating (Chapter 21). In the UK these arise

---

[1] A more detailed treatment of this subject can be found in *The Valuation of Public Sector Specialised Properties* by W. Britton and O. Connellan (The Centre of Research in the Built Environment, University of Glamorgan CPD Study Pack, 1997) and *Depreciated Replacement Cost* by W. Britton and O. Connellan (College of Estate Management CPD Study Pack, 1994).

mainly in no-market situations, that is where there is no actual market for the particular type of property, where the valuation is being performed in a hypothetical market or where the market cannot provide reliable direct guidance. In the USA, cost-based methods have been used for many years (originally mainly in order to support the housing market in the 1920s depression years) in parallel with market-based methods, but disillusionment with this practice is now widespread and appraisers are moving towards limiting their use to no-market situations.

Much time is frequently spent discussing whether or not a calculation based on cost can be described as a "valuation".[2] In most cases such a calculation could not be described as a market valuation but, in so far as it is intended to indicate the worth of a property to an individual or organisation, there seems no reason for it not to described as a valuation.

## 2. The Origins of Depreciated Replacement Cost

Depreciated Replacement Cost (DRC) appears to have been devised as a basis of valuation by the Assets Valuation Standards Committee of the RICS and promulgated in the first edition of the Red Book.[3] Neither the Red Book nor its companion, the White Book,[4] in their original versions gave guidance on the methods of valuation appropriate for determining value on a DRC basis[5] and it is perhaps not surprising that the profession fell back on the only cost-based method of valuation of any standing, the Contractor's Basis of rating, despite the swingeing criticisms of the method made in many quarters, including the Lands Tribunal. The latest edition of the Red Book[6] consolidates the Red and White Books. It is intended that separate guidance on method should be published but nothing has so far been provided on DRC.

In 1989 Kingston Polytechnic was commissioned by the RICS to research the Contractor's Basis of valuation as part of a programme of research into valuation methods started in 1975.

---

[2] See, for example *The Mallinson Report* (RICS, 1994).

[3] *Statements of Asset Valuation Practice and Guidance Notes* (RICS, 1974).

[4] *Manual of Valuation – Guidance Notes* (RICS, 1980).

[5] There were references to methods in a subsequent edition of the White Book but they were not particularly helpful.

[6] *RICS Appraisal and Valuation Manual* (RICS, 1995).

The original brief was to identify the present agreed and accepted method of valuation and to consider to what extent it was deficient in any respect but this brief was too narrow and the Research Report[7] published in 1991 addressed cost-based valuations generally and the DRC basis and methods of arriving at DRC in particular.

### 3. The DRC Basis of Valuation

The Red Book defines DRC as:

> The aggregate amount of the value of the land for the existing use or a notional replacement site in the same locality, and the gross replacement cost of the buildings and other site works, from which appropriate deductions may then be made to allow for the age, condition, economic or functional obsolescence and environmental factors etc: all of these might result in the existing property being worth less to the undertaking in occupation than would a new replacement.

It also requires all DRC valuations to be qualified as "subject to the adequate potential profitability of the business compared with the value of the total assets employed" or in the case of properties in public ownership "subject to the prospect and viability of the continuance of the occupation and use". DRC valuations of assets are mainly required for incorporation into company accounts in the private sector and for inclusion in financial statements and to provide a basis for the assessment of capital charges in the public sector.

### 4. Types of Property Involved

As was said above, the cost approach is particularly used in the UK for properties for which there is no market and therefore no evidence either of rents or capital values. It requires the valuer to assume that the capital cost of buying a site and the capital cost of erecting the buildings which are on it are, together, a guide to the value of the property as a whole. Properties for which there is no market are referred to as Specialised Properties and are defined in

---

[7] *The Cost Approach to Valuation* by W. Britton, O. Connellan and M. Crofts (RICS and Kingston Polytechnic, 1991).

the Red Book as "those which, due to their specialised nature are rarely, if ever, sold on the open market for single occupation for a continuance of their existing use, except as part of a sale of the business in occupation". In the private sector the most common examples are heavy industrial plants such as oil refineries, steelworks and chemical works but in the public sector the majority of buildings fall into this category including schools, hospitals, fire stations, police stations and museums.

## 5. Methods of Valuation

There are basically four methods of arriving at value on a DRC basis but all methods contain three elements in common:

(1) the current value of the land in its existing use;
(2) the current gross replacement cost (GRC) of the buildings; and
(3) the appropriate deductions from gross replacement cost for all types of obsolescence, to give net current replacement cost (NCRC).

The Red Book contains much detailed direction on land valuation but, in the majority of cases, it will be possible to value the land on the basis of yardsticks derived from market evidence. The general assumption is that the land value will not depreciate over time but the Red Book does envisage the possibility of such depreciation in what are assumed to be exceptional circumstances.

The GRC of the buildings is based on the yardstick of the current cost of erecting the buildings and site works with the normal plussages including fees and finance charges. Where buildings contain elements which are of no value to the occupier, such as surplus floor space or excessive ceiling heights, these may be excluded and in some cases the cost of erecting a modern substitute may be appropriate.

The main differences between the four methods of valuation arise in the methods of depreciating GRC to allow for all types of obsolescence. In the case of newly erected buildings it is probable that GRC and NCRC will be the same. In other cases, with the possible exception of protected buildings (which include listed buildings) where value does not decrease over time, the value of the existing building will be less than the current cost of its replacement due to economic and functional obsolescence and environmental factors.

Although the valuer is working in a no-market situation, the market is not entirely ignored and the Red Book requires that, in assessing depreciation, regard must be had "to the relative decline in value of buildings of similar age and type for which there is a market when compared with a new building".

The values referred to so far have been capital values but it is frequently necessary also to provide figures of annual value. An annual figure in this context is referred to as an asset rent which, like any other rent but explicitly, contains two elements, a charge for the use of capital and, if that capital is wasting, a charge for depreciation. An asset rent must be distinguished from a capital charge which is an accountancy concept.

## 6. Estimated Percentage Depreciation (EPD) Method of Valuation

This appears to derive from the Contractor's Basis and GRC is reduced to NCRC by way of either a single percentage deduction, or a series of percentages attributable to different causes of depreciation. This method relies purely on the valuer's experience of previous cases and makes no attempt at logical justification nor expressly revealed methodology. An asset rent can be derived from a straight percentage of DRC, that is building NCRC plus site value.

## 7. Straight-Line Depreciation (SLD) Method of Valuation

SLD is a technique for reflecting the wearing out of an asset from new until it has a negligible value. This method was adopted by the Inland Revenue Valuation Office Agency for the depreciation of hospital assets in its first valuation of the NHS Estate in 1988 but it is understood that a form of S curve has been adopted for the latest revaluation. At its simplest, this method assumes that an asset depreciates at a constant rate from new to the end of its life. Thus an asset with a life of 50 years wears out at a constant annual rate of 2% from its cost or value at the beginning of its life to nil at the end. To use the method, estimates of the total life of an asset must be made but the valuer should bear in mind that such estimates are very unlikely to prove realistic and the weight given to them should take cognisance of this. Revaluations at not more than five-yearly intervals, which are required for local authorities, seem likely to become the norm which should assist in due time in overcoming inaccurate life estimates, provided they do not go the way of the quinquennial revaluations of rating of the past.

The SLD method requires the valuer to establish the age of the property at the date of valuation and to estimate the remaining useful life. Added together, these figures give the total life of the building. The next step towards calculating the accumulated depreciation at the date of valuation is to determine the average annual percentage rate of depreciation which is simply 100/total life. Thus, with a building that is 10 years old and with an estimated 15 years of remaining life (total life 25 years) the average annual percentage rate of depreciation is (100/25), i.e. 4%. Applying this annual rate to a 10-year old building therefore gives an accumulated depreciation of 10 years @ 4% = 40%. This accumulated depreciation is deducted from GRC to give NCRC at the valuation date (10th year).

Table 1 shows the decreasing value on a straight-line basis of a building over 25 years, and the 6% annual return plus annual depreciation charge expected year by year by way of an asset rent. The cash flow or asset rent affords an internal rate of return of 6% weighed against the capital outlay (GRC). To interpret the table, it is assumed this assessment is done at the end of the 10th year, i.e. the building has an age of 10 years, and the estimate of future life is 15 years. The required NCRC is, therefore:

| | |
|---|---|
| GRC | £2,500,000 |

Annual Depreciation Rate
(100/25) = 4%

| | |
|---|---|
| *deduct* Accumulated Depreciation at 10th year = 10 years @ 4% = 40% = | <u>£1,000,000</u> |
| NCRC at 10th year | £1,500,000 |

The asset rent for the following year (11th year) is found by taking the sum of:

| | |
|---|---|
| £1,500,000 (NRC) @ 6% = | £90,000 |
| *plus* | |
| £1,500,000/15 yrs (future life) = | <u>£100,000</u> |
| Asset Rent | <u>£190,000</u> |

Note that the Table relates to the *building only*, i.e. no site value is included.

Table 1. Straight-line depreciation method

| Year | Cash flow asset rent £ | 6% NCRC £ | Depreciation £ | NCRC £ | Year |
|------|------|------|------|------|------|
| 0 | | | | 2500000 | 0 |
| 1 | 250000 | 150000 | 100000 | 2400000 | 1 |
| 2 | 244000 | 144000 | 100000 | 2300000 | 2 |
| 3 | 238000 | 138000 | 100000 | 2200000 | 3 |
| 4 | 232000 | 132000 | 100000 | 2100000 | 4 |
| 5 | 226000 | 126000 | 100000 | 2000000 | 5 |
| 6 | 220000 | 120000 | 100000 | 1900000 | 6 |
| 7 | 214000 | 114000 | 100000 | 1800000 | 7 |
| 8 | 208000 | 108000 | 100000 | 1700000 | 8 |
| 9 | 202000 | 102000 | 100000 | 1600000 | 9 |
| 10 | 196000 | 96000 | 100000 | 1500000 | 10 |
| 11 | 190000 | 90000 | 100000 | 1400000 | 11 |
| 12 | 184000 | 84000 | 100000 | 1300000 | 12 |
| 13 | 178000 | 78000 | 100000 | 1200000 | 13 |
| 14 | 172000 | 72000 | 100000 | 1100000 | 14 |
| 15 | 166000 | 66000 | 100000 | 1000000 | 15 |
| 16 | 160000 | 60000 | 100000 | 900000 | 16 |
| 17 | 154000 | 54000 | 100000 | 800000 | 17 |
| 18 | 148000 | 48000 | 100000 | 700000 | 18 |
| 19 | 142000 | 42000 | 100000 | 600000 | 19 |
| 20 | 136000 | 36000 | 100000 | 500000 | 20 |
| 21 | 130000 | 30000 | 100000 | 400000 | 21 |
| 22 | 124000 | 24000 | 100000 | 300000 | 22 |
| 23 | 118000 | 18000 | 100000 | 200000 | 23 |
| 24 | 112000 | 12000 | 100000 | 100000 | 24 |
| 25 | 106000 | 6000 | 100000 | 0 | 25 |

Apart from the criticism that it is wildly improbable that depreciation follows a straight-line, this method also suffers from the defect that the derived asset rent is "driven" from the yearly depreciation amount which is included every year at the same static figure, plus a percentage return on the decreasing NCRC. Thus, the valuer can exercise no judgement on fluctuations of the asset rent during the life-time of the asset.

It should be noted that the EPD and SLD methods are not mutually exclusive and some valuers use combinations of these methods to arrive at answers which satisfy their perceptions of value.

## 8. Discounted Asset Rent (DAR) Method of Valuation

This is a method first propounded in the Research Report *The Cost Approach to Valuation* referred to earlier and developed subsequently by two of the authors of the Report, Britton and Connellan, in co-operation with the now defunct Mid Glamorgan County Council into a user-friendly inspection and valuation computer program based on a shell expert system.

DAR is essentially a spreadsheet approach and the fixed inputs required for the calculation are:

(1) *The required rate of return on the invested capital* – essentially a choice for the organisation but public authorities tend to adopt the current Treasury borrowing rate.
(2) *The opening capital cost/value at year 0 (the beginning of the asset life)* – GRC plus the existing use value of the site.
(3) *The closing capital cost/value at the end of the building life* – existing use value of the site.
(4) *The closing asset rent at the end of the asset life* – required rate of return on the existing use value of the site. Usually the same rate as (1).
(5) *The age and estimated future life of the asset* – subject to the same considerations and qualifications as were noted for the SLD method.
(6) *The pattern of descending asset rents over the life of the asset* – this is at the valuer's discretion.

The cash flow or asset rents must be sufficient in total to satisfy the year by year requirements of an annual return on capital invested plus an annual amount for depreciation and the DRC of the asset descends over its life to the determined closing value at the end. If a required pattern is input into the cash flow, the various equations can eventually be solved through the spread-sheet, so that all values will correlate year by year over the life.

DAR has much in common with DCF in that the present value is determined by discounting future income flows. However, it was acknowledged by its authors to be subject to further research and that research (carried out by Connellan) has resulted in a further method of DRC valuation which is described next.

Table 2. S Curve Rental Method

S Curve valuation table over total useful life of building (at constant prices)

| Years | RV bldg | Building GRC bldg | Yield | Building plus site Capital Value bldg | Asset Rent | Capital Value | Years |
|---|---|---|---|---|---|---|---|
| 0 | £27,466 | £432,919 | 6.00% | £432,919 | £31,966 | £507,919 | 0 |
| 1 | £27,326 | | 6.00% | £429,227 | £31,826 | £504,227 | 1 |
| 2 | £27,143 | | 6.00% | £424,788 | £31,643 | £499,788 | 2 |
| 3 | £26,910 | | 6.00% | £419,501 | £31,410 | £494,501 | 3 |
| 4 | £26,622 | | 6.00% | £413,295 | £31,122 | £488,295 | 4 |
| 5 | £26,277 | | 6.00% | £406,125 | £30,777 | £481,125 | 5 |
| 6 | £25,871 | | 6.00% | £397,973 | £30,371 | £472,973 | 6 |
| 7 | £25,404 | | 6.00% | £388,842 | £29,904 | £463,842 | 7 |
| 8 | £24,878 | | 6.00% | £378,754 | £29,378 | £453,754 | 8 |
| 9 | £24,293 | | 6.00% | £367,750 | £28,793 | £442,750 | 9 |
| 10 | £23,653 | | 6.00% | £355,886 | £28,153 | £430,886 | 10 |
| 11 | £22,960 | | 6.00% | £343,231 | £27,460 | £418,231 | 11 |
| 12 | £22,219 | | 6.00% | £329,865 | £26,719 | £404,865 | 12 |
| 13 | £21,435 | | 6.00% | £315,876 | £25,935 | £390,876 | 13 |
| 14 | £20,612 | | 6.00% | £301,362 | £25,112 | £376,362 | 14 |
| 15 | £19,756 | | 6.00% | £286,422 | £24,256 | £361,422 | 15 |
| 16 | £18,873 | | 6.00% | £271,163 | £23,373 | £346,163 | 16 |
| 17 | £17,968 | | 6.00% | £255,691 | £22,468 | £330,691 | 17 |
| 18 | £17,048 | | 6.00% | £240,112 | £21,548 | £315,112 | 18 |
| 19 | £16,120 | | 6.00% | £224,532 | £20,620 | £299,532 | 19 |
| 20 | £15,188 | | 6.00% | £209,054 | £19,688 | £284,054 | 20 |
| 21 | £14,258 | | 6.00% | £193,776 | £18,758 | £268,776 | 21 |
| 22 | £13,337 | | 6.00% | £178,794 | £17,837 | £253,794 | 22 |
| 23 | £12,429 | | 6.00% | £164,193 | £16,929 | £239,193 | 23 |
| 24 | £11,540 | | 6.00% | £150,053 | £16,040 | £225,053 | 24 |
| 25 | £10,674 | | 6.00% | £136,448 | £15,174 | £211,448 | 25 |
| 26 | £9,836 | | 6.00% | £123,439 | £14,336 | £198,439 | 26 |
| 27 | £9,028 | | 6.00% | £111,081 | £13,528 | £186,081 | 27 |
| 28 | £8,256 | | 6.00% | £99,416 | £12,756 | £174,416 | 28 |
| 29 | £7,521 | | 6.00% | £88,478 | £12,021 | £163,478 | 29 |
| 30 | £6,826 | | 6.00% | £78,291 | £11,326 | £153,291 | 30 |
| 31 | £6,172 | | 6.00% | £68,865 | £10,672 | £143,865 | 31 |
| 32 | £5,560 | | 6.00% | £60,204 | £10,060 | £135,204 | 32 |
| 33 | £4,992 | | 6.00% | £52,298 | £9,492 | £127,298 | 33 |
| 34 | £4,466 | | 6.00% | £45,132 | £8,966 | £120,132 | 34 |
| 35 | £3,982 | | 6.00% | £38,678 | £8,482 | £113,678 | 35 |

| Years | RV bldg | GRC bldg | Building Capital Yield | Asset Value bldg | Building plus site Capital Rent | Value | Years |
|---|---|---|---|---|---|---|---|
| 36 | £3,540 | | 6.00% | £32,903 | £8,040 | £107,903 | 36 |
| 37 | £3,136 | | 6.00% | £27,766 | £7,636 | £102,766 | 37 |
| 38 | £2,770 | | 6.00% | £23,221 | £7,270 | £98,221 | 38 |
| 39 | £2,437 | | 6.00% | £19,219 | £6,937 | £94,219 | 39 |
| 40 | £2,134 | | 6.00% | £15,709 | £6,634 | £90,709 | 40 |
| 41 | £1,858 | | 6.00% | £12,641 | £6,358 | £87,641 | 41 |
| 42 | £1,605 | | 6.00% | £9,965 | £6,105 | £84,965 | 42 |
| 43 | £1,368 | | 6.00% | £7,638 | £5,868 | £82,638 | 43 |
| 44 | £1,144 | | 6.00% | £5,627 | £5,644 | £80,627 | 44 |
| 45 | £927 | | 6.00% | £3,906 | £5,427 | £78,906 | 45 |
| 46 | £712 | | 6.00% | £2,466 | £5,212 | £77,466 | 46 |
| 47 | £492 | | 6.00% | £1,314 | £4,992 | £76,314 | 47 |
| 48 | £261 | | 6.00% | £479 | £4,761 | £75,479 | 48 |
| 49 | £14 | | 6.00% | £14 | £4,514 | £75,014 | 49 |
| 50 | £0 | | 6.00% | £0 | £4,500 | £75,000 | 50 |

## 9. S Curve Rental Method of Valuation[8]

Whereas SLD uses a simplistic age/life fraction applied to GRC, the S Curve Rental method approaches the problem in terms of depreciating the rental value over the life of the building and capitalising that rent at the valuation date at an appropriate investment yield (or cap rate) for the unexpired useful life.

The opening rental value is derived from the GRC which is turned into an annual equivalent by dividing it by the years' purchase for the total useful life of the building at an appropriate investment yield. This process then gives the opening rental (termed 100% good) for a new building. This 100% good figure falls away and, assuming constant prices, will approach zero when the building reaches the end of its useful life.

The S Curve exercise involved a research project over five regional seminars in 1993/4 in the UK when valuers were asked to

---

[8] Copyright of the S Curve Rental Method is owned by Owen Connellan.

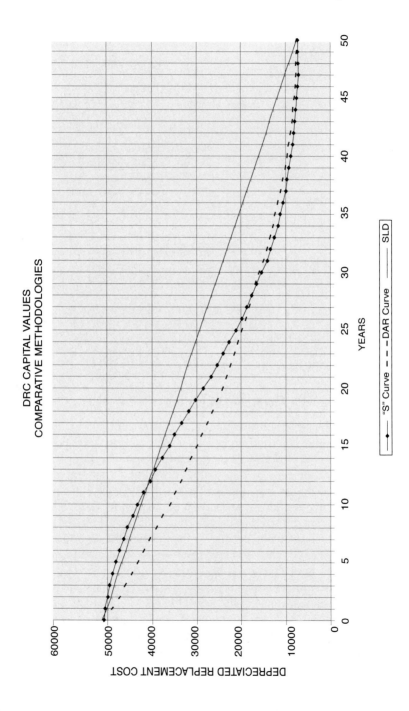

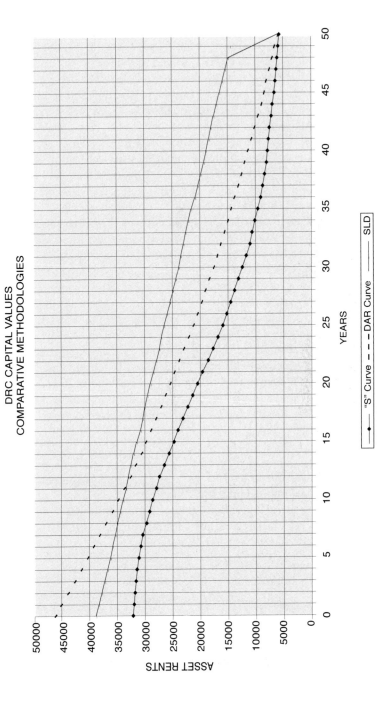

DRC CAPITAL VALUES
COMPARATIVE METHODOLOGIES

record their perceptions of the falling away of value over time. The example adopted was a specialised building with a total useful life of 60 years and each valuer recorded a "percentage good" rental value at every fifth year interval. In all, 389 valuers took part in the project and their results were data sorted and measures of central tendency for each fifth year interval were calculated in terms of the mean, median and mode. Each of these calculations showed "S" type patterns but the median was selected as the basis for the method. The valuers were required to assume that the opening rental for the combined building plus site was 100% good and that the closing rental (in the 60th year) was 20% good – the latter figure representing the site value element. The measures of central tendency were thus derived within the range of 100% to 20%, assuming constant prices. The application of these measures to the *building element* only can be made by a simple arithmetical adjustment for use in the S Curve method of valuation.

Having thus established the "S" type pattern over the life of a 60-year building it is possible to adapt the same pattern mathematically for use with buildings of different lives. Such a pattern can be applied to the opening rental, calculated as described above, and from this anchor point decreasing rental values follow that pattern in a downward "S" curve (all at constant prices as at the date of valuation). The relevant rental value (for the established age of the building) is then capitalised at the appropriate yield for the remaining life of the building as at that valuation date and the site value is then added to give the N.C.R.C.

*Example 20–1*

Taking the example of a building with a GRC of £432,919 and a site value of £75,000 with a total building life of 50 years and valuing at the 25th year using a yield of 6%, the following figures emerge for an S Curve calculation and, for comparison, the figures for DAR, SLD and EPD calculations are also shown:

|         | DRC       | Asset Rent |
|---------|-----------|------------|
| S Curve | £211,448  | £15,174    |
| DAR     | £198,249  | £20,581    |
| SLD     | £291,459  | £26,146    |
| EPD     | £233,719  | £20,633    |

Note that the EPD figure is derived from averaging the results from the other three methods but would require a percentage deduction from GRC of approximately 63%, i.e. GRC £432,919 less depreciation £274,200 plus site value equals £233,719.

The S Curve results can be checked from Table 2 and a comparison of the depreciation curves for SLD, DAR and the S Curve is shown on the two graphs.

## Mortgages and Loans

### 1. Borrowing against Property Assets

It is the nature of property that an interest in property tends to have a high value. As a result it is rare for a purchaser to be able to purchase an interest out of cashflow or accumulated cash savings. This is equally true at the top commercial end of property where a large property or a portfolio of properties may cost hundreds of millions of pounds as it is for the small residential property costing some thousands of pounds being purchased by a first time buyer. In all such cases the purchaser will borrow money to effect the purchase.

For the same reason the owner of a property who is seeking capital may prefer to borrow money, offering the interest in the property as a security, rather than to sell the interest which may realise far more than is required.

The largest market for such loan transactions by numbers is that for mortgages to finance the purchase of residential properties. The majority of lenders are either building societies or banks (frequently banks which were formerly building societies) and the loans take the form of mortgages.

A similar approach can be adopted for commercial property. However, when the sums become large various other forms of lending may be adopted although ultimately it is the underlying value of the asset which is the key element. (A loan made on the strength of covenant of the borrower is not a property loan.)

When any lender provides a loan secured against a property asset, a valuation of the property asset will be required. Before considering the approach to valuations for loan purposes it is helpful to be clear as to the nature of a mortgage and the rights and duties of the parties.

## 2. Nature of a Mortgage

A mortgage of freehold or leasehold property is a transaction whereby one party, the mortgagor, grants an interest in his property to another party, the mortgagee, as security for a loan.

The transaction is effected by means of a mortgage deed in which the mortgagor usually agrees to pay interest on the loan at a given rate per cent, and may also enter into express covenants as to the repair and insurance of the property. In some cases, the mortgage deed provides for periodical repayments of capital as well as interest (commonly in the case of a house mortgage).

The mortgagor retains the right to recover his property freed from the charge created by the mortgage deed on repayment of the amount due to the mortgagee. This is known as his "equity of redemption".

Since 1925 a legal mortgage of freeholds can only be made either (i) by the grant of a lease to the mortgagee for a long term of years, usually 3,000, with a provision for cesser on redemption, or (ii) by a charge expressed to be by way of legal mortgage.[9]

A legal mortgage of leaseholds can only be made either (i) by a sub-lease to the mortgagee of the whole term, less the last day or days, with a provision for cesser on redemption, or (ii) by a charge by way of legal mortgage.[10]

What is known as an "equitable mortgage" of land may be effected without a mortgage deed if it takes the form of (i) a signed written agreement acknowledging the loan and either stating that Specified Property is security for its repayment or promising to execute a legal mortgage if required or (ii) any mortgage of an equitable interest in land. A verbal agreement accompanied by deposit of the title deeds of the property can no longer suffice, since the Law of Property (Miscellaneous Provisions) Act 1989 – see *United Bank of Kuwait plc v Sahib*.[11]

As a general rule, a mortgage is a sound form of investment offering reasonable security and a fair rate of interest and, although

---

[9] Law of Property Act 1925, Sections 85 and 87.

[10] *Ibid.*, Section 86. A charge by way of legal mortgage places the mortgagee in the same position as though he had been granted a lease or sublease.

[11] [1996] 3 All ER 215. Such an agreement is a contract for a "disposition of an interest in land", and Section 2 of the 1989 Act provides that it must be *in writing* if it is to be valid.

the mortgage deed probably stipulates for repayment at the end of six months, it is usual for the loan to continue for a very much longer period.

So long as the mortgagor pays the interest regularly and observes the covenants of the mortgage deed, the mortgagee will usually be content to leave him in possession and control of the property. But if the interest is falling into arrears or the mortgagor is unable to meet a demand for repayment of the loan, the mortgagee must take steps to protect his security.

Property is also offered as a security for other types of lending, particularly development property where the sums borrowed are used to pay for the development and are repaid on successful completion and disposal of the development. It may also be used by companies to raise funding for expansion of the company's business generally, the loan being a secured loan or a debenture on the company's assets. Very large sums of money are raised in this way.

## 3. The Mortgagee's Security

The mortgagee's security for the money he has lent depends primarily upon the property and upon the sum it might be expected to realise if brought to sale at any time. His security for payment of interest on the loan at the agreed rate depends upon the net income the property is capable of producing in the case of commercial property which is commonly let, such as shops or offices. In the case of residential, the mortgagee typically takes account of the mortgagor's income when assessing the ability to pay. The underlying net rental value may be disproportionately low where the asset is valued on a vacant possession basis.

A common advance by way of mortgage is around two-thirds the estimated fair market value of the property, thus leaving the mortgagee a one-third margin of safety. Trustees do not usually advance more than two-thirds of a valuation of the property made by a skilled valuer for loans within the Trustee Act 1925. Other investors, such as building societies and banks, often make larger advances, particularly if some form of collateral security is offered or where provision is made for repayment of capital by instalments over a certain period. On the other hand, when there is a risk of future depreciation in value, an advance of three-fifths, or even one-half may be more satisfactory than the usual two-thirds. In all cases the mortgagee, or his adviser, should consider not only the value of the property in relation to the proposed loan, but also whether the net income from the property is sufficient to provide interest at the agreed rate. The strength of covenant of the borrower will also be a factor. It is the role of the mortgagee and not the valuer to determine the proportion of the value to be offered as a loan. An exception to that is in the case of loans made by trustees which are covered by the Trustee Act 1925. In such cases the "skilled valuer" employed by the trustees does have the duty to decide on a prudent level for the loan sought.

If the mortgagor defaults in payment of interest, observance of the covenants of the mortgage deed or repayment of the loan when legally demanded, the mortgagee has the following remedies against the property:

(i)   Under certain conditions[12] he may sell the mortgaged property and apply the proceeds to repayment of the loan and any arrears of interest together with the expenses of sale. Any surplus must be paid to the mortgagor.
(ii)  He may apply to the Court for a foreclosure order which will have the effect of extinguishing the mortgagor's equity of redemption.
(iii) He may at any time take personal possession of the income from the property, and after paying all necessary outgoings may apply the balance to paying interest on the mortgage debt, including any arrears. This surplus, if any, must be paid to the mortgagor or applied to reducing the mortgage debt.
(iv)  Under the same conditions as in (i) he may appoint a receiver to collect the income from the property and apply it to the purposes indicated in (iii), including payment of the receiver's commission.

It is evident that these remedies will only be fully effective where, in cases (i) and (ii), the market value of the property exceeds the amount due to the mortgagee, or where, in cases (iii) and (iv), the net income from the property, after paying all outgoings and annual charges having priority to the mortgage, is sufficient to discharge the annual interest on the loan with a margin to cover possible arrears of interest.

---

[12] Law of Property Act 1925, Section 101.

Although the property itself is the mortgagee's principal security, it is usual for the mortgage deed to include a personal covenant by the mortgagor to repay the loan. This may be reinforced by the personal guarantee of some third party, so that, in addition to the remedies mentioned above, there is that of action on the personal covenant.

The character and position of the borrower and his guarantor (if any) are therefore matters of considerable importance to the mortgagee, both as an additional security for the repayment of the loan and also as a guarantee for the regular payment of interest as it accrues due.

## 4. The Valuation

The Red Book, referred to before in Section 1 of this chapter, contains both Statements of Practice and also Guidance Notes on the valuation of property for mortgage and other loan purposes, with residential property considered separately from commercial property. In addition to this guidance a valuer needs to be aware of the several decisions on claims of negligence made against valuers which are considered later.

Any valuation will need to have regard to the requirements of the lender set out in the valuation instructions. In most cases what is required is the market value or open market value as described in Section 1. However, some lenders require the valuer to report the Estimated Realisation Price. This is defined in the Red Book as

> an opinion as to the amount of cash consideration before deduction of costs of sale which the Valuer considers, on the date of valuation, can reasonably be expected to be obtained on future completion of an unconditional sale of the interest in the subject property assuming:
>
> (a)  a willing seller;
> (b)  that completion will take place on a future date specified by the Valuer to allow a reasonable period for the proper marketing (having regard to the nature of the property and the state of the market);
> (c)  that no account is taken of any additional bid by a prospective purchaser with a special interest; and
> (d)  that both parties to the transaction will act knowledgeably, prudently and without compulsion.

The difference between this and open market value is the assumption that the property is to be sold in the future after an appropriate marketing period, instead of the assumption that the marketing period has expired with the valuation coinciding with the completion of the deal. It is in effect the answer that an agent would give to the question "if I put my property up for sale today what will I get for it?" This is not a forecast of future value except in so far as it reflects anticipated market changes during the marketing period. Given the relative lack of volativity in the property market there will normally be little or no difference between open market value and estimated realisation price, but this may not be the case for leaseholders where the lease term is short.

There are, however, certain aspects of a valuation for the purposes of a mortgage which will distinguish it from valuations for other purposes. The property is offered as a security for a loan so the valuation must reflect what can be realised if the lender will need to sell it to recoup the loan.

Hence the valuer will disregard the value of items which can be sold or removed by the borrower such as furniture and timber unless there are adequate safeguards against this happening. In the case of business premises, goodwill must be ignored.

The valuer must consider very carefully any factors such as probable action by the local authority which may unfavourably affect the value of the property in the future, particularly if it will require expenditure or will restrain the use of the property. In one negligence case the valuer ignored the absence of a fire certificate when valuing an hotel which could have led the local authority to close it down. He should fully consider any possible effects of the development plan for the area. Valuations should usually be on the basis of existing use unless planning permission has been granted for development. Development value should only be taken into account to the extent that buyers in the open market would take it into account when considering their bid for a property. Thus, for example, if a house is built on a double building plot with a tennis court occupying the second half, the market would recognise the development potential of the tennis court. The extent of any addition to EUV will depend on the certainty of obtaining planning permission. If an owner wishes to raise capital to carry out development then he will enter into a funding agreement which is different from a straightforward mortgage.

The valuer may also have to ask himself whether the existing market is unduly influenced by national or local conditions of a temporary nature which may have caused something like an artificial "boom" in prices. Any likely capital expenditure on the property, such as accrued dilapidations or the estimated cost of future development or reconstruction, must be allowed for as a deduction.

Any future element of value which is reasonably certain in its nature, such as reversion to full rental value on the expiration of an existing lease at a low rent, may properly be taken into account; but anything which is purely speculative in its nature should be disregarded. The valuer must be cautious when capitalising "full" rents and must consider whether they will be maintained or are likely to fall; this is particularly important where the property is not of a first-class type. The valuer must beware of including in his valuation elements of potential value which may never come about.

Information is sometimes tendered as to the price paid for the property by the mortgagor. If this was a recent actual or agreed purchase then it may provide a useful indication of the market value. However, the valuer will need to be sure that an excessive price was not paid or that the property was not purchased at a bargain price. This is particularly so where the valuer's opinion of value is significantly different from the price paid.

Notwithstanding the need for caution that the valuer must recognise when preparing a valuation for mortgage purposes, the valuation is not necessarily below a valuation say for advice on a sale or purchase. If there are no special factors as described then a valuation for mortgage purposes should be no different from a valuation for advice on a sale or purchase.

A lender may also require a valuer to give an opinion on "a forced sale value". This is no longer accepted terminology. It is the same as estimated realisation price but on the assumption that there is a stated time in which to achieve a sale which is less than the normal marketing period required to achieve full open market value. Clearly the shorter the time allowed to achieve a sale, the lower will be "the forced sale value". Given the probable absence of comparable sales on this basis the valuer will need to exercise his judgement as to the discount to apply to the normal market value.

## 5. Second or Subsequent Mortgages

It is possible for there to be more than one mortgage on a property, the second or subsequent mortgages being mortgages of the mortgagor's equity of redemption. Second or subsequent mortgages are also termed mezzanine finance. Provided they are all registered, the second and subsequent mortgagees will each have a claim on the property in their regular order after the first mortgagee's claim has been satisfied.

It is evident that there will be little security for such an advance unless care is taken to ensure that the total amount advanced, including the first mortgage, does not exceed what may reasonably be lent on the security of the property.

## Fire Insurance

### 1. Generally

When insuring against the loss of a property by fire, the sum insured is broadly the estimated cost of reinstatement of the property. Such reinstatement and cost assessments have traditionally been termed fire insurance valuations, and valuers regarded the preparation of such "valuations" as coming within their field of expertise.

These "valuations" frequently caused confusion to non-valuers since they are usually different from the open market value of the freehold interest, sometimes lower, other times higher.

As a result, the RICS has promoted the use of "cost of reinstatement" or "reinstatement cost assessment" and the abandonment of "valuations". Further, the work is more and more carried out by building or quantity surveyors who have detailed knowledge of the construction of buildings and the cost thereof. Where valuers still prepare cost estimates they are acting as cost advisers and not as valuers.

### 2. Cost of Reinstatement

The basis is the estimated cost of reinstating the asset damaged or destroyed to its former condition. Full reinstatement cost is determined on the basis of building costs prevailing at the time of reinstatement[13] and to this must be added the other incidental costs such as architect's fees, site clearance, loss of income to a landlord

---

[13] In *Glennifer Finance Corporation Ltd* v *Bamar Wood & Products Ltd* (1978) 37 P&CR 208, this point and the method of estimation were considered.

during the rebuilding period, cost of alternative accommodation during rebuilding in the case of a house, reductions in takings in a shop. So far as the estimation of reinstatement cost is concerned, many house insurance policies incorporate automatic updating of the sum insured on the basis of the RICS House Rebuilding Cost Index.

Although the basic proposition in relation to fire insurance cost estimates is comparatively simply put it is an area fraught with difficulties.

### 3. Effect of Under Insurance

It is essential that a building should be insured for the full current cost of re-building to whatever standard may be required at the time of re-building. In many cases a building will not be completely destroyed by fire, but if the insurance policy contains an average clause the insurers may pay only such proportion of the cost of repairs as the full cost of reinstatement bears to the insured value. Thus, if a building costing £40,000 to re-instate is insured for only £20,000 and there is damage by fire costing £5,000 to repair, the insurer may agree to bear only £2,500 of the cost. In cases such as churches, where the full cost of reinstatement results in an impossibly high premium and in any event the building once destroyed may not be re-instated, it may be possible to arrange a fire-loss policy to cover major repairs up to a particular percentage of the full cost of reinstatement.

### 4. Public Authority Requirements

It may not be realistic to assume that a building completely destroyed can be re-instated in its present form, quite apart from any new for old aspect. If, for example, a building does not meet current building regulations, the cost of providing a new building may be in excess of the cost of reinstatement. In the extreme case it may be that planning permission to rebuild would not be given and such a refusal of permission does not attract compensation. In the former case a sufficient addition should be made to the reinstatement cost to cover additional requirements but the latter case would probably be dealt with by a separate policy to cover the loss of site value.

## 5. Cost and Value

Since there is no definite relationship between cost and value it may be that a building which would cost say £150,000 to reinstate is worth on the market only say £100,000. In these circumstances, for the reasons given in (3) above, insurance cover cannot be confined to value although it may be possible to insure against the average clause being applied.

## 6. Incidental Costs

Mention was made in (2) of some of the incidental costs, such as architects' fees and loss of income which may arise. In most cases it is necessary to identify such costs separately from building costs and some may have to be the subject of separate insurance policies.

## 7. Value Added Tax on Building Costs

Complete re-building of a building destroyed by fire is treated as new work. The position regarding VAT depends on the circumstances. If the property is residential, for example, there are several possible situations. In the case of an owner occupier the owner may recover VAT on building materials under the DIY rules. If the property was tenanted the VAT position depends on whether the landlord will continue to let the replacement property on further periodic tenancies. If so, the VAT on building costs will not be recovered. On the other hand, if the intention is to sell the landlord's freehold interest in the replacement property then the building process will become a zero rated supply with no VAT on building costs. Whatever the VAT position on complete rebuilding, any repair following partial damage will attract VAT. The ability to recover VAT again depends on the circumstances.[14]

In preparing a cost estimate for fire insurance purposes, the manner in which VAT should be treated must be agreed with the insurer. As for the insured it would be prudent to establish that the policy does provide for the reimbursement of any VAT that may be incurred on building work and related fees.

---

[14] See Chapter 22.

# Chapter 21

# Valuations for Rating

## Introduction

This chapter is concerned with the practice and procedures applicable to the Rating List in respect of non-domestic rates with effect from 1 April 1990 and the Valuation List in respect of Council Tax with effect from 1 April 1993 in England and Wales. Those wishing to contrast the present legislation with that which applied until 31 March 1990 under the General Rate Act 1967 should refer to earlier editions of this book.

The law is stated as at 30 September 1998.

## Section A: Non-Domestic Properties

## 1. General

Non-domestic rates are a form of tax levied, as a general rule, upon the occupiers of property in respect of the annual value of their occupation in order to assist in defraying the expenses of government.

To ascertain the rate liability of a particular occupier, two things must be known. These are, first, the rating assessment of the occupied property expressed in terms of rateable value and, secondly, the national non-domestic rate or uniform business rate (UBR) applicable to all non-domestic hereditaments. This is fixed by central government and indexed to allow for inflation. As a result, once the rateable value is established, a business is able to budget with some degree of accuracy for its rate liabilities within the duration of a Revaluation cycle.

The making of assessments for rating purposes has, since 1950, been the responsibility of the Inland Revenue as provided originally by the Local Government Act 1948 and it is with these assessments that this chapter is primarily concerned. Whilst the main principles of procedure and rateable occupation will be considered, it is not proposed to deal here with the history and development of the rating system, with the law of rateable occupation in detail or with the minutiae of procedural details.

## 2. Procedure

The main statute governing assessment and valuation procedure is the Local Government Finance Act 1988, enlarged by a multitude of Statutory Instruments. However, this legislation maintains many of the principles of valuation and rating practice which have evolved over previous years.

In London, the Common Council of the City of London and the various London Borough Councils are the billing authorities for their respective areas. Outside London, the billing authorities are the district councils and Metropolitan councils. Before 1950, the billing authorities were responsible both for making rating assessments in the form of draft Valuation Lists, which were subsequently approved by Assessment Committees, and for the collection of rates. Since the Inland Revenue became responsible for the making of assessments in 1950, the responsibilities of billing authorities have been confined to rate collection.

The rating assessments of all rateable properties, known as "hereditaments", within a rating area are entered in the Rating List for that area. The intended procedure is for new Rating Lists to come into force every five years. The 1990 List came into force on 1 April 1990, the 1995 List on 1 April 1995 and the next List on 1 April 2000. By virtue of the New Valuation Lists (Time and Class of Hereditaments) Order 1987, the List is restricted to non-domestic properties. For the 1990 List, 1 April 1988 was the "antecedent valuation date" specified in that Order by reference to which rateable values were to be ascertained. A valuation date of 1 April 1993 applies for the 1995 Rating List and 1st April 1998 for the 2000 List. The Bayliss Report, produced by the RICS in 1997, advocated making the antecedent valuation date only one year prior to the commencement date of the new List. It was thought that this would help to reduce the risk of there being wide variations in rental values between the antecedent valuation date and the commencement date of the List, as occurred with the 1990 and 1995 Lists and may occur with the 2000 List in certain areas, and for certain types of property. The RICS proposal has a wide degree of support.

Briefly, the procedure for the preparation of a new Rating List is as follows. The Valuation Officer of the Inland Revenue sends out return forms, for completion by owners and occupiers, asking for details of such matters as the rent paid, length and date of grant of lease, repairing covenants, etc. Hereditaments are then inspected and valuations prepared. The new List is drawn up and transmitted to the billing authority, at whose offices it is available for inspection by the end of December, before coming into force on the following 1 April.

Once a List has come into force, the Valuation Officer has a statutory duty to maintain it in a correct form. This he does by altering it when any defect, such as an incorrect value in respect of an existing hereditament or the need to include a new hereditament, is brought to his attention. The Valuation Officer is required to serve notice on the ratepayer and billing authority stating the effect of any such alteration within six weeks of altering the List.

Challenges to a rateable value assessed by the Valuation Officer can only be made by certain persons and within certain time limits, having regard to certain criteria.

Only an "interested person" can make a Proposal for the alteration of a Rating List. An interested person means primarily the occupier of the property in question, any other person having a legal estate or an equitable interest entitling that person to future possession, but not a mortgagee not in possession, or anybody having a qualifying connection with either of these. A proposal can be made at any time against the assessment that first appears in a Rating List with effect from the start of the List, known as a "compiled list" entry. Otherwise the main criteria that restrict the service of Proposals are that any Proposal must be served within six months of:

- the service of a Notice by the Valuation Officer of an alteration of the List;
- a material change of circumstances;
- a Valuation Tribunal or Lands Tribunal decision that is relevant to the hereditament in respect of which the proposal is being served;
- a change of occupation provided that there has been no previous appeal on the same grounds determined by the Valuation Tribunal or the Lands Tribunal.

Exceptionally, a Proposal can be made at any time where a single assessment needs to be split because the hereditament has become occupied in parts, or vice versa, or where there has been a combination of the foregoing so as to reconstitute the occupation of two or more former hereditaments.

If the Valuation Officer cannot accept that the Proposal is well founded or a settlement cannot be agreed with the proposer, the dispute is referred to the Valuation Tribunal as an appeal.

The Valuation Tribunal is empowered to hear all aspects of an appeal and give a decision. If either the Valuation Officer or the proposer are dissatisfied with the decision, they may further appeal to the Lands Tribunal provided that they were represented at the Valuation Tribunal hearing. On questions of fact and valuation, the decision of the Lands Tribunal is final, but on matters of law, a right of appeal lies to the Court of Appeal and thence to the House of Lords.

## 3. Exemptions and Reliefs

Until 1963 only a very few persons were required to pay rates on the basis of the full assessment for their property appearing in the, then, Valuation List. The occupiers of agricultural land and buildings were (and still are) entirely exempt. Many other occupiers were relieved by way of a statutory adjustment to their assessment which reduced their rate liability.

Some order and reason for reliefs and exemptions was first introduced by the Rating and Valuation Act 1961, the provisions of which are generally incorporated in the Local Government Finance Act 1988, Schedule 5. Although a detailed analysis of exemptions and reliefs is not appropriate for inclusion in this book, classes of property that are exempt include agricultural land and buildings, fish farms, places of religious worship, parks, and property used for disabled persons. Reliefs are available to, *inter alia*, charities and similar organisations and for empty and partially empty property.

In the case of exemptions, no rateable value will appear in the Rating List; in the case of reliefs, specific relief will be given from the calculated rate bill usually by a percentage allowance determined by statute.

## 4. Elements of Rateable Occupation

Although not strictly connected with the methodology of rating valuation, this chapter would not be complete without some reference to the rateable occupier. Before any rating valuation can be undertaken, it is necessary to establish the extent of the hereditament and, to some extent, the ultimate establishment of that hereditament is a combination of the identification of both the unit of rateable occupation and the rateable occupier.

For an occupier to be rateable it is necessary to establish four essential elements. There must be:

(i)   Actual occupation.
(ii)  Exclusive occupation.
(iii) Beneficial occupation.
(iv)  The occupation must not be too transient.

## (i) Actual Occupation

In general terms, whether or not there is actual occupation will generally be self-evident. However, a number of points should be noted.

Title does not limit occupation and single occupation can extend over differing Titles. It is not necessary for there to be legal entitlement to possession for there to be rateable occupation; a squatter has been held liable for rates. Finally, there can be more than one rateable occupation in respect of the same area of land. An area of woodland might be let to A for paintball games on Monday to Friday and to B for rough shooting on Saturday and Sunday, both lettings on five-year leases. Both lettings would be rateable despite being in respect of the same area of land.

It should also be noted that where there is an intention to occupy or re-occupy at some date in the future, a person may be held to be in actual possession.[1]

## (ii) Exclusive Occupation

The occupation has to be exclusive. In cases where it would appear that a number of persons have a right of occupation, it will be necessary to ascertain who has the paramount control in relation to the land under consideration.

In *Westminster City Council* v *Southern Railway Co*[2] Victoria Railway Station was under consideration. Here various kiosks, bookstalls, etc., were let to individual traders, but the railway company exercised a degree of control over their occupations to the

---

[1] See *R* v *Melladew* [1907] 1 KB 192.
[2] [1936] AC 511.

extent that at certain hours of the day they were physically prevented from occupying their premises due to the fact that the station was closed. The House of Lords held that notwithstanding that they were denied possession at certain times of the day, the kiosks, etc., were in the occupation of the traders and their occupation was exclusive.

### (iii) Beneficial Occupation

The occupation must be of value or be of some benefit to the occupier. However, this does not necessarily mean that the occupier will make a profit.[3]

Similarly, where a local authority is under a statutory duty to provide some benefit to the community, it will still be regarded as having beneficial occupation even though clearly it operates at a loss.[4] This situation, however, must be contrasted with properties such as public parks which have been held not to be in the beneficial occupation of the local authority that holds the Title to the land or is required to maintain them.[5]

### (iv) Transience of Occupation

Transience of occupation really has two elements to it: transience in the context of time and transience in the context of the degree of use made of the land.

Out of *London County Council* v *Wilkins (VO)*,[6] concerning the rateability of contractors' site huts, a rule of thumb arose that an occupation had to exist for longer than 12 months to be rateable. However in *Dick Hampton (Earth Moving) Ltd* v *Lewis (VO)*,[7] concerning quarrying operations only lasting six months, the 12-month rule of thumb was dismissed and emphasis was placed more on the nature of the occupation and use of the land itself.

---

[3] See *London County Council* v *Erith Churchwardens and Overseers* [1893] AC 562.
[4] *Governors of the Poor of Bristol* v *Wait* (1836) A&E 1 (the subject property was a workhouse).
[5] See *Lambeth Overseers* v *LCC* [1897] AC 625.
[6] [1957] AC 362.
[7] [1975] 3 All ER 946.

In summary, therefore, once a person has been identified who can satisfy the four elements set out above in respect of an occupation of land, the valuer has then identified both the rateable occupier and the next step is to establish the rateable hereditament in respect of which the valuation has to be made.

## 5. Principles of Assessment

It is of prime importance that the extent of the hereditament to be valued is established at the outset. Without first determining the hereditament it is impossible to ensure that the correct unit of property is valued. Section 64, Local Government Finance Act 1988 maintains the definition of "hereditament" found in Section 115, General Rate Act 1967, namely "property which is or may become liable to a rate being a unit of such property which is, or would fall to be, shown as a separate item in the rating (*previously valuation*) list".

In general, for a property to be a hereditament there must be a single rateable occupier and a single geographic unit. Secondly, there are certain general principles to be observed in preparing valuations for rating purposes, the most important of which are:

(i) The hereditament being valued must be assumed to be vacant and to let. The statutory definition assumes a tenancy and it is, therefore, necessary to assume that the hereditament is available so that the bid of the hypothetical tenant on the statutory terms can be ascertained. However, a local authority scheme which would involve disturbance at a sufficiently early date might affect the bid of a hypothetical tenant from year to year.

(ii) The hereditament must be valued *rebus sic stantibus* which means in its actual existing physical state, "as it stands and as used and occupied when the assessment is made".[8] This rule does not, however, prevent the assumption of changes of a non-structural nature, e.g. a change of use, provided planning permission would be forthcoming for that change of use. However, the mode or category of occupation by the hypothetical tenant must be conceived to be the same mode or

---

[8] Lord Parmoor – *Great Western & Metropolitan Rail Companies* v *Hammersmith Assessment Committee* [1916] 1 AC 23.

category as that of the actual occupier; thus a shop must be valued as a shop rather than any particular type of shop.

The principle of *rebus sic stantibus* has been the subject of considerable judicial development over the years. Reference may be made to *Midland Bank Ltd* v *Lanham (VO)*[9] for a comprehensive résumé of the point.

(iii) Prior to *Garton* v *Hunter (VO)*,[10] if a property was let at what was plainly a rack rent, then that was the only permissible evidence. However, this case decided that "the actual rent is no criterion unless it indeed happens to be the rent the imaginary tenant might reasonably be expected to pay" and as a result other evidence of value may now be examined. This includes rents passing on comparable properties.

(iv) Assessments on comparable properties may be considered in the absence of better evidence once the basis of value of a new List has been established.

(v) A question of considerable importance is the extent of the rateable hereditament and whether separate assessments should be made in respect of different parts. The rateable unit will normally be taken as the whole of the land and buildings in the occupation of an occupier within a single curtilage. However, the Court of Appeal in *Gilbert (VO)* v *(S) Hickinbottom & Sons Ltd*,[11] ruled that a bakery and a building used for repairs essential for its efficient operation could constitute a single hereditament, although separated by a highway, as they were functionally essential to each other. Similarly in *Harris Graphics Ltd* v *Williams (VO)*[12] a proven functional connection between two geographically separate buildings on the same industrial estate was held to justify a single assessment.

It is not necessary for parts to be structurally severed to be capable of separate assessment if the parts are physically capable of separate occupation.[13] From the valuation viewpoint, separate assessments usually, but not necessarily, result in a higher total

---

[9] [1978] 1 EGLR 189.
[10] [1969] 2 QB 37.
[11] [1956] 2 QB 240.
[12] [1989] RA 211.
[13] See *Moffat (VO)* v *Venus Packaging Ltd* [1977] 2 EGLR 177.

rateable value. However, if parts are separately assessed and one part ceases to be occupied then there might be no liability for rates on that part.

## 6. Statutory Basis of Assessment

All non-domestic hereditaments are assessed to rateable value in accordance with the statutory provision set down in Schedule 6, Local Government Finance Act 1988. This states that:

> the rateable value of a non-domestic hereditament none of which consists of domestic property and none of which is exempt from local non-domestic rating shall be taken to be an amount equal to the rent at which it is estimated the hereditament might reasonably be expected to let from year to year if the tenant undertook to pay all the tenant's rates and taxes and to bear the cost of repairs and insurance and other expenses (if any) necessary to maintain the hereditament in a state to command the rent.

The "rent" is not necessarily the rent actually paid, but the rent which a hypothetical tenant might reasonably be expected to pay. Hypothetical tenants would include all possible occupiers including the actual occupier.

A tenant from "year to year" can occupy the hereditament indefinitely, having a tenancy which will probably continue but which can be determined by notice. There is therefore a reasonable prospect of the tenancy continuing and no implied lack of security of tenure which would adversely affect the expected rent.[14]

"Usual tenants' rates and taxes" include general rates, water rates and occupiers drainage rate but not Schedule A Income Tax and owner's drainage rate, which are landlords' taxes.

The cost of repairs, insurance and other expenses necessary to maintain the hereditament are deemed to be the responsibility of the tenant under the statutory definition of rateable value.

Although the question of lack of repair had been before the Courts on a number of occasions in recent years, this was usually in relation to the hypothetical landlord's liability in the context of the former Gross Value. The definition of rateable value assumes

---

[14] See *Humber Ltd* v *Jones (VO) & Rugby Rural District Council* (1960) 53 R&IT 293.

that the tenant will undertake to bear the cost of repairs and consequently the Valuation Office maintained that it could be further assumed, as was the established principle from Gross Value, that the tenant would be liable first to put the hereditament into repair. However, since rateable value is based on rents, ratepayers' surveyors argued that the extent to which the direct rental evidence reflected disrepair would determine whether or not the assessment under consideration should have an allowance for actual disrepair.

The issue was decided by the Lands Tribunal in *Benjamin (VO) v Anston Properties Ltd*[15] where the Tribunal determined that the state of repair was indeed a factor to be taken into account in assessing rateable value and reduced the rateable value relating to an office building in disrepair to a nominal £1. However, the decision presented government with something of a dilemma. It was clearly felt that the decision was correct according to the wording of the statute, since the decision was not appealed to the Court of Appeal. However, acceptance of the decision might result in a serious erosion of the countrywide rateable value base with a consequent overall drop in government revenue.

The government response was swift and sure. Shortly after the Lands Tribunal decision the local government minister Hilary Armstrong announced that the law was to be amended with retrospective effect to 1 April 1990 to ensure that assessments are made on the basis of the property being in a good state of repair.

As has already been stated, the hereditament has to be valued having regard to the level of rental values established as at the antecedent valuation date; 1 April 1988 in respect of the 1990 Rating List, 1 April 1993 in respect of the 1995 List, and 1 April 1998 in respect of the 2000 List. However, the physical circumstances of the hereditament are taken as at the date when the new Rating List comes into force (1 April 1990, 1 April 1995 and 1 April 2000 respectively), or at any other "material day", primarily the date of any alteration affecting the hereditament, as set down in The Non-Domestic Rating (Material Day for List Alterations) Regulations 1992.

In general the rating concept of rateable value is almost identical to the valuation practitioners concept of a modern commercial lease.

---

[15] [1998] 2 EGLR 147.

## 7. Rating of Unoccupied Properties

Despite the absence of a beneficial occupier, unoccupied properties are rateable under general rating law. However, although under previous legislation, billing authorities had discretion with regard to levying rates on unoccupied property, which meant ratepayers across the country suffered different treatment, unoccupied rates are now dealt with primarily by statute.

Under the provisions of the Non-Domestic Rating (Unoccupied Property) Regulations 1989 fully unoccupied property will be exempt from rates for a period of three months in general; thereafter rate liability will be at 50% of full liability. However, certain hereditaments enjoy total exemption from rates. These include "qualifying industrial hereditaments", e.g. factories, warehouses, etc.; listed buildings; hereditaments where occupation is prohibited by law; hereditaments with a Rateable Value below defined levels, currently £1,000, and certain other categories of hereditament.

Partly occupied property can also receive rate relief under the provisions of Section 44A, Local Government Finance Act 1988 where the part not occupied is only vacant short-term. This statutory provision is still applied at the discretion of the billing authority.

## 8. Methods of Assessment

There are different methods of assessment but it is vital to bear in mind that, whichever method may be used, the end product must be the rent on the basis of the statutory definition of rateable value. Although most case law relating to the statutory definition was established under the former legislation of the General Rate Act 1967 and so related to Gross Value and Net Annual Value, it is submitted that it is still relevant except where it would obviously be inapplicable, e.g. the landlords notional liability to repair under the former definition of Gross Value.

The principal valuation methods of assessment used are:

(a)  by reference to rents paid, which is usually known as the rental method;
(b)  the contractor's test;
(c)  the profits basis; and
(d)  by formula.

The choice of method to be used is not necessarily a free one. For some types of hereditament a statutory formula is laid down or a Minister is empowered to make an order laying down the method of determining the assessment. Examples of these are usually found amongst the statutory undertakers.

In cases where there is no such restriction, the first three methods are appropriate, but the general rule is that if the rental method can be used, it should be used in preference to the contractor's test or the profits test. However, all these methods are means to the same end, namely, to arrive at a rental value.

Their application to particular types of property is dealt with in Section 9 of this chapter, but a brief description of the first three methods is given here.

### (a) Valuation by Reference to Rents Paid

In many cases the rent paid for the property being valued may be the best guide to a rating assessment. However, Rateable Value can only correspond with rents actually paid provided that (i) the rent represents the fair annual value of the premises at the relevant date and (ii) the terms on which the property is let are the same as those assumed in the statutory definition. The rent paid for any particular property will almost certainly not be the true rental value in cases where property has been let for some time and values in the district have since changed.

Again, properties are often let at a low rent in consideration of the lessee paying a premium on entry, or surrendering the unexpired term of an existing lease, or undertaking improvements or alterations to the premises. In the case of premiums or such equivalent sums, the rent must be adjusted for assessment purposes by adding to it the annual equivalent of the consideration given by the tenant for the lease. This would not, however, apply if the premium was paid in respect of furniture or goodwill. A rent of business premises may also be low if a tenant's voluntary improvement under an earlier lease has been ignored under Section 34 of the Landlord and Tenant Act 1954, as amended by the Law of Property Act 1969. If the tenant has carried out non-contractual improvements, the rental value of these improvements must also be added to the rent paid.

By contrast, a rent may be high where a tenant has received inducements to take the lease such as a reverse premium or a long rent-free period.

The rent paid may not be equal to the rental value because there is a relationship between the lessor and the lessee. A typical example is where the property is occupied by a limited liability company on lease from the freeholder who is a director of that company. Rents reserved in respect of lettings between associated companies are often not equal to the proper rental value. Sale and leaseback rents may also differ from rental value.

Adjustments may be necessary in cases where the terms of the tenancy differ from those of the statutory definition of Rateable Value. A typical example is that of a shop let on a lease, under which the lessee is responsible for internal repairs only. In such a case, the rent will be higher than it would have been had the tenant been responsible for all repairs. A deduction, in respect of the repairs, must therefore be made to the rent to obtain the Rateable Value. As a rule of thumb it is common practice to deduct 5% from the reserved rent where the tenant has the responsibility for internal repairs only and 10% where there is no responsibility for repairs.

The relationship between rents in accordance with the statutory definition and lease rents has been a subject of contention. It has been held by the Courts in a number of cases that rents on leases for up to 21 years did not differ materially from rents from year to year. However, these cases are not recent ones and now it is most likely that a rent fixed for 21 years without rent review clauses would be held to differ from a rent from year to year, because the rent reserved over the earlier years of the lease would be high to balance a later expected profit rent.[16] This point is of some importance when comparing and analysing rents paid under leases which appear to be above other market evidence on account of the security of lengthy terms without review, particularly in a rising rental market.

In *Dawkins* v *Ash Brothers & Heaton Ltd*,[17] the House of Lords by a 3–2 majority decision held that the prospect of the demolition of part of a hereditament within a year was a factor which a hypothetical tenant would consider in making his rental bid. This resulted in a lower assessment but involved complex argument as to the point at which actual facts displaced the hypothetical circumstances to be considered in rating law.

---

[16] See *Humber Ltd* v *Jones (VO) & Rugby Rural District Council* (1960) 53 R&IT 293 and *Baker Britt & Co Ltd* v *Hampsher (VO)* [1976] 2 EGLR 87.
[17] [1969] 2 All ER 246.

Where the rent actually paid is considered to be above or below the prevailing rental value, or where a property is owner-occupied, a valuation must be made by comparing the property being valued with other similar properties let at true rental values. Sometimes a proper rent can be fixed by direct comparison with other similar properties. In other cases it may be necessary to compare their relative size and to use some convenient unit of comparison, such as the rent per square metre.

With certain special types of property a unit of accommodation rather than of measurement may be used, such as a figure per seat for theatres and cinemas, or per car parking space for car parks.

When using the rents paid for other similar properties as a basis of assessment regard must be had to the terms on which those properties are let, and the need to make any necessary adjustments to those rents so that they conform with the statutory definition.

Mention may also be made of the principle of "equation of rents". This is a theory based on the assumption that a tenant would pay only a fixed sum by way of rent and rates. If the rates burden increases substantially, as it did on retail property particularly in areas of Central London, as a result of the 1990 Revaluation, then the rent should fall. This, however, may be very difficult to prove. In a case concerning the assessment of a large factory following the abolition of industrial derating for the 1963 Revaluation, the Tribunal decided that there was no evidence of industrial rents falling after that Revaluation. It was thought that this issue of equation could well be of some interest in the 1990 revaluation, which was based on rental evidence as at 1 April 1988. A possible point of argument and negotiation could be whether or not those 1988 rental levels were agreed in the full recognition of the potentially significant changes in future rate burdens as a result of the forthcoming revaluation. However, the argument never really materialised, probably due to the fact that it was impossible to isolate the element of any rent theoretically affected in a rising rental market.

## (b) The Contractor's Test

This method is used in cases where rental evidence, either direct or indirect, is not available or is inconclusive. The theory behind the contractor's test is that the owners, as hypothetical tenants, in arriving at their bid would have regard to the yearly cost to them of acquiring the ownership of the hereditament by taking a rate of interest on capital cost or capital value. Examples of properties for which the contractor's basis is appropriate are public libraries, town halls, sewage disposal works, fire stations, colleges, municipal baths, schools, crematoria, specialised industrial properties and most rateable plant and machinery. The method has also been used in the case of a motor racing track where no licence was required and there was no quasi-monopoly. If a quasi-monopoly exists, as for example with a licensed hotel, then the profits basis based on available accounts would be more appropriate.

The contractor's method was described in outline in Chapter 2 and is applied by taking a percentage of the effective capital value of the land and buildings to arrive at the annual value.

There are two possible methods of determining the effective capital value of buildings for this purpose:

(i) by reference to the known or estimated cost, inclusive of fees, of reconstructing the existing building, known as the replacement cost approach. In the case of a new building, the effective capital value might well be the actual cost of construction unless there was some form of surplusage, such as excess capacity, or other disability which would justify a reduction in that cost. With an older property, however, the cost of reconstruction would have to be written down to take account of disabilities of the existing building, i.e. reflecting its age and obsolescence. Excessive embellishment and ornamentation would normally have to be ignored, as it is not something for which the hypothetical tenant would pay more rent.

(ii) by reference to the cost, including fees, of constructing a simple modern building capable of performing the functions of the existing building. This is a more refined approach known as the "simple substitute building" solution to the same problem of valuing obsolete or older properties.

In either case, the effective capital value which is required is the cost of a modern building less the necessary deductions in respect of disabilities, so that the final figure obtained represents as nearly as practicable the value of the old building with all its disadvantages. To this figure must be added the effective capital value of the land, the value being assessed having regard to the existing buildings *rebus sic stantibus* and ignoring any prospective development value.

The final stage in the valuation is the conversion of effective capital value into a rental value by the application of an appropriate percentage which theoretically could be argued to be the market borrowing rate for funds. Over the years the Lands Tribunal provided some judicial guidance on this point and a range of decapitalisation rates built up.

In *Coppin (VO)* v *East Midlands Airport Joint Committee*,[18] concerning the assessment of the East Midlands Airport, Castle Donington, the Lands Tribunal in their decision referred to the fact that both valuers agreed that 5% was the generally accepted rate to be applied to the effective capital value to arrive at Net Annual Value in the normal type of case. In *Shrewsbury School (Governors)* v *Hudd (VO)*,[19] concerning the rating of a public school, 3.5% of the cost of a substitute building, fees and site, less 70% for age and obsolescence, formed the basis of the Lands Tribunal's decision in arriving at Gross Value, while in a cemetery and crematorium case, *Gudgion (VO)* v *Croydon London Borough Council*,[20] the Lands Tribunal adopted 3% on replacement cost, less surplusage and disabilities, to arrive at Net Annual Value.

Given these variations in practice, it is perhaps not surprising that statutory decapitalisation rates were introduced for the 1990 Revaluation. The Non-Domestic Rating (Miscellaneous Provisions) (No. 2) Regulations 1989 set down rates of 4% for hospitals and educational establishments and 6% for other properties, thus effectively putting an end to the debate on the appropriate rate to use. For the 1995 Revaluation the rates were altered to 3.67% and 5.5% respectively.

To reduce areas of contention further, the 1990 List led, for the first time, to guidance notes agreed between the Valuation Office Agency and many of the various bodies that generally occupy property to which the contractors method of valuation applies.

A typical contractor's method of valuation of a college in the 1990 List might be as shown in the following example.

It is possible to incorporate an end allowance in a contractor's method valuation to reflect a disability not otherwise dealt with.

---

[18] (1970) RA 503.

[19] (1966) RA 439.

[20] (1970) RA 341.

| Accommodation | Date Built | Floor Area GIA $m^2$ | Building Cost $£/m^2$ | Obsolescence allowance | £ |
|---|---|---|---|---|---|
| **Main School "Classical"** | | | | | |
| Façade | 1893 | 1,090 | 1,000 | 50% | 545,000 |
| Standard Façade | 1893 | 2,154 | 540 | 50% | 581,580 |
| Extension | 1990 | 521 | 540 | 0% | 281,340 |
| Music Block | 1951 | 356 | 540 | 25% | 144,180 |
| Language Block | 1986 | 228 | 540 | 15% | 104,652 |
| Tennis Court | | | | | 5,000 |
| | | | | | 1,661,752 |

Factor Adjustment for building cost (location: Outer London)    1.15

   1,911,015

Site Value (Outer London: 15% of costs)    1.15

   2,197,667

Sports Field 1.5 ha @ £15,000/ha    22,500

Effective Capital Value    2,220,167

Decapitalisation Rate @ 4%    £88,806 RV

Say    £88,800 RV

For example, an end allowance for piecemeal development, i.e., the buildings spread apart thus being inefficient to operate, would be justified. However, an allowance for poor natural light within classrooms would not be justified, as this should be reflected in the obsolescence allowance.

Finally, it should not be overlooked that the hypothetical tenant's ability to pay the resultant figure of rent arrived at by the contractor's method valuation is a material factor. Once this mathematical valuation exercise has been completed, it is always necessary to "stand back" and consider whether or not the resultant answer looks sensible.

### (c) The Profits Basis

This method is also used where rental evidence is absent or inconclusive. The theory is that the hypothetical tenant would relate his rental bid to the profits he would be likely to make from the business he would conduct on the hereditament. It must be emphasised that the profits themselves are not rateable but that they are used as a basis for estimating the rent that a tenant would pay for the premises vacant and to let. Examples of properties to which the profits basis may be applied are racecourses, licensed premises, caravan sites and leisure centres including pleasure piers. The general rule is that the method is particularly applicable where there is some degree of monopoly, either statutory, as in the case of public utility undertakings, or factual, as in the foregoing examples.

The method involves the use of the accounts of the actual occupier, but if the actual occupier's management skill is below or above the normal level of competence assumed for the hypothetical tenant, then consequential adjustment to those accounts will be necessary. From the gross receipts, purchases are deducted, after they have been adjusted for variation of stock in hand at the beginning and end of the year, thus leaving the gross profit. Working expenses are then deducted. The residue, known as the divisible balance, is divided between the tenant, for his remuneration and interest on his capital invested in the business, leaving the remainder for the rates and the landlord by way of rent. The tenant's share must be sufficient to induce him to take the tenancy, irrespective of the balance for rent and rates.

Traditionally, the actual rates were not deducted as an outgoing in making the initial valuation since the year in which the valuation was being prepared was not necessarily the statutory date of valuation. However, as a result of the 1990 Rating List, and subsequent Rating Lists, having an antecedent valuation date there is an argument that the actual rates payable in that year should be deducted to arrive directly at the Rateable Value.

A profits valuation for the assessment of a licensed hotel is set out in outline opposite.

## 9. Assessment of Various Types of Property

### (a) Agricultural properties

Agricultural land and buildings are wholly exempt from rating under Schedule 5, Local Government Finance Act 1988. "Agricultural land" is defined as:

> land used as arable, meadow or pasture ground only; land used for a plantation or a wood or for the growth of saleable

## Outline Profits Valuation

|  | | £ |  |
|---|---|---|---|
|  |  | Receipts |  |
| *less* |  | Purchases | (adjusted for stock position) |
|  |  | Gross Profit |  |

*less*
Working expenses, from the trading
figures, which would be appropriate
to a tenant from year to year
occupying under the definition of
Rateable Value. The working expenses
deducted may be adjusted, if
necessary, to reflect those which the
hypothetical tenant would allow
and need not necessarily
be those actually incurred.

Net Profit / Divisible Balance

*less*
Tenant's share consisting of:

(i) interest on his capital invested
in furniture, equipment, stock
and working capital.

(ii) remuneration for running the
business including risk, etc., on
capital (an assessed figure) – or,
a percentage of tenant's capital
(or of the Divisible Balance)

Available for rent and rates

*less* Rates paid in the antecedent
valuation date year

RV

*Note*: A shortened version of the profits test is frequently used, such
as taking a percentage of gross turnover or receipts to arrive at
Rateable Value or, in the case of petrol filling stations, adjusting
throughput to arrive at Rateable Value.

underwood; land exceeding 0.10 hectare and used for the purpose of poultry farming; a market garden, nursery ground, orchard or allotment, including allotment gardens within the meaning of the Allotments Act 1922.

However, the definition does not include "land occupied together with a house as a park; gardens (other than as aforesaid); pleasure grounds; land used mainly or exclusively for purposes of sport or recreation; land used as a racecourse".

It will be noticed that the definition includes land used for a plantation or a wood or for the growth of saleable underwood, but if such land is kept or preserved mainly or exclusively for the purposes of sport or recreation or is otherwise excluded from the definition of "agricultural land" (for example, if it is "occupied together with a house as a park"), the exemption will not apply.

"Agricultural buildings" are also defined in Schedule 5 to the Act. A building is an agricultural building if it is not a dwelling and:

(a) it is occupied together with agricultural land and is used solely in connection with agricultural operations on the land, or

(b) it is or forms part of a market garden and is used solely in connection with agricultural operations at the market garden.

The test of whether a building is occupied with land is not a geographic test but rather a question of whether or not the buildings are occupied at the same time as the land, see *Handley (VO)* v *Bernard Matthews plc*.[21]

Bee-keeping and fish farming, excluding ornamental fish, are similarly exempted under Schedule 5. Certain buildings occupied by farmers co-operatives are also entitled to relief under Schedule 5.

## (b) Offices

The analysis of actual rents paid, checked by comparison with rents paid for other similar premises, will usually provide the basis of valuation. Most properties of this type will be let under a lease which makes the tenant responsible for insurance, all repairs, etc.

---

[21] [1988] RA 222.

and so may well equate to the definition of Rateable Value. However, some leases will provide for the tenant to pay towards these costs through a fixed or variable service charge. Whatever the individual arrangements, the actual terms must be adjusted as appropriate to accord with the definition of Rateable Value. This is particularly important if an analysis of the rent of the property in question is to be used as the basis of valuation for other properties.

Office rents are usually analysed to a price per square metre and the basis of measurement tends to be Net Internal Area, as defined in the RICS and ISVA Code of Measuring Practice. Factors which influence the value of office accommodation are:

(i)   external factors such as geographic location, namely central to the perceived office location core; proximity to transport connections; proximity to staff facilities, e.g. shopping, lunch and leisure opportunities.

(ii)  internal factors such as feature reception area, lifts, air conditioning, raised floors, double glazing.

*(c)  Shops*

The assessment of shops is governed mainly by consideration of the rents actually paid. In principle, if a property is let at a market rent at 1 April 1993, then that rent, adjusted if necessary to conform to the definition of Rateable Value, should form the basis of the assessment in the 1995 List. Indeed, as shops are usually let on full repairing and insuring leases, little or no further rental adjustment is usually necessary, but if the tenant is responsible for internal repairs, the rent is usually decreased by 5%. However, such a general practice may be considered too arbitrary and should not be adopted if other figures can be justified. An addition to the rent must be made for any tenant's expenditure such as the installation of a shop front, particularly in the case of a new shop where the tenant leases and pays rent for only a "shell" and covenants to complete and fit out the shop. It might be thought that the allowance for such expenditure should be obtained by decapitalising at dual rates with allowance for income tax on the sinking fund. The position is by no means clear. In *Caltex Trading and Transport Co Ltd* v *Cane (VO)*[22] the Lands Tribunal considered

---

[22] (1962) 182 EG 159.

that only simple interest on expenditure was correct, while in 1967 they agreed with the use of dual rates but excluded allowing for the effect of income tax on the sinking fund as being a factor which the tenant would not consider.[23] In the later case of *F W Woolworth & Co Ltd* v *Moore (VO)*,[24] the 1967 principle was upheld and finishing costs to the hereditament were decapitalised over a 35-year period using a 6% and 3% dual rate table (excluding income tax). The use of simple interest gives the lowest addition to the rent paid, while the use of dual rates with allowance for tax gives the highest. Any necessary adjustment for repairs not undertaken by the tenant must be applied to the rent after it has been increased by allowing for tenant's expenditure. The final figure obtained, by whatever approach, should be the rent which the hypothetical tenant would pay for the completed shop on the basis of the statutory definition.

Shops only a short distance apart can vary considerably in rental value because of vital differences in their positions. Any factor influencing pedestrian flow past the shop has its effect on value, such as proximity to "multiples", adequate and easy car parking or public transport, the position of shopping "breaks" such as intersecting roads, bus stops and traffic lights and widths of pavements and streets.

When the time comes either to analyse adjusted rents or to value a particular shop, certain recognised methods of valuation are used with the object of promoting uniformity and facilitating comparison between different properties.

A method in general use for valuing the ground floor of a shop, which is usually a very much more valuable part than the rest of the premises, is the "zoning" method. The front area of the shop is the most valuable part, as it provides the most prominent selling space, and the value per unit of area decreases as the distance away from the front of the shop increases until a point is reached beyond which any further reduction would not be sensible. The zoning method allows for this progressive decrease in value from the front to the rear.

The first step in the analysis of a rent paid for premises including a ground floor shop is to deduct the rent attributable to all ancillary accommodation, leaving that paid for the ground floor selling space only. This rent is then broken down on the assumption that

---

[23] See *F W Woolworth & Co Ltd* v *Peck (VO)* (1967) 204 EG 423.
[24] [1978] 2 EGLR 213.

the front part of the shop, back to a certain depth, is worth the maximum figure, say £x per unit of area, the next portion £x/2 per unit of area, and so on. This process, known as "halving back", is used most extensively and can be varied, but usually there will be up to three zones and a remainder.

The depth of each zone varies according to circumstances and valuers have different opinions as to the most suitable depths to be used. The choice of the depth of zone should, however, be logical and governed by the depth of the shops under review, as it might be pointless to halve back after 7 metres if the standard of shop under consideration is 10 metres. Historically, valuers took a depth of 15 feet for the first zone and 25 feet for the second and subsequent zones, but more recently it has become more common practice to divide the shop into consistent zones of 6.1 metres (20 feet) deep.

The first zone (termed Zone A) is measured backwards from the front of the shop, which is usually the rear pavement line of the demise. This zone is clearly the most valuable and may consist entirely of shop front, arcades and show cases. Subsequent zones are measured from the end of the previous zone. The benefits of a return frontage on a corner shop can be reflected by a percentage addition to the value of the shop. As far as sales space on upper or basement floors is concerned, this may be valued as a fraction of the Zone A value in comparative terms. Other ancillary space is usually taken at a unit value applied to the area concerned.

The Lands Tribunal have indicated that, in their opinion, the number and depths of the zones to be adopted is not so important, provided that the zoning method is applied correctly and consistently according to the actual depths of the shops whose rents are being devalued and that the same method is applied subsequently when valuing.[25]

When a large shop is being valued on the basis of rental information derived from smaller shops, a "quantity allowance" is sometimes made in the form of a percentage deduction from the resulting valuation of the large shop. This procedure may be justified on the grounds that there is a limited demand for large shops and that therefore a landlord may have to accept a rather lower rent per unit of area than he would for a small shop. This

---

[25] See *Marks & Spencer v Collier (VO)* (1966) 197 EG 1015.

may be valid when a large shop is in what is basically a small shop position, but it is probably invalid when it is in part of a shopping area where large chain stores have established themselves. For this reason, the Lands Tribunal rejected quantity allowances on two shops in Brighton in 1958, in assessments for the 1956 List.[26] Later, in a 1963 List reference concerning a new store built for Marks & Spencer in Peterborough and their old store, which they had sold to Boots, the Lands Tribunal refused quantity allowances on the ground that the limited demand and limited supply of large shops was in balance.[27] Furthermore, in *Trevail (VO) v C & A Modes Ltd* and *Trevail (VO) v Marks & Spencer Ltd*,[28] the Lands Tribunal expressed doubts as to whether the zoning method was not being stretched beyond its capabilities in the valuation of these large walk-round stores. The problems of valuing large stores for rating purposes was comprehensively considered in the Lands Tribunal's decision on the John Lewis's store in Oxford Street.[29]

An alternative approach to quantity is to value such large stores on an overall pricing method, but this can only be supported where there is appropriate and reliable rental evidence which has been tested in the market.

Apart from any possible quantity allowance, a disability allowance may be an appropriate deduction in the case of shops with disadvantages such as obsolete layout, inconvenient steps, changes in floor level, excessive or inadequate height, lack of rear access, inadequacy of toilets or large and obstructive structural columns, if the valuation basis has been derived from rental evidence on shops without such disabilities.

In order to be able to check an assessment of a particular shop to see if it is fair by comparison with the assessments of other neighbouring shops, it is essential to be at least reasonably conversant with the shopping locality.

The following example illustrates the method of analysing the rent of premises comprising a ground floor shop and two upper floors occupied together with the shop for residential purposes.

---

[26] See *British Home Stores Ltd v Brighton Corporation and Burton; Fine Fare Ltd v Burton and Brighton Corporation* [1958] 51 R&IT 665.

[27] See *Marks & Spencer Ltd v Collier (VO)* [1966] RA 107.

[28] (1967) 202 EG 1175.

[29] See *John Lewis & Co Ltd v Goodwin (VO) and Westminster City Council* (1979) 252 EG 499.

The analysis is made to ascertain the rental value per unit of area of the front zone of the shop so that it can be compared with the rents of other shops. It is not in this case considered likely that there would be any doubt about the correctness of the figure used for valuing the upper floors.

*Example 21–1*

Shop premises comprising a ground floor "shell" and two upper floors situated in a good position in a large town were let in April 1998 at £5,000 p.a. on a full repairing and insuring lease for 15 years with a review in the fifth and 10th years. The tenant spent £4,500 on a shop front and other finishings. Analyse this rent for the purpose of rating assessments. The shop has a frontage of 6 metres and a depth of 30 metres. The net floor areas of the residential first and second floors are 50 square metres each.

*Analysis*

| | | |
|---|---|---|
| Rent reserved | | £5,000 |
| *Add* | | |
| Annual equivalent of cost of shop front and finishings provided by the tenant | | |
| $\dfrac{\text{£4,500}}{\text{YP 15 yrs at 6\% and 3\%}} = \dfrac{4,500}{8.7899} =$ | | 512 |
| Full rental value in terms of RV | | £5,512 |
| *Deduct* | | |
| Rental value of upper floors | | |
| 100 sq metres at £7 per sq metre | | £700 |
| Therefore leaving for shop rental value in terms of RV | | £4,812 |

Measurement shows that the effective net area of the shop is as follows:

| | | |
|---|---|---|
| Zone 'A' | (first 6.1 metres) | 36.6 sq metres |
| Zone 'B' | (second 6.1 metres) | 36.6 sq metres |
| Zone 'C' | (third 6.1 metres) | 36.6 sq metres |
| Zone 'D' | (remaining depth) | 70.2 sq metres |

Let £x per square metre represent the rental value of the first zone and adopting "halving back":

then       $36.6 x + \dfrac{36.6x}{2} + \dfrac{36.6x}{4} + \dfrac{70.2x}{8}$        = £4,812

$$72.825x \qquad\qquad = £4,812$$
$$x \qquad\qquad = £66.08$$

∴ Zone A rental value is approximately £66 per square metre in terms of Rateable Value.

The above analysed figures can then be compared with other Zone A rents based on the letting of similar shops in the vicinity to assist in obtaining a rent which represents a fair and consistent basis for the assessment of the shops under consideration.

A more convenient approach is that of reducing areas to "equivalent Zone A" areas. Using the above facts, the total area of 180 square metres is expressed as:

Zone A                 36.6

Zone B                 18.3           $\left\{ \dfrac{36.6}{2} \right\}$

Zone C                 9.15           $\left\{ \dfrac{36.6}{4} \right\}$

Zone R                 8.775          $\left\{ \dfrac{70.2}{8} \right\}$

Equivalent Zone A area = 72.825 square metres.

Analysis then becomes $\dfrac{£\,4,812}{72.825}$ = £66 approximately as above.

The whole area has been converted to the area in terms of Zone A, and the figure of 72.825 square metres would be expressed as "72.825 square metres ITZA".

It must be emphasised that the zoning method, although used very commonly in practice, is merely a means to an end in order to find the true rental value. The fixing of zone depths and the allocation of values to the zones are arbitrary processes and strict adherence to an arbitrary pattern may result in absurd answers in particular cases.

On the other hand, the method is based on a sound practical principle, namely, that the front portion of a shop, including windows, is the part which attracts the most customers and is therefore the most valuable part as selling space. Consequently, a shop with moderate but adequate depth is likely to have a higher overall rental per unit of floor space than one with the same frontage but a much greater depth. Similarly, a shop with a frontage of 7 metres and a depth of 15 metres may be worth £6,000 p.a. but an adjoining shop with the same frontage of 7 metres and otherwise identical but with double the depth, i.e. 30 metres, will be found to be worth considerably less than £12,000 p.a.

Reference may usefully be made to *WH Smith & Son Ltd* v *Clee (VO)*[30] and *Hiltons Footwear Ltd* v *Leicester City Council and Culverwell (VO)*.[31]

### (d) Factories and Warehouses

Both factories and warehouses are usually assessed by reference to Gross Internal Area, the value per square metre being fixed by comparison with other similar properties in the neighbourhood. In rare cases a flat rate per unit of floor space may be used throughout the building, but usually it is more appropriate to vary the rate applied to each floor and/or possible use for the available floor space.

Properties of this type differ considerably in such matters as construction, accommodation, planning and situation. Any special advantages or disadvantages attaching to the property under consideration must be very carefully examined when applying rents of other properties. Particular factors affecting rental values are the proximity of motorways or rail facilities, availability of labour, internal layout, clear height of the main warehouse area, nature of access to upper floors, loading and unloading facilities, natural lighting, heating, ventilation, adequacy of toilets, nature of ancillary offices, canteen facilities, yard areas, availability of public transport and car parking for staff.

There may be no general demand for certain industrial properties of highly specialised types in some areas and therefore no evidence of rental values which could be used as a basis for comparison, particularly since most are owner-occupied in any event. With such cases as, for example, oil refineries, chemical

---

[30] (1977) 243 EG 677.

[31] (1977) 242 EG 213.

plants, cement works and steelworks, it will probably be necessary to resort to the contractor's test by taking an appropriate percentage of the effective capital value of the land and buildings as a method of obtaining the Rateable Value.

Plant and machinery is only to be taken into account in the rating assessment if it is deemed to be a part of the hereditament. For the 1990 List, the Valuation for Rating (Plant and Machinery) Regulations 1989 specified that five classes of plant and machinery were deemed to be a part of the hereditament and therefore rateable.

During the course of negotiation of many 1990 Rating List assessments it became apparent, as if it had not already been so, that there were references to obsolescent items, incompatibilities with Scottish rating practice and other deficiencies that had been carried over into the 1990 List. A list of plant and machinery had been adopted that had essentially been first prepared in 1960, with only relatively minor amendments in the intervening years.

The Wood Committee was charged with a review of the law relating to the rateability of plant and machinery and out of its Report was born the Valuation for Rating (Plant and Machinery) Regulations 1994 which apply to valuations for rating for the 1995 Rating List. These now provide for rateability of plant and machinery in the following classes:

Class 1. Relates to the generation, storage, primary transformation or main transmission of power in the hereditament.

Class 2. Relates to "services" to the hereditament, i.e. heating; cooling; ventilating; lighting; draining; supplying of water; protection from trespass, criminal damage, theft, fire or other hazard.

This class excludes items used to provide "services" that are in the hereditament mainly to provide "services" as part of manufacturing or trade processes.

Class 3. Has a variety of items in various categories, including railway lines, etc.; lifts, elevators, etc.; cables, pylons, etc., used for transmission, etc. of electricity which are part of an electricity hereditament; cables, masts, etc. used for the transmission of communications; pipelines.

Class 4. Plant or machinery identifiable by name, either specific or generic, in the class, except any item which is not, or is not in the nature of, a building or structure.

The effect of these provisions is that the rent to be ascertained is that which a tenant would pay for the land and buildings, including rateable plant and machinery, in accordance with the statutory definition.

There may be no difficulty in assessing the rental value, on a unit of floor space basis, of premises including certain rateable plant and machinery such as sprinkler and heating systems. Other rateable plant and machinery may have to be valued separately from the land and buildings on the contractor's test, applying the statutory decapitalisation rate to its effective capital value to arrive at Rateable Value.

Three further points should be noted. They are:

(1) Rateable plant and machinery is not restricted to that found on industrial hereditaments; for example, a passenger lift in a department store is rateable. However, an exception to the general statement applies to pipelines.

(2) Whilst the statutory provisions regarding plant and machinery did not apply to a valuation made on a profits basis for the 1990 Rating List, they do apply for the 1995 and 2000 Rating List.

(3) The Valuation Officer, if asked to do so by the occupier, must provide written particulars of what items of plant and machinery are deemed to be part of the hereditament and have been included in his valuation.

### (e) Cinemas and Theatres

A method of valuation which has been approved by the Lands Tribunal is a quasi-profits or takings method. It involves finding the gross receipts from the sale of seats plus the takings from bingo, screen advertising and the sale of ice-cream, cigarettes and sweets, etc. A percentage of this total figure is then taken to obtain the Rateable Value. In *Thorn EMI Cinemas Ltd* v *Harrison (VO)*,[32] a percentage of 8.5% was applied to gross receipts (adjusted to 1973 levels as the 1973 Valuation List was in force). An alternative method of making the assessment is to relate it to a price per seat, but very wide variations can be found in analysing assessments by this method.

---

[32] [1986] 2 EGLR 223.

### (f) Licensed Premises

Until the 1990 Rating List, the method of valuation was a shortened profit basis which assumed that the brewer was the hypothetical tenant who would sub-let to a tied tenant.

For the 1990 Rating List the Valuation Office adopted the overbid method in respect of brewery pubs. This method still assumed that the brewer was the hypothetical tenant but that he would pay to the hypothetical landlord 100% of the tied rent he would expect to receive. In addition, he would pay a percentage of the profit that the brewer would make from both brewing and wine and spirit wholesaling. However, it was also recognised in the 1990 List that few brewers would be in the market for many rural pubs and in such cases the overbid for brewing/wine and spirit wholesaling was not appropriate. In respect of freehold pubs a normal profit basis valuation was applied.

For the 1995 List the VO approached all valuations on a profit basis.

### (g) Canal, Electricity Supply, Gas, Railway, Telecommunications, Water Supply and Long Distance Pipeline Hereditaments

Section 52 of the Local Government Finance Act 1988 provides for the compilation of central rating lists and Section 53 enables the Secretary of State to prescribe Regulations designating the hereditaments to be dealt therein. The Central Rating Lists Regulations 1994 apply for the 1995 Rating List and designated the above hereditaments accordingly. All these hereditaments were rateable in accordance with statutory formulae, but gradually they are being removed from formula rating to a normal rating valuation approach.

### (h) Caravan Sites

A caravan used as a sole or main residence will be liable to Council Tax (see Section B below). In the case of other caravans and sites, Statute[33] provides that the whole geographic unit comprising caravan pitches, together with any caravans thereon, and common facilities shall be treated as a single assessment. The site operator

---

[33] Non-Domestic Rating (Caravan Sites) Regulations 1990, SI 1990 No 673.

is made liable for payment of rates in respect of any such assessment.

In the absence of good rental evidence, caravan sites are normally assessed by a profits test. A Lands Tribunal decision on a 1963 List reference, *Garton v Hunter (VO)*,[34] established that various methods of valuation were admissible but that it was up to the then Local Valuation Court (now Valuation Tribunal) to decide the most appropriate.

### (i) Garages and Service Stations

These are normally valued on a direct profits basis. The rental value depends primarily on throughput, being the quantity of petrol sold. A rental value for a certain throughput (per thousand litres) is taken to represent the tenant's bid in terms of Rateable Value. The figure varies on a sliding scale which increases with throughput, as the tenant's overheads would be proportionately less as the trade increases. The scale also has regard to varying revenue from credit card sales, petrol accounts, opening hours, incentives offered, etc. since these all affect profitability. The Rateable Value thus arrived at allows for the petrol sales area and is taken to include any small sales kiosk but not a shop as such. It also assumes a canopy over the pump sales area and an allowance will be appropriate if there is no canopy. Additions to the valuation would need to be made for other buildings and land such as storerooms, showrooms, lubrication bays, workshops, and lock-up garages and other facilities.

### (j) Minerals

Mineral hereditaments are valued according to their component parts. The mineral itself is valued on a royalty basis applied to the volume extracted. Plant, machinery and buildings are valued on an estimated capital value basis. Surface land may be valued by reference to its value to the overall operation. 50% relief is applied to the mineral value and the value placed on any land used for actual quarrying operations. Any value for waste disposal purposes should be taken into account.

Mineral hereditament valuations are subject to annual review.

---

[34] [1969] 2 QB 37.

## Section B:  Domestic Properties

### 1.  General

Prior to 1 April 1990, domestic properties were liable for local rates. All domestic properties had rating assessments entered in the then Valuation List. These domestic hereditaments were assessed to Gross Value which was broadly defined as the rental value of the property on the assumption that the landlord undertook the repair liabilities. The Rateable Value was derived from the Gross Value by a statutory formula and the property's rate liability was determined by multiplying the Rateable Value by the locally set rate in the £.

On 1 April 1990 the Community Charge was introduced. This sought to raise local revenue by taxing persons resident in the local charging authority area – a poll tax. However this proved difficult to administer and collect and additionally became politically unacceptable.

The Local Government Finance Act 1992 made provision for the abolition of the Community Charge and the introduction of the Council Tax.

### 2.  Procedures

The Act provides that the Valuation Office will compile and maintain a Valuation List for each Billing Authority in England and Wales. Billing Authorities are the same authorities that exist for non-domestic rates.

The Valuation List will show every dwelling in the billing authority area, together with a band of value applicable to that dwelling. A dwelling is defined as any property which would have been a hereditament for the purposes of the General Rate Act 1967, does not fall to be shown in a non-domestic Rating List and is not exempt from local non-domestic rating.

The bands for England are as follows:

| | |
|---|---|
| Band A | Value not over £40,000 |
| Band B | Value over £40,000 but not over £52,000 |
| Band C | Value over £52,000 but not over £68,000 |
| Band D | Value over £68,000 but not over £88,000 |
| Band E | Value over £88,000 but not over £120,000 |
| Band F | Value over £120,000 but not over £160,000 |
| Bank G | Value over £160,000 but not over £320,000 |
| Band H | Value over £320,000 |

The bands for Wales are lower, with Band H, for example, being value above £240,000.

## 3. Basis of Assessment

The appropriate banding will be decided by determining the value the property might reasonably have been expected to realise if it had been sold on the open market by a willing vendor on 1 April 1991 subject to the following assumptions:

(a) that the sale was with vacant possession;
(b) that the interest sold was the freehold or, in the case of a flat, a lease for 99 years at a nominal rent;
(c) that the size, character and layout of the property, and the physical state of its locality were the same as at the time the valuation of the property was made;
(d) that the property was in a state of reasonable repair.

The actual liability to Council Tax is determined by the banding, i.e. value, of the property and also by the number of adults living in it.

The legislation provides for appeals to be made under certain circumstances where it is considered that the banding is incorrect. Appeals are determined by the Valuation Tribunal whose decisions can be appealed to the High Court on points of law only.

## 4. Compilation and Maintenance of New Lists

The Act provides that the Secretary of State may make an Order varying the valuation bands and requiring the preparation of new Valuation Lists.

# Chapter 22

# Taxation

## Capital Gains Tax

### 1. Taxation of Capital Gains – Capital Assets and Stock in Trade

Before 1962 a capital gain was generally free from taxation. In the Finance Act 1962 provisions were introduced to treat gains made in a short period, normally one year but three years for land, as income to be taxed accordingly. The Finance Act 1965 introduced a comprehensive system for the taxation of long-term capital gains and the 1962 Act provisions were abolished finally in 1971. Thus, since 1965 there has existed a scheme for the taxation of capital gains, by which is meant gains realised when a capital asset is disposed of and the proceeds exceed the costs incurred in acquiring the asset.

These provisions apply to assets of a capital nature. In the case of property these would be the land and buildings owned by a person or company which they occupy or let as an investment. It is important to note that in some instances such property will not be regarded as capital but as stock in trade which is not subject to the provisions applicable to capital gains.

The determination of whether or not property is held as stock in trade is not always clear. Broadly, stock in trade refers to those instances where a property was purchased with the purpose of making a profit. A clear example is land purchased by a company engaged in house building. The company buys land on which to build houses and the land is as much a raw material as the bricks and timber which go into the houses themselves. Such land is clearly not a capital asset since the company will sell the houses with the land as soon as possible and so move on to the next site. Thus, the cost of buying the land is one of the general costs of the business, whilst the sale proceeds from the disposal of the houses constitute its income. Any gain in respect of the land is therefore a part of the company's profits to be treated as income.

On the other hand, if the company owns an office building wherein its staff are housed, such a building is a capital asset since the offices were not purchased with a view to selling on for a profit. If the company does choose to sell its offices at any time, any gain realised would be subject to capital gains tax legislation. This legislation is examined below.

## 2. Taxation of Chargeable Gains Act 1992 (TCGA 1992)

*(a) Occasions of Charge*

Capital gains tax is payable on all chargeable gains accruing after 6 April 1965, on the disposal of all forms of assets including freehold and leasehold property, options and incorporeal hereditaments.

An asset may be disposed of by sale, exchange or gift. Disposal also includes part disposals and circumstances where a capital sum is derived from an asset. The transfer of ownership following a death is not a disposal.

A part-disposal arises not only when part of a property is sold but also when a lease is granted at a premium. By accepting a premium, the landlord is selling a part of his interest and the lessee is purchasing a profit rent. Premiums may be liable to both income tax under the Income and Corporation Taxes Act 1988, which deals with taxation of premiums, and capital gains tax under the 1992 Act. The Act contains provisions for avoiding double taxation on any part of the premium.

A capital sum is derived from an asset when the owner obtains a capital sum such as a payment received to release another owner from a restrictive covenant or to release him from the burden of an easement and would also include compensation received for injurious affection on compulsory purchase.

It also includes capital sums received under a fire insurance policy, although if the money is wholly or substantially applied to restore the asset no liability will arise.

The transfer of an asset out of fixed assets into trading stock is deemed to be a disposal. This would arise, for example, if an owner occupier of commercial premises decided to move elsewhere and to carry out a redevelopment with a view to selling the completed new property.

Where a disposal is preceded by a contract to make the disposal, the disposal is taken to occur when the contract becomes binding. This typically occurs on the sale of property where the parties enter into a contract with completion at a later date. If the contract is conditional, for example subject to grant of planning permission, it becomes binding on the date when the planning permission is granted which will be the date of disposal.

## (b) Computation of Gains

The general rule applicable to the disposal of non-wasting assets, which includes freeholds, and leases with more than 50 years unexpired, is that in arriving at the net chargeable gain certain items of allowable expenditure are deductible from the consideration received on disposal. These items are:

(1) Expenditure wholly and exclusively incurred in disposing of the asset. This includes legal and agents' fees, advertising costs and any valuation fees incurred in preparing a valuation required to assess the gain. It does not include fees incurred for a valuer to negotiate and agree the values with the District Valuer.[1]

(2) The price paid for the asset, together with incidental costs wholly and exclusively incurred for the acquisition, for example legal and agents' fees and stamp duty.

(3) Additional expenditure to enhance the value of the asset and reflected in the asset at the disposal date, including expenditure on establishing, preserving or defending title. A typical example would be improvements to the property carried out by the taxpayer such as extensions.

Any item allowed for revenue taxation cannot be allowed again and thus only items of a capital nature are deductible. Special rules apply where expenditure has been met out of public funds.

Certain of the above-mentioned items will need to be apportioned in the case of a part-disposal and further modifications will be required to meet special circumstances, as described later.

### Example 22–1

A purchased a freehold interest in a shop in 1982 for £200,000. In 1987 A sold his interest for £400,000.

---

[1] *Couch (Inspector of Taxes)* v *Administrators of the Estate of Caton, The Times,* 16 July 1997; Sub nom. *Administrators of the Estate of Caton* v *Couch (Inspector of Taxes)* [1997] STC 970.

| | | | |
|---|---|---:|---:|
| Proceeds of Disposal | | | £400,000 |
| *Less* | | | |
| Agents' fees on sale, say | | £ 4,000 | |
| Advertising for sale, say | | 800 | |
| Legal fees on sale, say | | 2,000 | 6,800 |
| Net Proceeds of Disposal | | | 393,200 |
| *Less* | | | |
| Acquisition Price | | 200,000 | |
| *Add* | | | |
| Agents' fees on purchase, say | 2,000 | | |
| Legal Fees on Purchase,  say | 1,000 | | |
| Stamp Duty | 2,000 | 5,000 | 205,000 |
| Gain | | | £188,200 |

## (c)  Disposals after 5 April 1988

The Finance Act 1988 introduced a significant reform of the capital gains tax rules in respect of disposals on or after 6 April 1988. The general provisions are noted elsewhere but one change requires specific attention. This concerns cases where the asset, the subject of the disposal, was acquired before 1 April 1982. In such cases the taxpayer may elect (irrevocably) for capital gains and losses on all assets held at 31 March 1982 to be calculated by reference to values at that date. This relieves those who make the election of the need to maintain records going back beyond that date.

The gain is determined by deducting from the net proceeds of disposal the market value of the asset as at 31 March 1982 in lieu of the actual price paid at the actual time of acquisition. The effect of this is to charge to CGT only the gain arising after 31 March 1982. The Act states that the taxpayer is deemed to have sold the asset and immediately to have re-acquired it at market value. No allowance is given for the notional fees that would have been incurred on the notional purchase. Further, any capital expenditure incurred before 31 March 1982 is excluded from the calculation of the gain since it took place before the asset was (notionally) purchased.

## (d)  Indexation

Provisions were introduced in 1982 to make some allowance for the effects of inflation on the amount of gains realised. This is achieved by deducting from the actual gain arising after 31 March 1982 the gain which would have arisen if the growth in value had merely been in line with inflation. If the inflation gain is less than the actual gain then only the balance of the gain attracts CGT. Where the inflation gain exceeds the actual gain, so turning the actual gain into a notional loss, the asset is deemed to have produced no gain and no loss. If the inflation gain produces a greater notional loss than an actual loss, the sale of the asset is deemed to have produced a loss equal to the actual loss. The importance of this is that, where a disposal produces a loss rather than a gain, the loss may be set against gains realised in the same year or carried forward to be set against gains in future years.

Inflation is measured by changes in the Retail Price Index. So the indexation allowance is calculated by applying the percentage increase in the Retail Price Index (RPI) between 31 March 1982 and the date of disposal to the market value on 31 March 1982, or, if acquired after 31 March 1982, the percentage increase between the date of acquisition and disposal to the acquisition price.

RPI figures are calculated and published each month by the Government in relation to the R.P.I. Index. The Index was re-based in January 1987 to 100. However, the earlier figures of RPI have been re-calculated against the 1987 base so that, for example, the RPI figure for March 1982 is 79.44 in the 1987 Index (313.4 in the previous Index).

Thus, if an asset acquired before 31 March 1982 was sold in January 1995, the indexation allowance would be:

$$\frac{\text{RPI (Jan '95)} - \text{RPI (Mar '82)}}{\text{RPI (Mar '82)}}$$

RPI (Jan '95) = 146.00 : RPI (Mar '82) = 79.44

$\therefore \dfrac{146.00 - 79.44}{79.44}$ = 83.79% of Market Value on 31 March 1982

(i.e. general inflation between March 1982 and January 1995 was 83.79%).

*Example 22–2*

A purchased a freehold interest in a shop in January 1981 for £200,000 and sold the interest in January 1995 for £500,000.

| Proceeds of Disposal | | £500,000 |
|---|---|---|
| *Less* | | |
| Agents' fees on sale, say | £7,500 | |
| Advertising for sale, say | 1,000 | |
| Legal fees on sale, say | 4,500 | |
| Valuation fees to prepare 1982 valuation, say | 1,000 | 14,000 |
| Net Proceeds of Disposal | | 486,000 |
| *Less* | | |
| Acquisition Costs, say | | 205,000 |
| Gain | | £281,000 |
| *Less* | | |
| Indexation Allowance: | | |
| Market Value 31 March 1982, say | 220,000 | |
| RPI increase | 0.8379 | 184,338 |
| Gain Chargeable to CGT | | £96,662 |

If the taxpayer elects to adopt the Market Value at 31 March 1982 as the purchase price, the result is:

| Net Proceeds of Disposal (as before) | £486,000 |
|---|---|
| *Less* | |
| Acquisition Price (OMV 31.8.82) | 220,000 |
| Gain | £266,000 |
| *Less* | |
| Indexation Allowance (as before) | 184,338 |
| Chargeable Gain | £81,662 |

The taxpayer would benefit by an election for 1982 value to be adopted.

It should be remembered that the date of sale is taken to be the date of a binding contract if, as is usual, one precedes a conveyance, and the RPI on disposal would be that for the month of the binding contract.


*(e) Taper Relief*

In the case of individuals and trusts taper relief was substituted for indexation relief as from 6 April 1998. Taper relief operates by reducing the amount of the gain by a percentage dependent on the type of, and the amount of time that the taxpayer has owned the asset.

The percentages are:

|  | Gains on business assets | Gains on non-business assets |
| --- | --- | --- |
| Number of complete years after 5.4.98 for which assets are held | Percentage of gain chargeable | Percentage of gain chargeable |
| 0 | 100 | 100 |
| 1 | 92.5 | 100 |
| 2 | 85 | 100 |
| 3 | 77.5 | 95 |
| 4 | 70 | 90 |
| 5 | 62.5 | 85 |
| 6 | 55 | 80 |
| 7 | 47.5 | 75 |
| 8 | 40 | 70 |
| 9 | 32.5 | 65 |
| 10 or more | 25 | 60 |

A business asset is extensively defined, but broadly it is one used by the taxpayer in his business (a trade, profession or vocation) or held for the purpose of a qualifying office or employment or comprises shares in a qualifying company (5% of voting rights if he works full time, or otherwise 25%).

Where an asset was owned on 17 March 1998 and 5 April 1998 then one year is added to the period of ownership after 5 April – the "bonus year".

Where an asset is held as a business asset for a period of ownership and as a non-business asset for the balance, any gain is apportioned on a time basis before applying the appropriate relief. For example, if an asset is owned for five whole years, three of them as a business asset, and a gain of £100,000 is realised on a disposal then taper relief is:

$$3/5 \times 100,000 \times 62.5\% : 2/5 \times 100,000 \times 85\%.$$

*Example 22–3*

Assume facts as in Example 22–2 but A sold the interest in May 1999. The RPI in April 1998 was 162.6. A occupied the property as a business asset.

Adopt Market Value at 31 March 1982.

| | | |
|---|---|---|
| Gain (as before) | | £266,000 |
| *Less* | | |
| Indexation Allowance: | | |
| Market Value 31 March 1982 | 220,000 | |
| × $\dfrac{162.6 - 79.44}{79.44}$ = | 1.05 | £231,000 |
| Indexed Gain | | £35,000 |

| | | |
|---|---|---|
| *Less* | | |
| Taper Relief | | |
| Asset owned on 17 March 1998 = | 1 year | |
| Whole years since 6 April 1998 = | 1 year | |
| | 2 years | |
| Relief for 2 years = 85% = | | 0.85 |
| Chargeable Gain | | £29.750 |

*Note*: If A had let the property so that it was not a business asset there would be no taper relief for two years of ownership.

*(f) Rates of Tax*

In the years before 1987 the rate of tax was 30% for all taxpayers. In the Finance Act 1987 the rate of tax for companies changed to the same rate of tax which applied to their general profits (corporation tax rate). In 1988, this approach was extended to individuals, the rate of tax being the taxpayer's marginal income tax rate.

Hence, if an individual realises a chargeable gain and his income tax rate is the basic rate or a reduced rate, then those rates were applied to the gain as if it were income. If his income tax rate is already at the higher rate then that rate will be applied. In cases where the gain, if added to other income, would take the taxpayer into the higher rate, the rate of tax from 6 April 1999 is 20% on the amount equal to the difference between other taxable income and the higher rate threshold, with the higher rate on the balance.

The first part of a gain is exempt for individuals, with half that amount for trustees. This varies from year to year. For example, in the tax years 1994/1995 to 1999/2000 the exempt amount rose from £5,800 to £6,000 to £6,300 to £6,500 to £6,800 to £7,100. There is no exempt amount for companies.

## (g) Exemptions and Reliefs

### 1. Exempted bodies
Certain bodies are exempt from liability, including charities, local authorities, friendly societies, scientific research associations, and pension funds, as are non-resident owners.

### 2. Owner-occupied houses
Exemption from capital gains tax is given to an owner in respect of a gain accruing from the disposal of a dwelling-house (including a flat) which has been his only or main residence, together with garden and grounds up to 0.5 hectare or such larger area as may be appropriate to the particular house ("required for the reasonable enjoyment of the house").

Where, in addition to the main building, there are outbuildings such as garages, greenhouses, sheds, and even staff accommodation in the grounds of the house, the first step is to determine which buildings are part of the house. The test is whether the buildings are within the curtilage of the house, which requires that they must be close to the house (*Lewis* v *Rook*).[2] In that case a gardener's cottage 190 yards from the house was not part of the house.

The test of which buildings are included with the house is settled between the taxpayer and the Inspector of Taxes, whilst it is the District Valuer who negotiates with the taxpayer as to the area of land required for the reasonable enjoyment of the house.

Where part of the garden of a house is sold but the house is retained, the exemption applies if the land sold is within a garden of less than 0.5 hectares or, if larger, is within the area required for the reasonable enjoyment of the house. This will not be so if the house is sold first without the land in question, and the land is sold later (*Varty* v *Lynes*).[3] This can prove to be a valuable exemption if part of the garden is sold for a high development value.

The exemption does not apply, however, if the house was purchased with a view to making a gain or was not used as a home – which implies that the taxpayer can show a degree of permanence and expectation of continuity. In *Goodwin* v *Curtis*[4] the taxpayer purchased a farmhouse on 1 April 1985 and sold it on 3 May 1985

---

[2] [1992] STC 171.
[3] [1976] STC 508.
[4] *The Times*, 2 March 1998.

at a considerably enhanced price. It was held that the private residence relief did not apply.

The degree of exemption is proportionate to the period of owner-occupation during ownership. Full exemption is given if the owner has lived in the house throughout the whole of his period of ownership. To cover the circumstances where a house is vacant after the owner has moved and is looking for a purchaser, the exemption includes the last three years of ownership. Certain other periods of absence have also to be disregarded under the Act, such as where the taxpayer has to live elsewhere temporarily as a condition of his employment.

Thus, if a property is sold and was not owner-occupied for the whole period of ownership, the degree of exemption depends upon the ratio between the period of owner-occupation (including in any event the last three years of ownership) and the total period of ownership.

If, therefore, X buys a house on 1 April 1989, and lives in it for two years until 1 April 1991, when he lets it, subsequently occupying the house on 1 April 1995, and selling it on 1 April 1998, the proportion of any capital gain which is exempt from capital gains tax is:

$$\frac{2+3}{9} = \frac{5}{9}$$

Where premises are used partly as a dwelling and partly for other purposes, the exemption applies only to the residential part and it will be necessary to apportion any capital gain on the whole property between the two portions. This apportionment should be made on a value basis. Thus, if a house is half let and half owner-occupied, the gain on disposal is not necessarily apportioned on a 50:50 basis. It is likely that more value is attributable to the owner-occupied part and that the growth in value is greater for that part. The apportionment should reflect these circumstances.

The exemption also applies to one other house which is occupied rent free by a dependent relative if the house was bought before 6 April 1988, and also where trustees dispose of a house which has been occupied by the beneficiary under the trust.[5]

---

[5] For further details see CGT4, issued by The Board of Inland Revenue and available from Tax Offices free of charge.

## 3. *Replacement of business assets*

Where trade assets, including buildings, fixed plant and machinery, and goodwill are sold and the whole of the proceeds are devoted to the replacement of the assets with other trade assets, the trader may claim to defer any capital gains tax which ordinarily may have been payable on the sale. He may choose, instead of paying the tax which arises, to have the actual purchase price of the replacement written down by the amount of any chargeable gain on the sale of the original assets. This process may be repeated on subsequent similar transactions so that tax on the accumulated capital gain will not be paid until the assets are sold and not replaced. To qualify for this relief, the acquisition of the new assets must be within a period commencing one year before and ending three years after the sale of the old assets. The Act also deals with circumstances where not all the proceeds of sale are re-invested.

### *Example 22–4*

X purchased a freehold shop in June 1991 for £15,000, the expenses of acquisition being £500, and occupied and commenced trading as a grocer. In September 1997, he sold this shop for £40,000, his expenses of sale being £900. He immediately purchased a new shop for £45,000, incurring expenses of £1,000 and continues in business in the same trade.

X claimed relief from the payment of tax in 1997 on the sale for £40,000 so the cost of acquisition of the new shop will be dealt with as follows:

*Chargeable Gain on sale in 1997*

| | | |
|---|---:|---:|
| Proceeds of Disposal | | £40,000 |
| *Less* Expenses | | 900 |
| Net Proceeds of Disposal | | £39,100 |
| *Less* Acquisition Price | £15,000 | |
| *Add* Costs | 500 | 15,500 |
| Gain | | £23,600 |
| *Less* Indexation Relief | | |
| $\dfrac{159.3 - 134.1}{134.1} \times 15,500 \ =$ | | 2,900 |
| Chargeable Gain | | £20,700 |

The chargeable gain is deducted from the purchase price of the new shop so that, on a subsequent sale, £24,300 (£45,000 – 20,700) and not £45,000 will be deducted as the purchase price. In this way, the chargeable gain is said to be "rolled over" into the new asset – hence rollover relief.

| | |
|---|---:|
| Cost of Replacement Asset | 45,000 |
| *Less* Gain on Sale of Old Asset | 20,700 |
| Notional Cost of Replacement Asset | £24,300 |

If the replacement shop had been sold in 1999 for £80,000, with the expenses of sale being £2,000 and the owner then ceasing to continue in business, liability to capital gains tax would then arise as follows (ignoring indexation and taper relief):

| | | |
|---|---:|---:|
| Proceeds of Disposal | | £80,000 |
| *Less* Expenses | | 2,000 |
| Net Proceeds of Disposal | | 78,000 |
| *Less* | | |
| Notional Acquisition Price | £24,300 | |
| Expenses of Acquisition | 1,000 | 25,300 |
| Gain | | £52,700 |

Thus, the actual net gain in respect of the new asset of £78,000 – (£45,000 + £1,000) = £32,000 is boosted by the previous gain of £20,700 carried (or "rolled") over to give a terminal gain of £32,000 + £20,700 = £52,700.

X may, however, be eligible for relief from this charge if he retired on the disposal (see 4 below).

Where the sum re-invested in the replacement asset is less than the net proceeds realised on the sale of the old asset, the difference is subject to capital gains tax, since it is part of the gain itself which is not being rolled over.

Where rollover relief was claimed between 31 March 1982 and 6 April 1988 and part of the gain accrued before 31 March 1982, any gain arising on the disposal of the replacement asset is halved.

It should be emphasised that rollover relief applies to properties occupied for business purposes but not to properties held as investments. The exception to this is where an interest in an investment property is acquired under compulsory purchase powers when the owner has 12 months to roll over the compensation proceeds into another investment property.

### 4. *Retirement relief*

On retirement at age 50 or later, an owner may be exempted from capital gains tax on gains which accrue to him from the disposal by way of sale or gift of a family business or of shares or securities in a family trading company. In the years to 1999, chargeable gains up to £250,000 are exempt whilst, for gains above £250,000, only half of the next £750,000 of gains is brought into tax. In the tax year 1999/2000 these amounts are reduced by £50,000 and £150,000 respectively and also in succeeding years so that the relief eventually is abolished. The owner must have been in the business for ten years or more. If the period is less than 10 years, the maximum relief is reduced by 10% for each year below 10 years, e.g. in business for four years, the relief is 40% of maximum. No relief is given if the owner has been in business for less than one year. A similar relief is given to persons who are forced to retire before the age of 50 due to ill-health.

### (h) *Losses*

A capital loss might arise as a result of a disposal, say where the asset is sold for less than the price paid for it. Such a loss is allowable if a gain on the same transaction would have been chargeable. The loss would normally be set off against gains accruing in the same year of assessment but if, in a particular year, losses exceed gains, the net loss can be carried forward and set off against gains accruing in the following or subsequent years.

### (i) *Part Disposals*

The rule for calculating any gain or loss on a part disposal is that only that proportion of the allowable expenditure which the value of the part disposed of bears at the date of disposal to the value of the whole asset, can be set against the consideration received for the part. Thus, it is necessary to value the whole asset, and this necessitates a valuation of the retained part, since for these purposes the value of the whole asset is treated as being the aggregate of the values of the part disposed of and the part retained. The proportion disposed of is then:

$$\frac{A}{A + B} = \frac{\text{Consideration on part disposal (A)}}{\text{Consideration on part disposal (A) + value of retained part (B)}}$$

*Example 22–4*

X bought an asset in 1982 for £100,000, including costs, and later sells part of it for £60,000, no disposal costs being involved. The first step is to value the part retained, say £90,000. The proportion of the asset disposed of is:

$$\frac{£60,000}{£60,000 + £90,000} \qquad \text{or } 40\%$$

i.e. X is realising 40% of the value of the asset. (£60,000 out of asset worth £150,000.)

Allowable expenditure of costs of acquisition and enhancement expenditure is therefore taken to be 40%. Hence the gain realised, ignoring indexation, is:

| | | |
|---|---|---|
| Proceeds of Disposal | | £60,000 |
| *Less* | | |
| Acquisition Price | £100,000 | |
| $\dfrac{\times A}{A + B} \quad = \quad 40\% \ =$ | 0.4 | |
| Apportioned Acquisition Price | | 40,000 |
| Gain | | £20,000 |

Where enhancement expenditure relates solely to the part sold, it is fully allowable; where it relates solely to the part retained, then none is allowable on the part disposal. In other cases it is similarly apportioned by A/A+B.

Indexation relief is applied to the apportioned acquisition price.

An alternative basis of calculating the cost of a part disposal of an estate for capital gains tax purposes avoids having to value the unsold part of the estate. Under this alternative basis the part disposed of will be treated as a separate asset and any fair and reasonable method of apportioning part of the total cost to it will be accepted, e.g. a reasonable valuation of that part at the acquisition date. Where the market value at 31 March 1982 is to be taken as cost of acquisition, a reasonable valuation of the part at that date will be accepted.[6]

---

[6] This alternative basis is an extra-statutory concession introduced by the Inland Revenue; see Statement of Practice D1 published on 22 April 1971.

The cost of the part disposed of will be deducted from the total cost of the estate (or from the balance) to determine the cost of the part retained. This avoids the total of the separate parts ever exceeding the whole. Thus, in the above example, the total costs were £100,000. Of this, £40,000 was treated as the cost of the part disposed; the remaining £60,000 is carried forward as the cost of the retained interest.

The taxpayer can always require the general rule to be applied, except in cases already settled on the alternative basis. However, if he does so, it will normally be necessary to adhere to the choice for subsequent part disposals.

Where a part of a holding is disposed of for £20,000 or less, such consideration being not more than 20% of the value of the whole holding, it is termed a small part disposal[7] and the taxpayer can elect to pay no capital gain on such a part disposal. This is effected by deducting the consideration from the acquisition costs. Thus, in the above example, if the value of the whole is taken to be £150,000 at the time of disposal, a disposal of part at £20,000 would be less than 20% of the whole. No tax would arise, but the acquisition cost of the retained land would be £100,000 − £20,000 = £80,000. Where other land has been sold in the same tax year the aggregate consideration must not exceed £20,000.

### (j) The Market Value Rule on Certain Disposals

In a normal case the actual sales price will be treated as the consideration received on disposal. On certain occasions, however, the Act provides for a deemed sale at market value at the date of disposal. This rule on market value operates in the case of gifts. It also applies in the case of other dispositions not by bargain at arm's length, such as transfers between closely related or associated persons. It applies as well where an asset is transferred from being a fixed asset into trading stock. The rule applies to both transferor and transferee so that if X gives a property to his son worth £10,000, he will be deemed to have sold for that sum and will be assessed for capital gains tax accordingly. The son can then adopt £10,000 as his cost of acquisition on a subsequent disposition.

Market value is the price which the asset "might reasonably be expected to fetch on a sale in the open market".[8]

---

[7] TCGA 1992, Section 242.
[8] TCGA 1992, Section 272(1).

## (k)  Sale of Leasehold Interests

A long lease, that is a lease with more than 50 years unexpired, is not treated as a wasting asset and the whole of the original acquisition cost and other expenditure can be set against the consideration received on the sale of the lease.

A lease with 50 years or less to run, however, is treated for capital gains tax purposes as a wasting asset and on the sale of such an interest the whole of the acquisition cost and other expenditure cannot be deducted. Instead, it is deemed to waste away at a rate which is shown in a table of percentages in Schedule 8 to TCGA 1992. Only the residue of the expenditure which remains at the date of disposition can be set against the consideration received.

The table of percentages is derived from the Years' Purchase 6% Single Rate table. The YP for 50 years is taken to be 100. Thus, against 40 years the figure is 95.457 and against 25 years, 81.100. This implies that the value of the lease with 40 years to run should be 95.457% of its value when it had 50 years to run, and 81.100% with 25 years to run. Similarly, the value of the lease with 25 years to run should be 81.100/95.457% of the value when it had 40 years to run.

The Act provides a formula for writing down acquisition costs whereby there is deducted $P1 - P3/P1$ from the costs where $P1 =$ figure from the Table in Schedule 8 relating to years to run at time of purchase and $P3 =$ figure from Table relating to years to run at time of sale.

It is suggested that a more convenient approach is to multiply the acquisition cost by $P3/P1$.

### Example 22–5

A lease with 40 years to run was purchased for £50,000 and sold when it had 25 years to run for £55,000. The gain, ignoring costs and indexation, is:

| | | | |
|---|---|---|---|
| Proceeds of Disposal | | | £55,000 |
| *Less* | | | |
| Acquisition Price | | £50,000 | |
| $\times \dfrac{P3}{P1} = \dfrac{81.100}{95.457} =$ | | 0.850 | 42.500 |
| Gain | | | £12,500 |

If a lease which has more than 50 years to run is purchased, the appropriate percentage at acquisition will be 100 and it will commence to be a wasting asset when it has less than 50 years unexpired.

Where an election is made to adopt market value at 31 March 1982, it should be remembered that this is a (deemed) acquisition cost which is also subject to writing down by applying P3/P1. Indexation relief is applied to the written down value.

If any value of P3 or P1 is required, it can easily be calculated by dividing YP @ 6% for the years unexpired by YP 50 yrs @ 6% for 50 years multiplied by 100.

e.g. P3 when 25 years to run

$$= \quad \frac{\text{YP 25 yrs @ 6\%}}{\text{YP 50 yrs @ 6\%}} \quad \times 100 = \quad \frac{12.783}{15.762} \times 100 = 81.100$$

### (l) Premiums for Leases Granted out of Freeholds or Long Leases

The receipt of a premium on the grant of a lease is treated as a part disposal of the larger interest out of which the lease is granted. Against the premium may be set the appropriate part of the price paid for the larger interest and any other allowable expenditure as determined by the formula applicable to part disposals.[9]

Where a lease for a period of 50 years or less is granted at a premium, a part, at least, of the premium will be liable for income taxation under Schedule A. To avoid double taxation the part of a premium for a lease which is thus chargeable to income tax is excluded from liability to capital gains tax. This necessitates a modification to the normal formula applicable to ascertain allowable expenditure.

### Example 22–7

X purchased a freehold shop for £50,000 (inclusive of costs) in 1986. In 1999, he grants a 25-year lease at £6,000 per annum net, taking a premium of £75,000.

*Step 1.*
Calculate the amount taxable under Schedule A.

---

[9] See Section (j) of this chapter.

The amount taxable is the whole premium less 2% of the premium for each complete year of the term except the first year. Hence:

| | |
|---|---|
| Whole premium | £75,000 |
| *Less* 2 (25 – 1)% of £75,000 | £36,000 |
| Part subject to Schedule A | £39,000 |

£39,000 is excluded from any capital gains tax liability, leaving £36,000 as the amount of premium on which to base capital gains tax.

*Step 2.*

Allowable expenditure is then to be considered in accordance with the part disposal formula[10] but there is to be excluded from the consideration in the numerator of the fraction (but not in the denominator) that part of the premium taxable under Schedule A.

The application of the formula involves a valuation of the interest retained, which includes the capitalised value of the rent reserved under the lease:

| | | | |
|---|---|---|---|
| Rent reserved | £6,000 p.a. | | |
| YP 25 years at 6% | 12.78 | £76,680 | |
| Reversion to estimated full rental value (net) | £12,000 p.a. | | |
| YP in perp deferred 25 years at 6% | 3.88 | £46,560 | £123,240 |
| Value of Retained Interest, say | | | £123,000 |

Using the modified part disposal formula, the gain (ignoring costs and indexation) is:

$$\text{Gain} = £36,000 - \left\{ £50,000 \times \frac{£36,000}{£75,000 + £123,000} \right\}$$

$$= £36,000 - £9,090$$

$$= £26,910 \text{ (subject to indexation and taper relief).}$$

---

[10] See Section (j) of this chapter.

## (m) Premiums for Sub-leases Granted out of Short Leases

As the larger interest is a wasting asset, the normal part disposal formula does not apply. Only that part of the expenditure on the larger interest that will waste away over the period of the sub-lease in accordance with the table of percentages[11] can be set against the premium received.

Thus, if a person acquired a 40-year lease for £5,000 and when the lease had 30 years to run, he granted a sublease for 7 years at the head lease rent, taking a premium of £6,000, the following percentages from the table are required:

40 years 95.457 (percentage when interest acquired);
30 years 87.330 (percentage on grant of sub-lease);
23 years 78.055 (percentage on expiration of sub-lease).

The percentage for the 7 years of sub-lease is:

$$87.330 - 78.055 = 9.275$$

and the expenditure which is allowable against the premium of £6,000 is:

$$£5,000 \times \frac{9.275}{95.457} = £486$$

The capital gain is therefore £6,000 − £486 = £5,514.
(*Note*: The allowable expenditure has to be written down if the sub-lease rent is higher than the head lease rent.)

In this case 2(7 − 1)% = 12% of the premium = 12% of £6,000 = £720, leaving £5,280 liable to income tax under Schedule A. The method of avoiding double taxation is different in this case. The amount of the premium chargeable to income tax is deducted from the capital gain of £5,514 and only the balance of £5,514 − £5,280 = £234, is subject to capital gains tax. If the income tax proportion exceeds the capital gain then no capital gains tax is payable.

## Inheritance Tax

### 1. General

The Finance Act 1975 repealed the long established provisions for taxing the value of a deceased's estate at death by the imposition of

---

[11] See Section (k) of this chapter.

estate duty. The Act replaced estate duty with a new tax, Capital Transfer Tax. The provisions relating to this tax were incorporated in the Capital Transfer Tax Act 1984. The Finance Act 1986 abolished capital transfer tax and replaced it with Inheritance Tax. Since many of the features of inheritance tax are the same as capital transfer tax, the 1984 Act was renamed the Inheritance Tax Act 1984.

## 2. Taxable Transfers

Inheritance Tax (IHT) applies when there is a transfer of value. This will arise where a person makes a gift in his lifetime or his estate passes in its entirety on his death. A gift may be an absolute gift, whereby an asset is passed to another without any charge, or where an asset is sold to another at an "under value", by which is meant the sale price is deliberately low so as to confer an element of gift in the price, i.e. there is an intention to confer a gratuitous benefit.

## 3. Basis of Assessment

Since IHT relates to the transfer of value, the value on which the tax is assessed is not the value of the asset itself but the diminution in the value of the donor's estate caused by the transfer. This will commonly be the same figure as the value, but not necessarily so. For example, suppose that a person owns a home which is vacant and which he lets to his son at a nominal rent of £100 p.a. for 20 years. The grant of the lease at a low rent – under value – is a gift to his son. The value of the house before the gift, with vacant possession, is say £30,000, but after the gift when subject to the lease is, say, £8,000. The value of the lease is, say, £12,000. Although the subject of the gift, the lease, is worth £12,000, he has actually made a transfer of value of £30,000 (pre-gift) – £8,000 (post-gift) = £22,000 which will be subject to IHT.

Sums which may be set against the transfer of value are any incidental costs incurred in making the transfer and any capital gains tax payable on the gift. Note that it is the capital gains tax itself and not the whole gain so that there is an element of double taxation.

Since it is the transfer of value which is to be taxed it is important to establish who is paying the tax, as IHT may be paid by the donor or donee in the case of a lifetime gift. If the donee pays, then the IHT is assessed on the value transferred. If, for example, the value transferred were £12,000 and the tax rate was 40%, the donee would pay 40% of £12,000 = £4,800. On the other hand, if the donor pays, the transfer value is not only the asset but also the IHT payable out of the estate. Hence it is necessary to gross up the value transferred before determining the tax so that the grossed up value less IHT equals the value transferred. Following the above example, where the value transferred is £12,000, this is grossed up by:

$$\frac{100}{100 - 40} \times 12,000 = £20,000$$

Thus, IHT at 40% of £20,000 = £8,000, leaving £12,000 as the value transferred.

#### 4. Rates of Tax

Prior to 1988 there were several rates of tax on bands of value, the rates and bands changing yearly. The Finance Act 1988 introduced a greatly simplified system whereby the first £110,000 of value were exempt from IHT, a rate of 40% being the rate on all value in excess of this sum. The exempt band has been increased since then by changes made in successive Finance Acts. For example, the band for the tax year 1996/1997 was £200,000, for 1997/1998 it rose to £215,000, to £223,000 for 1998/1999 and to £231,000 for 1999/2000. Lifetime gifts are subject to Capital Gains Tax calculated on the basis that they are transferred at their full value. There is no CGT on death.

#### 5. Exemptions and Reliefs

*(i) General*

There are several exemptions and reliefs of a general nature such as exemptions for transfers between husband and wife, certain gifts to charities and other bodies, lifetime gifts up to £3,000 per annum, and wedding gifts up to £5,000.

*(ii) Potentially Exempt Transfers*

Where a gift is made by an individual to another individual or into certain trusts such as an accumulation and maintenance trust, no IHT liability will arise if the donor lives for seven years after making the gift. In such cases the gift is an exempt transfer. However, if the donor dies within seven years IHT becomes payable although there is tapering relief whereby a proportion only of the tax is charged. The tapering relief is:

| *Years between Gift and Death* | *% age of full IHT payable* |
|---|---|
| 0–3 | 100 |
| 3–4 | 80 |
| 4–5 | 60 |
| 5–6 | 40 |
| 6–7 | 20 |

IHT is charged at the rate of tax applicable at the time of death but on the value at the time of making the gift. Thus, if an asset was given to another individual in June 1994 with a value of £100,000 and the donor dies in September 1998 when the asset has a value of £150,000, the IHT liability will be 60% of 40% of £100,000. The executors can elect to adopt the value at death if this is less than the value at the time of the gift.

Because of these provisions, only gifts made within the period of seven years prior to death normally attrach IHT liability.

### (iii) Business Assets

Where a person owns a business as the sole proprietor, or otherwise has a controlling interest (husband and wife being treated as one), on the transfer of relevant business assets including shareholdings in unlisted companies, or in listed companies where the person has a controlling interest, the relief is 100%. In the case of land and buildings used by a partnership of which the person was a partner or by a company which he controlled, the relief is 50%. Business does not include dealing in land or investment so that property held for such purposes is not included. On the other hand, development is a business for this purpose.

### (iv) Agricultural Property

Where a person owns or occupies property for agricultural purposes and is engaged in agriculture – the "working farmer" – he is entitled to relief on most transfers of that property of 100% of the agricultural value of the property.

Agricultural value is the value assuming it will always be used for agricultural purposes, so any value attributable to non-agricultural use is ignored, particularly development value or hope value. Hence, assume two hectares of agricultural land were bequeathed by a farmer to his son which had a possibility of residential development. The residential value is £1,000,000 but the value as farmland is £20,000. The residential value would be taken into account in arriving at the value of the estate, but the relief would be 100% of £20,000. Similarly, when farm cottages are the subject of the transfer, the agricultural value is determined by assuming that they could only be occupied by farm workers, and so ignoring any enhanced value if they were to be sold as say "weekend cottages".

### (v) Woodlands

Where on death the deceased's estate includes woodlands, an election may be made to exclude the value of the timber from the value of the estate. Thus, no IHT will be payable on the value of the timber at that time. If there is a subsequent disposal or transfer of the timber in the new owner's lifetime, then IHT becomes payable at the deceased's marginal rate of IHT on the disposal proceeds. If the new owner retains the timber to his death, then no IHT will be payable on the original owner's estate at death in respect of the timber.

In determining the value of the deceased's estate where this relief is claimed, the value of the land on which the woodlands grow is included at what is often termed "prairie value". Since it is land covered with trees, the value will be relatively low.

Where a person owns and runs woodlands as a commercial enterprise, the business relief referred to above may be claimed. Where the woodlands are ancillary to an agricultural holding, then agricultural relief may be claimed. Both of these reliefs are more favourable.

### 6. Valuation

The general principle for determining the value of a person's estate which is the subject of a transfer is to assume a sale at market value being the price which might reasonably be expected to be fetched if sold in the open market at the time of the transfer. No reduction is to be made because such a sale would have "flooded the market". Further, it is to be assumed that the assets would be sold to achieve the best price, which means that if a higher overall price would be achieved by selling parts rather than as a single lot, then it is to be assumed that the sale would follow that approach.[12]

## Income Tax on Property

### 1. Schedule A

Income tax is charged under Schedule A to the Income and Corporation Taxes Act 1988, on income from rents and certain other receipts from property after deductions have been made for allowable expenses.

Liability under Schedule A extends to rent under leases of land and certain other receipts such as rent charges. Premiums receivable on the grant of a lease for a term not exceeding 50 years are also subject to taxation under Schedule A. The amount of the premium taxable is the whole premium less 2% for each complete year of the lease, except for the first year[13] $(2(n-1)\%$ of the premium where n is the number of complete years of the lease).

Deductions from the gross rents receivable may be made in respect of allowable expenses incurred, including cost of maintenance, repairs, insurance, management, services, rates and any rent payable to a superior landlord.

Rents receivable from furnished lettings are similarly chargeable to income tax on the gross rents less allowable expenses. The whole of the rent will be charged under Case VI of Schedule D, but the landlord may elect for the profit arising from the property itself to be assessed under Schedule A, with the profit from the furniture continuing to be assessed under Case VI. A writing-down allowance may be claimed in respect of furniture and any other items provided. As an alternative to this, the taxpayer may choose to deduct the expenses of renewing these items.

---

[12] *Duke of Buccleuch* v *Inland Revenue Commissioners* [1967] 1 AC 506.
[13] See Part 1(l) of this chapter for Capital Gains Tax provisions concerning premiums.

## 2. Capital Allowances

Items of a capital nature, such as expenditure on improvements, additions or alterations, cannot be deducted except for writing down allowances of the cost of certain specific categories. These capital allowances include expenditure on such items as the provision or improvement of agricultural or industrial buildings and the provision of certain plant and machinery. These allowances generally take the form of annual allowances such as 4% of the cost of industrial or agricultural buildings, and 25% of plant and machinery on a reducing cost basis. 100% allowances are available in Enterprise Zones. The provisions which govern the availability and calculation of capital allowances are contained in the Capital Allowances Act 1990.

## Value Added Tax

### 1. VAT and Property

Value Added Tax (VAT) was introduced in 1973 when the UK joined the European Community (EC). The rules for membership of the EC are contained in the Treaty of Rome and associated Directives, and include a rule that member states must operate a VAT system of taxation. VAT is thus a eurotax and the detailed provisions of VAT legislation are consequently capable of challenge in the European Court. That this is so, is well illustrated by the fact that VAT in respect of property was radically changed in 1989 following a decision of the European Court in 1988.

Prior to 1989 most property activities were effectively outside the VAT net but, since the Finance Act 1989, many activities are now subject to VAT and the number is growing. The main legislation is contained in the Value Added Tax Act 1994.

### 2. VAT – The General Scheme

#### (a) Added Value

VAT is chargeable on added value when a taxable supply is made by a registered person in the course or furtherance of a business. Some definitions are required before this statement can be understood.

(i)  *Supply*

This refers to the sale of goods or the provision of services. They are expressed as being the supply of goods or the supply of services.

(ii)  *Registered Person*

Where a company, individual, partnership or other body makes taxable supplies above a prescribed level over a prescribed period that supplier is required to register for VAT. The result of that is that each supplier is allotted a unique VAT registration number which is used in respect of all VAT activities. VAT returns are then made at the end of successive three-monthly periods when the supplier provides the details required by Her Majesty's Customs and Excise ("HMCE") who administer the tax.

(iii)  *In the course or furtherance of business*

There is no detailed definition of "business". It "includes any trade, profession or vocation" and also the provision of facilities to subscribing members of a club and admission of persons to any premises for a consideration.[14] The meaning of business is fairly widely interpreted but clearly excludes activities by individuals acting in their personal capacity, such as when they sell their own car or house, or assist organisations in a voluntary capacity.

(iv)  *Added Value*

The VAT system operates by applying tax to the value added to goods as they pass through the business chain.

*Example 22–8*

A produces raw materials which are sold to B for £100. B converts the raw materials into parts which are sold to C for £240. C assembles the parts to produce an item which is sold to D, a person who will use the item in his home, for £400.

---

[14] VAT Act 1994, Section 94.

| Transaction | Price Paid (a) | Sale Price (b) | Value Added (b)–(a) |
|---|---|---|---|
| A to B | — | £100 | £100 |
| B to C | £100 | £240 | £140 |
| C to D | £240 | £400 | £160 |
| | | | £400 |

Assuming the rate of tax is 10% then the tax payable to HMCE will be:

| Transaction | Value Added | Tax (at 10%) |
|---|---|---|
| A to B | £100 | 10 |
| B to C | £140 | 14 |
| C to D | £160 | 16 |
| | | 40 |

In practice, this result is achieved by each party in the chain acting in the course or furtherance of a business charging tax on the whole sale proceeds and deducting any tax paid on the purchase costs.

| Party | Tax Charged on Sale (x) | Tax Paid on Purchase (y) | Tax Paid to HMCE (x − y) |
|---|---|---|---|
| A | 100 @ 10% = 10 | — | 10 |
| B | 240 @ 10% = 24 | 10 | 14 |
| C | 400 @ 10% = 40 | 24 | 16 |
| | | | 40 |

As can be seen, the one person who bears the tax, since it is a real irrecoverable cost, is D, the final consumer, whereas it is collected and passed over to HMCE by the parties in the chain of production as it passes down the chain.

Tax charged on a sale price is termed output VAT and that paid on the purchase price is termed input VAT. The quarterly tax returns referred to above require the taxpayer to total output and input taxes over the quarter and to pay the excess of output VAT over input VAT to HMCE. If input VAT exceeds output VAT then HMCE reimburse the difference to the taxpayer.

*(b) Taxable Supplies*

The VAT scheme applies to persons making taxable supplies in the course or furtherance of a business. As has been shown, the scheme operates by taxpayers making VAT returns whereby they pay to HMCE the excess of output VAT which they have collected from the input VAT which they have paid out. This is described as the recovery of input VAT. The effect is that VAT is a neutral tax for such persons (apart from the costs of managing their VAT records).

However, the VAT legislation divides supplies of goods or services into three categories – two of which are of taxable supplies and one that is not. Hence, where a person makes supplies in the non-taxable supply category, the scheme does not apply as described so far for such a person, who is treated as a consumer; any input VAT paid out is not recoverable.

The two categories of taxable supplies are standard rated supplies and zero rated supplies.

(i)  *Standard Rated Supplies*

Standard rated supplies are all supplies which are not found in the detailed list of supplies coming within the other two categories. They are known as standard rated supplies since VAT is applied at the prevailing standard rate of VAT – 17.5% at the time of writing. Hence, if goods are sold for £100 and they constitute a standard rated supply, then VAT of £17.50 is added to the sale price. (The supply of domestic fuel has imposed a rate below the standard rate but is more by accident than design.)

(ii)  *Zero Rated Supplies*

These supplies are also taxable supplies. However, the rate of VAT charged is zero so that no VAT is actually added to the price. Zero rated goods sold for £100 will have the sale price of £100 with nothing added. Nonetheless, the supplier of zero rated goods or services is entitled to recover input VAT paid out on purchases. If the supplier makes wholly zero rated supplies, the quarterly return would show output VAT of nil so that any input VAT paid out would be fully refunded by HMCE.

Zero rating exists within the VAT scheme to keep down the price of goods and services to consumers where it is considered socially desirable to do so. This is demonstrated by the types of supplies which are zero rated, including food, children's clothing, most public transport, and products designed for use by disabled persons. A full and detailed list of zero rated supplies is contained in Schedule 8 to the VAT Act 1994. Those relating to property are considered below.

### (c) Non-Taxable Supplies

The third category of supplies of goods or services is termed Exempt Supplies. Where a person makes an exempt supply, no VAT is added to the sale price. To this extent it is like a zero rated supply. However, wholly unlike a zero rated supply, the goods or services are not subject to any VAT charge at all, neither 17.5% nor 0%. A person making wholly exempt supplies does not make taxable supplies, does not register for VAT, makes no VAT returns, and does not recover any input VAT paid out on purchases of goods or services. For such a person any VAT paid out on purchases of goods or services is a charge which increases the cost of purchases, as it does for consumers generally. Where a person makes some taxable supplies and some exempt supplies ("makes partially exempt supplies"), any input VAT incurred on goods or services attributable to an onward exempt supply by that person cannot be recovered notwithstanding that the person be registered for VAT.

Those items which comprise exempt supplies are fully set out in Schedule 9 to the VAT Act 1994. They include supplies by the Post Office of postal services, financial services provided by banks, products of insurance companies, education provided by schools and universities, and health services provided by doctors, dentists and opticians. It is important to be familiar with those items constituting exempt supplies since exempt suppliers are likely to be VAT-averse and this could condition their response to proposals put to them. Those supplies related to property are considered below.

### 3. VAT and the Property Market

In order to consider the impact of VAT on the property market, it is essential first to identify the VAT treatment of the various property activities.

## (a) Residential Property

Zero rated supplies are mainly confined to the provision of new residential accommodation. For this purpose, residential encompasses the provision of accommodation with a view to long-term residence such as care homes for the aged or infirm, or children's homes, student hostels, hospices, even nunneries. It does not include hotels or prisons.

Thus, the construction of new houses or flats, including most of the building materials, are zero rated supplies, as are the final sale or letting, other than when they are let for less than 21 years. This applies also to similar sales or lettings following the conversion of non-residential property and to certain types of significant work to listed buildings so as to produce residential property which is to be sold or let for not less than 21 years.

It follows that the sale of a new house or flat has no actual VAT added to the sale price, but is subject to the notional VAT at nil %. All other transactions in relation to residential property are either exempt supplies or are supplies not in the course or furtherance of a business – which is the situation for the great majority of transactions. As a result, no VAT is chargeable on any transactions involving residential property.

On the other hand, work carried out to residential property such as repairs or improvements is a standard rated supply. Since the supply is in respect of property for which no taxable supply can be made, VAT increases the cost of such work to the property owner. Hence the valuation of a residential property where a deduction is made for repairs requires the cost to be included gross of VAT (unless the work is of a small scale which may be carried out by a small builder – see (b)(ii) below).

## (b) Non-Residential Property

Consideration of the VAT implications of activities in respect of non-residential properties will be considered in respect of sales and lettings, and development and building works.

(i) *Sales and Lettings*

All sales and lettings are generally exempt supplies. However, there are specific exceptions to this general rule. Most important are:

(a) The sale of the freehold interest in a new or partly completed building. "New" means a building completed less than three years ago. Such a sale is a standard rated supply so that VAT must be added to the sale price.

(b) The sale of the freehold interest in a new or partly completed civil engineering work. This is also a standard rated supply. In practice, this will usually relate to infrastructure. For example, if an estate road and some services have been built into a development site and completed within the past three years then, on a sale of the freehold interest in the site, VAT is added to the sale price, or that part of the price reflecting the infrastructure. This could require an apportionment of the price between the standard rated supply, on which VAT is charged, and the balance which is an exempt supply where no VAT arises.

(c) The provision of space for car parking is a standard rated supply. This applies to licences, a fact familiar to all car drivers who pay VAT on the charge for parking their car.

There are various other supplies set out in Group 1 of Schedule 9 which are also standard rated, many of them related to sports and leisure activities.

(ii) *Development and Building Works*
The supply of building works of all kind is a standard rated supply. Hence, demolition, new construction, refurbishment, and repairs are all standard rated supplies on which VAT is added at 17.5% to the cost of the work.

The exception would be if a builder has a low turnover below the VAT registration threshold. Such 'small builders' are not required to register and so do not add VAT to their charges.

## 4. Option to Tax

As is clear from the above, most transactions are exempt supplies, whereas physical works are standard rated. This creates a VAT burden where the owner of a property carries out work to the property on the cost of which VAT will be charged. Where the owner will make exempt supplies if the property is subsequently let or sold, the VAT on the cost of the work is irrecoverable.

However, the Act provides that an owner of property may elect to have all supplies of the property treated as standard rated supplies – the option to tax. The incentive to elect is that, because supplies become taxable supplies, input VAT is recoverable. So an owner carrying out work to a property will be able to recover the input VAT on the cost of the work if the election is made. If the cost of the work is significant, with a large amount of input VAT payable, the consequence of an election could be highly beneficial.

*Example 22–9*

X owns the freehold interest in office premises which have recently been vacated on the expiry of a lease. The property can be let in its present condition for £100,000 p.a. If X spends a total sum of £300,000 on refurbishment, the property will be let at £120,000 p.a. and the yield will fall from 9% to 8%.

(a) Value without refurbishment

| | | |
|---|---|---|
| Rental Value | | £100,000 p.a. |
| YP in perp at 9% | | 11.1 |
| Capital Value | | £1,110,000 |

(b) Value with refurbishment (no election to Tax)

| | | |
|---|---|---|
| Rental Value | | £120,000 p.a. |
| YP in perp @ 8% | | 12.5 |
| | | £1,500,000 |
| *Less* | | |
| Refurbishment Cost | £300,000 | |
| *Add* | | |
| VAT @ 17.5% | 52,500 | 352,500 |
| Capital Value | | £1,147,500 |

Increase in Capital Value £1,147,500 – £1,110,000 = £37,500
Return on Costs = 10.64%

(c) Value with refurbishment (X elects to Tax)

| | | |
|---|---|---|
| Rental Value | | £120,000 p.a. |
| YP in perp @ 8% | | 12.5 |
| | | £1,500,000 |
| *Less* | | |
| Refurbishment Cost | | 300,000 |
| Capital Value | | £1,200,000 |

Increase in Capital Value £1,200,000 – £1,110,000 = £90,000
Return on Costs = 30.00%

*Note*

(i)   As a result of the election, the return on the additional investment is 30%, which would justify carrying out the work. If no election is made, the return of 10.64% would render the additional investment of marginal benefit.

(ii)  VAT would be payable on the rent by the tenant but this would be "passed on" to HMCE so is ignored.

(iii) If the offices are in an area where potential tenants making exempt supplies form a dominant part of the market the yield might need to be increased to say 8.25% if X elects. This would still justify an election.

(iv)  If the cost of the work were £350,000, the value would increase by £40,000 if the election were made, but would be £21,250 lower if no election were made.

The effect of opting to tax a property is that all supplies become standard rated so that, if the property is let, VAT at 17.5% is added to the rent, or, if sold, to the sale price.

If the tenant or purchaser makes taxable supplies, then input VAT on the rent or price is of little or no consequence since they would recover the input VAT. However, if they make exempt supplies, then the imposition of input VAT would be a deterrent to them in taking a lease or purchasing the property. This might have an adverse effect when marketing the property, particularly if exempt suppliers are the most likely tenants or purchasers. This is a factor that an owner would need to take into account when deciding whether to elect.

When an election is made, if the property is a building then the election will apply to the whole building (but does not apply to owners of other interests in the building). In the case of land, the owner can define the area of land to which the option shall apply.

If a building is part residential, the option will not apply to the residential part. Hence, if a shop with a flat over is let in a single lease at £20,000 p.a. and the landlord opts to tax, the rent would have to be apportioned between the shop and the flat (assume £18,000 p.a. and £2,000 p.a. respectively), when VAT will be payable on £18,000 p.a. only.

An election is irrevocable apart from an initial three-month period or after 20 years. The election ceases to be of effect if a building to which it applies is demolished. It also ceases when ownership ceases following a sale. It does not bind a purchaser. However, since the purchaser will have paid VAT on the purchase price there will be a strong incentive on the purchaser to elect in order to recover the input VAT. This will not be the case if the property is vacant and the purchaser proposes to occupy for business purposes making taxable supplies: in these circumstances the purchaser can recover the input VAT on the purchase price as normal input VAT of the business.

The option to tax may be ineffective if the building is one designed to be occupied by a business making wholly or mainly exempt supplies. Hence, a bank building offices for its own occupation may not opt to tax the building. However, if a developer built the offices and let them to the bank, the bank making no financial contribution to the development, then an option would be effective.

## 5. Transfer of a Going Concern

Under the VAT legislation, the transfer of a going concern (TOGC) is not a supply for VAT purposes. Hence no VAT charge can arise when a transaction is a TOGC.

HMCE take the view that the letting of property can be a going concern. If so, the change of ownership of a property investment may be treated as a TOGC. Whether this is the case is of no importance if the owner of such a property has not opted to tax, since the sale of the landlord's interest will be either an exempt supply or a TOGC. In either case, no VAT charge will arise. An exception to this would be if the landlord's interest is the freehold and the building is less than three years old, as the sale of the freehold interest would be of a new building and so standard rated if not a TOGC.

Whether a sale of the landlord's interest is a TOGC or not is of crucial importance where the landlord has opted to tax (or if it is the sale of the freehold interest in a new building) since VAT is chargeable if it is not a TOGC but is not chargeable if it is a TOGC. The position affects both vendor and purchaser.

The view of HMCE is that, if the purchaser elects to tax the building prior to completion of the purchase, the sale will be a TOGC but not otherwise. The reasoning behind this is that all circumstances, including the VAT position, will be the same both before and after the change of ownership and VAT will thus be payable on the rents from the lettings without a break.

## 6. Value and VAT

In nearly all cases VAT is charged on the consideration passing. In some cases the consideration may not consist wholly or partly of money, in which case the value may need to be determined. This can also be the case where there is a transaction between connected persons at less than open market value.

Valuers may be called upon to provide the value required. There is no clear guide in the legislation as to valuation assumptions that the valuer may or may not make, save that it must be assumed that the parties are, effectively, at arm's length.

## 7. Valuation and VAT

As a general principle, the inclusion of VAT on the elements of a valuation will arise where VAT is a cost to the interest being valued because it is not recoverable.

For example, in preparing a valuation where an allowance for expenditure has to be made and VAT is payable on that expenditure, the valuation will allow for the gross of VAT cost if the input VAT cannot be recovered. So, if the valuation is of a residential property, any allowance for expenditure on matters such as repairs or improvements will be included gross of VAT. An exception would be the valuation of a listed building where the expenditure is of a nature which is within the zero-rated provisions of Schedule 8 to the VAT Act 1994.

In the case of residual valuations, building costs and fees will normally be included net of VAT. This is so because, if the development is residential, it will probably result in a zero rated supply; if not residential, it is probable that the developer (if not selling the freehold interest of the new building on completion – a standard rated supply) will opt to tax in order to recover VAT on the development costs and so will make a standard rated supply of the building.

An exception would be the residual valuation of a building to be developed and used to make exempt supplies. For example, the residual valuation of a hospital or doctor's clinic development, or of a university teaching block, would include costs and fees gross of VAT.

The valuation of an occupier's leasehold interest in a building where the landlord has opted to tax and the occupier makes exempt supplies, and the use under the lease is limited to the making of exempt supplies, is a further example where expenditure, in this case the rents payable, should be included gross of VAT.

As is clear, the inclusion of VAT in a valuation is something which turns on the facts of the case, with the test being, can the input VAT be recovered?

## 8. Special Cases

### (a) Incentives

When the market is weak, property owners are prone to offer incentives to tenants to take a lease of their premises. These take various forms, with the most common being long rent-free periods and capital payments to the tenant (reverse premiums).

In the case of rent-free periods, this is not treated as a supply by the tenant so that no VAT is chargeable. The exception is where the rent-free period is in return for some act by the tenant which benefits the landlord when it is, in effect, a reward or payment to the tenant. In such cases, the value of the rent-free period has to be determined and the tenant charges the landlord VAT on that value. A typical example is where a tenant takes a lease of property which requires substantial work to it which the tenant agrees to carry out in return for the rent-free period.

As to reverse premiums it was held in the Neville Russell case[15] that a payment to the tenant must be in return for services by the tenant which are a standard rated supply. Hence the tenant should charge the landlord VAT on the amount of the premium. Failure to do so brings into play the general rule that, if VAT should have been charged but no explicit addition for VAT has been made, the payment is deemed to include the VAT element. So, for example, a tenant receiving £100,000 as a reverse premium should charge the landlord £17,500 VAT (at standard rate of 17.5%). Failure to do so would result in HMCE treating the payment as £85,106.33 plus VAT (at 17.5%) of £14,893.62, a total of £100,000. The VAT element would be payable by the tenant to HMCE.

---

[15] *Neville Russell* v *HMCE* (1987) VAT TR 194.

The Neville Russell decision is contradicted by the VAT treatment of other incentives and similar situations and it would be no surprise if it were overturned. It remains in effect, however, at the time of writing.

### (b) Statutory Compensation

Payments to agricultural or business tenants under the relevant statutes following the termination of their leases are not a supply for VAT purposes. Indeed, such payments are tax free for the tenants.

### (c) Mixed Supplies

Where a mixed supply is made of two or more elements which would be different types of supply if made separately, the type of supply is determined by the dominant purpose if one supply is ancillary to the other. For example, a letting of offices is an exempt supply (unless the landlord has opted to tax), whereas a separate letting of car parking spaces in the building to the same tenant would be a standard rated supply. However, a letting of offices with car spaces in a single lease would be an exempt supply since the dominant purpose is the letting of the offices to which the car spaces are an ancillary matter.

Similarly, if a building is opted to tax and is compulsorily purchased, VAT would be payable by the acquiring authority on the purchase price of the building and also on any payment of compensation for disturbance which is an ancillary matter. However, no VAT is payable on compensation for injurious affection since no supply of the land affected is being made.

Where each supply is complementary to the other, then any consideration would have to be apportioned between them. So, the letting of a shop with a flat above in a single lease would probably be treated as two separate supplies.

## Stamp Duty

### 1. Purchases

Stamp Duty is payable on certain deeds. In the case of property, a sale is effected by a deed. So stamp duty is payable on a sale by the purchaser. The amount of duty is dependent on the price. The rate of duty remained fixed for many years but was increased sharply in successive years from 1997. The rate fixed for sales after 5 April 1998 was nil on prices up to £60,000, 1% above £60,000 and up to £250,000, 2% above £250,000 up to £500,000, and 3% above £500,000, the rates of 2% and 3% being increased to 2.5% and 3.5% respectively from 9 March 1999. Note that the duty is charged on the whole price so that, for example, a purchaser paying £70,000 pays 1% of £70,000 = £700 whereas no duty is payable on a price of £60,000. The duty at £70,000 is, in effect, at the rate of 7% on the marginal sum. Purchasers will tend to resist prices close to the changeover point because of this approach.

## 2. Leases

Further stamp duty is payable by lessees when taking a lease. The amount of duty depends on the length of the lease and the amount of rent. The duty is a percentage of the rent which increases when the term goes beyond seven years, 35 years and 100 years. As a result, it is common for a term of building lease agreed to be for 125 years to take the form of a lease for 99 years and an option for a further 26 years. This halves the duty otherwise payable on a lease of 125 years. It also explains why the traditional building lease is for the curious term of 99 years.

# Chapter 23

# Compensation Under the Town and Country Planning Acts I
# Revocation, Modification and Discontinuance Orders, Etc.

Until 1991 the law on this subject was in its essentials a direct survival from the immediate post-war period when the Town and Country Planning Act 1947 was passed. But the Planning and Compensation Act 1991 swept away the remains of the post-war system of compensation for planning restrictions – one of the two pillars of the "compensation betterment" question – except for cases of revocation and discontinuance (which occur but rarely in practice). There is now no trace remaining of the concept which emerged with the Uthwatt Report during the Second World War, namely that some kind of national balance should be achieved between: (i) a proportion of prospective development value which should be transferred to the state and not left to landowners in the event of a grant of planning permission; and (ii) a proportion of that value which should be conceded to landowners as compensation in the event of a refusal of planning permission.

## 1. Revocation, Modification and Discontinuance Cases

The survival of planning compensation in revocation, modification and discontinuance cases represents something quite different, which is the loss not of *prospective* development value but of *authorised* development value. In revocation cases (though not discontinuance cases) this distinction is probably more apparent than real; but it can be traced back to the Town and Country Planning Act 1954 which gave effect to the then Government's belief that some planning compensation should be available within modest limits.

The real practical distinction is that local planning authorities cannot avoid making decisions on a vast number of planning applications each year, many of which must inevitably be refused (or granted subject to burdensome conditions), but they can certainly avoid making more than a very few decisions to revoke planning permissions already granted (though the Secretary of State can do it for them and saddle them with the consequences). The initiative lies with them in revocation cases but most certainly not in cases of refusal. The former continue to involve a heavy burden of compensation; the latter none.

## 2. Revocation and Modification Orders

Under Sections 97–100 of the 1990 Act a local planning authority may, by means of an order which must be confirmed by the Secretary of State if it is opposed by the owners or occupiers of the land in question or anyone else with any property interest in the land, but does not require such confirmation in other cases, revoke or modify a planning permission already given; but a revocation or modification order can only be made (a) before building or other operations authorised by the permission have been completed, in which case work already done will not be affected by the order, or (b) before any change of use authorised by the permission has taken place.

If the proposed Order is unopposed, the local planning authority need not submit it to the Secretary of State for confirmation. Alternatively, the Secretary of State may himself make a revocation order, provided that he first consults the local planning authority.

Not less than 28 days' notice of the proposed Order must be given to the owner and occupier of the land affected and to any other person who, in the opinion of the local planning authority will be affected by the Order, to enable them to apply to be heard by a person appointed by the Secretary of State.

When the Order takes effect, Section 107 of the 1990 Act provides for the payment by the local planning authority of compensation under the following heads:

(i)  Expenditure on work which is rendered abortive by the revocation or modification – including the cost of preparing plans in connection with such work or other similar matters preparatory thereto.

But no compensation will be paid in respect of work done before the grant of the relevant planning permission, e.g. in anticipation of such permission being granted.

(ii)  Any other loss or damage directly attributable to the revocation or modification, which includes depreciation of the land value in consequence of the total or partial revocation of planning permission.

No compensation of this kind was payable before 1954 (i.e. under the 1947 Act) for depreciation in the value of land unless either (a) a development charge had been paid, or (b) the land was exempt from the charge under Part VIII of that Act.

Claims for compensation must be served (in writing) on the local planning authority within 12 months of the revocation or modification order, or of the adverse decision under Section 108 (below).[1] In accordance with the provisions of Section 117 of the 1990 Act, compensation for depreciation in the value of an interest in land will be calculated on a market value basis as prescribed by Section 5 of the Land Compensation Act 1961, ignoring any mortgage to which the interest may be subject. The measure of compensation will be the amount by which the value of the claimant's interest with the benefit of the planning permission exceeded the value of that interest with the planning permission revoked or modified.

In calculating such depreciation it is to be assumed that planning permission would be granted for development within paragraphs 1 and 2 of Schedule 3 of the 1990 Act. These comprise development that would involve:

(i) *Rebuilding* any building built on or before 1 July 1948 which was demolished after 7 January 1937, or any more recently built building "in existence at a material date", provided that the cubic content "is not substantially exceeded", and

(ii) *Conversion* of a single dwelling into two or more dwellings.

The gross floor space to be used for a given purpose must not be increased by more than 10% in the rebuilt building, if the previous building was original to the site or built before 1 July 1948, and must not be increased at all in other cases (1990 Act, Schedule 10).

Thus, if planning permission for actual development within Schedule 3 is revoked or modified, compensation must be assessed on the assumption that development value within Schedule 3 is enjoyed by the claimant, whereas of course the opposite is true. In the absence of compulsory purchase (by purchase notice or

---

[1] Town and Country Planning General Regulations 1992 (SI 1992 No 1492), Regulation 12.

otherwise) there can be no redress. The same unreal assumption applies also when permission for new development is revoked.

In *Canterbury City Council* v *Colley*,[2] the House of Lords dealt with the question of compensation for revocation. The facts were that in 1961 planning permission had been granted to demolish a house at Whitstable and build a new and larger one on the same site. The demolition occurred but no new house was built. The planning permission was revoked in 1987. The owners of the land claimed compensation under what is now Section 107 of the 1990 Act. The existing use value of the vacant site was then £8,250. Full development value with the benefit of the 1961 planning permission (had it not been revoked) was £115,000. The difference was the measure of loss of land value, i.e. £106,750. But on the assumption that the site must be treated as having the benefit of planning permission to rebuild, under what is now Schedule 3 of the 1990 Act, the value of the site was £70,000, reducing the depreciation from £106,750 to £45,000. Yet no such permission had been granted. The House of Lords held that on the plain meaning of the statutory wording the compensation must be restricted to £45,000. This is clearly an unjust situation. It does not apply to assessment of compensation under Section 115 (see below: Discontinuance Orders).

Section 108 of the 1990 Act applies the above provisions as to compensation where planning permission given by a Development Order is withdrawn – whether by revoking or amending the Order or by making a direction under it[3] – and, on an express planning application being then made, permission is refused or is granted subject to conditions other than those imposed by the Development Order. In cases where the Development Order is itself revoked or modified, the express planning application must be made within a year thereafter.[4]

Disputes as to the compensation payable under Sections 107 and 108 are referable to the Lands Tribunal (Section 118). Special provisions are laid down for compensation in cases of revocation or modification of planning permission for mineral development,

---

[2] [1993] 1 EGLR 182.

[3] For example, a direction made by the Secretary of State or the local planning authority under Article 4 of the General Permitted Development Order 1995.

[4] Town and Country Planning Act 1990, Section 108(2).

whether under Section 97 of the 1990 Act or by way of various mineral orders under Schedule 9 of the Act (Section 116).

Revocation or modification of planning permission may entitle the owner to serve a purchase notice under Part VI of the Act.[5]

Provision is made[6] for the apportionment of compensation for depreciation in value exceeding £20 between different parts of the land, where practicable, and for the registration of the "compensation notice" as a land charge. Compensation so registered will be repayable in whole or in part if permission is subsequently given for development which consists (wholly or largely) of the construction of residential, commercial or industrial buildings or any combination thereof, or of any other activity to which the Secretary of State believes these provisions should apply by reason of the likely development value. This applies whether the development occurs on the whole or only part of the land affected by the order, but not where the registered compensation was paid in respect of a local planning authority's order modifying planning permission previously given and the subsequent development is in accordance with that permission as modified.[7]

## 3. Discontinuance Orders

Under Sections 102–104 of the Act of 1990 a local planning authority may, by means of an order confirmed by the Secretary of State, at any time require the discontinuance or modification of an authorised use of land (as in *Blow* v *Norfolk County Council*,[8] in which the authorised use of land as a caravan site was required to be discontinued) or the alteration or removal of authorised buildings or works.[9]

Section 115 provides that any person who suffers loss in consequence of a discontinuance order, either through depreciation

---

[5] Section 137(1)(b). See Chapter 25.

[6] Sections 109–110.

[7] Sections 111–112.

[8] [1966] 3 All ER 579.

[9] Where, on the other hand, uses or works are "unauthorised", i.e. have been begun or carried out without a grant of planning permission or contrary to conditions attached to such permission, the local planning authority will proceed by means of an "enforcement notice" under Part VII of the T&CP Act 1990, so that no compensation is payable. On this, see Chapter 14.

in the value of his land, or by disturbance, or by expense incurred in complying with the order, is entitled to compensation provided his claim is made within 12 months.[10] An example of this type of case is given later in this chapter. Interest at a prescribed statutory rate is payable on the compensation from the date that the discontinuance takes effect.[11] As an alternative to claiming compensation for depreciation in the value of his land, the owner of the interest may be able to serve a purchase notice under Part VI of the Act.[12] Special provisions apply to certain discontinuance orders relating to mineral development (see Section 116).

## 4. Other Cases

Compensation on account of planning requirements may also be payable in connection with the following matters under the Town and Country Planning Act 1990:

(i)   Refusals, or restrictive grants of consents; or directions for replanting trees without financial assistance from the Forestry Commission under tree preservation orders (Sections 203–4).
(ii)  Cost of removal of advertisements in certain special and limited circumstances (Section 223).
(iii) Stop notices (Section 186).
(iv)  Pedestrian precinct orders (Section 249 in Part X).

Compensation is payable in these cases, as in those mentioned earlier in this chapter, by the local planning authority to whom application has to be made within six months of the official decision giving rise to the claim (12 months in the case of stop notices and tree preservation orders).[13]

The making of pedestrian precinct orders (in so far as they cause properties to lose vehicular access to highways), entitles owners to

---

[10] Town and Country Planning General Regulations 1992 (SI 1992 No 1492), Regulation 12.
[11] Planning and Compensation Act 1991, Schedule 18, Part I.
[12] Section 137(1)(c).
[13] Town and Country Planning General Regulations 1992 (SI 1992 No 1492), Regulation 12 and 14; Town and Country Planning (Control of Advertisements) Regulation 1992 (SI 1992 No 666), Regulation 20; Town and Country Planning (Tree Preservation Orders) Regulations 1969 (SI 1969 No 17), Model Order.

compensation for depreciation in value of the properties affected. Stop notices which fail (in effect) for lack of justification entitle persons affected by them to compensation for resultant "loss or damage". So do restrictions imposed on applications for consents under tree preservation order, except where the refusal, or restrictive grant, of a consent under a tree preservation order is accompanied by a certificate (not quashed on appeal) that specific trees are to be preserved "in the interests of good forestry" or by reason of their "outstanding or special amenity value". In *Bell* v *Canterbury City Council* (1988),[14] the Court of Appeal held that "loss or damage" in relation to the refusal of consent to remove protected trees included not only the value of the timber but also the reduction in the existing use value of the land on which the trees stood. Compensation in respect of advertisements is payable only "in respect of any expenses reasonably incurred" in carrying out works (i) to remove advertisements, or (ii) to discontinue use of any site for advertising, dating back (in either case) to 1 August 1948.

## 5. Listed Buildings and Ancient Monuments

In the case of listed buildings of special architectural or historic interest, Section 28 of the Planning (Listed Buildings and Conservation Areas) Act 1990 provides for payment of compensation for orders revoking or modifying listed building consents (except for unopposed orders) on the same basis as revocation or modification order compensation under the Town and Country Planning Act 1990 Section 107, described above.

Section 29 authorises payment of compensation for loss caused by the issue of a building preservation notice, on a similar basis to compensation for stop notices (above).

Claims must be made within six months of the making of the order.[15]

Part I of the Ancient Monuments and Archaeological Areas Act 1979 empowers the Secretary of State to make a "schedule of monuments" (rather on the lines of listed buildings) in order to protect ancient monuments. Consents are then required for carrying out works affecting ancient monuments and the Secretary

---

[14] [1988] 1 EGLR 205.
[15] Planning (Listed Buildings and Conservation Areas) Regulations 1990 (SI 1990 No 1519), Regulation 9.

of State must pay compensation to anyone who "incurs expenditure or otherwise sustains any loss or damage" in consequence of a refusal, or conditional or short-term grant, of consent by him (Sections 7–9).

## 6. Examples

*Example 23–1*

*Claim following revocation of a planning permission*

The owner of the freehold interest in a large Victorian house on the outskirts of a town obtained planning permission two years ago to redevelop the site with a new office building with a gross internal area of 700 m². The house was demolished last year at a cost of £2,000.

The owner recently invited tenders to carry out the development.

The local planning authority has changed its policy and has determined that commercial development should be mainly confined to the town centre. In consequence, it has made an Order under the Town and Country Planning Act 1990 to revoke the planning permission granted two years ago.

The owner has incurred professional fees of £50,000 to date, of which £15,000 relate to advice leading to the grant of the planning permission.

Compensation for the revocation is payable under Section 107 of the 1990 Act.

*Calculation of Compensation*

(a) Depreciation in value of freehold interest

| | |
|---|---:|
| Value with planning permission for offices. 700 m² @ say £300 per m² = | £210,000 |
| *Less* Value following revocation. | |
| Assume planning permission would be granted for development under Part I of Schedule 3 = right to rebuild former house: | |
| Value as residential site to build a large detached house say | <u>£100,000</u> |

| Depreciation in value | £110,000 |
|---|---|

(b) Abortive Expenditure

| (i) | Professional fees | £50,000 |
|---|---|---|
| (ii) | Demolition | £2,000 |
| | | £52,000 |

*Note*:
(a) If demolition had occurred before the grant of planning permission it could not be claimed as being work done in anticipation of the grant of permission. Fees incurred before the grant of permission are not excluded (*Burlin* v *Manchester City Council*).[16]
(b) A surveyor's fee under Ryde's Scale is payable in respect of the claim.

*Example 23–2*

*Claim for discontinuance of an existing use*

A factory has been established in its present building for many years. It is occupied by the freeholder. The building was constructed 60 years ago and the accommodation is on two floors, each having a gross internal area of 300 m². There are no offices other than "works offices". There is a small yard, and the total site area is 0.04 hectares.

The property is located in a predominantly residential area and is allocated for this use in the local plan. The local authority, following complaints about lorry traffic, smells and noise by local residents have ordered that it shall cease to be used for its present purpose.

It is required to assess the compensation payable under Section 107 of the Town and Country Planning Act 1990, in respect of the discontinuance of the authorised use. No other premises are available in this area.

The order gives planning consent for the erection of a single dwelling-house on the site.

---

[16] [1976] 1 EGLR 205.

*Calculation of Compensation*

Value of premises as a factory (existing use)

| | | |
|---|---|---|
| Ground floor | 300 m$^2$ at £40 per m$^2$ = | £12,000 p.a. |
| First floor | 300 m$^2$ at £30 per m$^2$ = | £9,000 p.a. |
| Estimated net rental value | | £21,000 p.a. |
| YP in perp | | 7 |
| | | £147,000 |

| | | |
|---|---|---|
| Value of site with planning consent to erect a single dwelling-house, say | £40,000 | |
| *Less*: | | |
| Cost of demolition of existing buildings, say | £5,000 | £35,000 |
| Depreciation in value of freehold interest | | £112,000 |

In addition, the freeholder would be entitled to claim for incidental losses due to "disturbance", such as are discussed in Chapter 28, and the payment of surveyor's fees under Ryde's Scale for agreeing the compensation.

# Chapter 24

# Compensation Under the Town and Country Planning Acts II Purchase Notices

## 1. Compulsory Purchase Instigated by Land Owners

A remedy provided for landowners who suffer from adverse planning decisions or proposals is afforded by the provisions of Part VI of the Town and Country Planning Act 1990. This involves procedures under which local authorities and other public bodies are, in certain circumstances, compelled to buy land affected by such decisions or proposals.

Planning compensation as described in previous editions of this book was, of course, claimable only on the basis that the claimant retained the land in question and should be reimbursed for depreciation or other loss relating to such land. The Planning and Compensation Act 1991 abolished the previously subsisting rights to such compensation, except in cases of revocation or modification orders, or discontinuance orders.

But where a purchase notice takes effect, the land is transferred and not retained; in other words, there is a compulsory purchase, and the compensation payable is, in fact, compulsory purchase compensation. This, of course, does not in principle prevent the claimant from receiving lost development value in terms of money, just as he does in planning compensation for revocation, modification and discontinuance orders, but he will only do so to the extent that the principles of compulsory purchase compensation allow. There is in addition a special type of case in which compensation is payable even though a purchase notice is prevented from taking effect.

## 2. Purchase Notices

(Sections 137–148 – Town and Country Planning Act 1990 – i.e. Part VI, Chapter I, of the Act).

If, in any case, planning permission is refused, either by the local planning authority or by the Secretary of State, or is granted subject to conditions, then if the owner[1] of the land claims that it has become incapable of reasonably beneficial use:

(i)   in its existing state; or
(ii)  even if developed in accordance with such conditions (if any); or
(iii) even if developed in any other way for which permission has been, or is deemed to be, granted under the Act, or for which the Secretary of State or the local planning authority have undertaken to grant permission;

he may, within twelve months[2] of the planning decision in question, serve a "purchase notice" on the district council or (as the case may be) the London Borough Council or the Common Council of the City of London, depending on where the land is situated, requiring them to purchase his interest in the land.

The Council on whom a purchase notice is served shall, within three months of such service, themselves serve a counter-notice on the owner to the effect:

(a)   that they are willing to comply with the purchase notice; or
(b)   that another specified local authority, or statutory undertakers, have agreed to comply with it in their place; or
(c)   that, for reasons specified in the notice, neither the council nor any other local authority or statutory undertakers are willing to comply with the purchase notice, and that a copy of the purchase notice has been sent to the Secretary of State on a date specified in the counter-notice, together with a statement of the reasons for their refusal to purchase.

In cases (a) and (b) above, the Council on whom the purchase notice was served, or the specified local authority or statutory undertakers, as the case may be, will be deemed to be authorised to acquire the owners' interest in the land and to have served a notice to treat on the date of the service of the counter-notice, as if the land were being acquired under a compulsory purchase order.

---

[1] "Owner" means the person entitled to receive the rack rent of the land, or who would be entitled to receive it if the land were so let – Section 336(1). (See *London Corporation* v *Cusack-Smith* [1955] AC 337.)
[2] Town and Country Planning General Regulations 1992, Regulation 12.

In case (c), where the purchase notice is forwarded to the Secretary of State, the following courses of action are open to him:

(i) to confirm the notice, if satisfied that the land is in fact incapable of reasonably beneficial use in the circumstances specified in Section 137;[3]

(ii) to confirm the notice but to substitute another local authority or statutory undertaker;

(iii) not to confirm the notice but to grant permission for the required development or to revoke or amend any conditions imposed;

(iv) not to confirm the notice but to direct that permission shall be given for some other form of development;

(v) to refuse to confirm the notice, or to take any action, if satisfied that the land has not in fact been rendered incapable of reasonably beneficial use.

"Reasonably beneficial use" is to be judged in relation to the present use of the land.

In considering whether or not the use of any land is "reasonably beneficial", the Secretary of State must not take account of the possibility of any development outside those types specified in Schedule 3 to the 1990 Act.[4]

"Incapable of reasonably beneficial use" signifies that the land in question must be virtually useless and not merely less useful, because practically any planning restriction will result in land being less useful (*R v Minister of Housing and Local Government, ex parte Chichester Rural District Council*).[5]

---

[3] But even in this case, the Secretary of State is not obliged to confirm under certain special circumstances relating to land, forming part of a larger area, which has a restricted amenity use by virtue of a previous planning permission – Town and Country Planning Act 1990 – Section 142. Thus, if planning permission is given to build houses on a field, subject to a condition that for amenity reasons trees are to be planted along the road frontage, the developers cannot, after building and selling the houses, claim that the road frontage land which they retain is "incapable of reasonably beneficial use" so as to justify serving a purchase notice (for the previous law, see *Adams & Wade v Minister of Housing and Local Government* (1965) 18 P&CR 60).

[4] As amended by the Planning and Compensation Act 1991, Section 31 and Schedule 6.

[5] [1960] 2 All ER 407.

Before taking any of the steps enumerated above – including refusal to confirm – the Secretary of State must give notice of his proposed action to the person who served the purchase notice, the local authority on whom it was served, the local planning authority and any other local authority or statutory undertaker, who may have been substituted for the authority on whom the notice was served. He must also give such person or bodies an opportunity of a hearing if they so desire. If after such hearing it appears to the Secretary of State to be expedient to take some action under the Section other than that specified in his notice, he may do so without any further hearing.

If the Secretary of State confirms the purchase notice, or has not taken any action within nine months from the date when it was served on the local authority, or within six months of its transmission to him by the local authority (whichever period is the shorter), the local authority on whom it was served will be deemed to be authorised to acquire the land compulsorily. The authority is then deemed to have served a notice to treat on such date as the Secretary of State may direct, or at the end of the period of six months as the case may be. The notice to treat is deemed to be authorised as if the acquiring authority required the land "for planning purposes" under Part IX of the 1990 Act.

Any party aggrieved by the decision of the Secretary of State on the purchase notice may, within six weeks, make an application to the High Court to quash the decision on the grounds that either (i) the decision is not within the powers of the Town and Country Planning Act 1990, or (ii) the interests of the applicant have been substantially prejudiced by a failure to comply with any relevant requirements (1990 Act, Part XII, Section 288) but cannot otherwise challenge the validity of his decision (Section 284).

If the Secretary of State's decision is quashed, the purchase notice is treated as cancelled; but the owner may serve a further purchase notice in its place as if the date of the High Court's decision to quash were the date of the adverse decision (Section 143).

Where a purchase notice takes effect, i.e. the authority are deemed to have served a notice to treat, the compensation payable to the owner will be assessed, as in any case of compulsory acquisition, under the provisions of the Land Compensation Act 1961 (as described in the following chapters).

Where the Secretary of State does not confirm the purchase notice but directs instead that permission shall be given for some alternative development, then if the value of the land with permission for such alternative development is less than its value with permission for development within Schedule 3 to the 1990 Act[6] the owner is entitled to compensation equal to the difference between the two values (Section 144). But this is subject to any direction by the Secretary of State excluding compensation in respect of any conditions as to design, external appearance, size or height of buildings, or number of buildings to the acre.

*Example 24–1*

*Example of compensation following refusal of Secretary of State to confirm a purchase notice.*

A site having a frontage of 30 ft and a depth of 75 ft stands on the inside of a sharp bend in a main road carrying heavy traffic. A shop with upper part on three floors was erected on the land in 1910 and was entirely destroyed by fire last year. The freeholder immediately applied for planning permission to build a replacement shop and upper part on the site, but permission was refused on the grounds that the re-erection of the premises would obstruct the view of traffic approaching the corner.

The freeholder served a purchase notice on the local authority. They referred it to the Secretary of State who refused to confirm it. He indicated in his decision that he would permit a single-storey lock-up shop to be erected on the corner to be set back so that the effective site depth would be only 45 ft.

If the original premises were to be reinstated on the site, they would have a net rack rental value of £20,000 p.a. producing a site value of £75,000. The lock-up shop for which permission would be granted would be worth only £10,000 p.a. producing a site value of £40,000.

Assess the compensation payable under Section 144.

*Compensation:*
"Existing use" development value, i.e. value of
site with permission to rebuild the former shop
and upper part worth £20,000 p.a.:

---

6 See *supra*, n.4.

| Site Value | £75,000 |
|---|---|

Value of site for erection of lock-up shop
worth £10,000 p.a.:

| Site Value | £40,000 |
|---|---|
| Compensation | £35,000 |

Purchase notices may also be served when land has been rendered "incapable of reasonably beneficial use" in consequence of a revocation or modification or discontinuance order (Section 144 of the 1990 Act), or a refusal (or restrictive grant) or revocation or modification of a "listed building" consent (Sections 32–36 of the Planning (Listed Buildings and Conservation Areas) Act 1990). In theory they may be made available under advertisement regulations or tree preservation orders but only if appropriate provisions were to be included in the Regulations issued, which has not so far happened.

Purchase notices are also met within housing legislation. Under Section 287 of the Housing Act 1985, the owner of an "obstructive building", being a building which, although it is not (if a house) unfit for habitation, the local authority wish to have demolished because its nearness to other buildings makes it injurious to health (i.e. it is a sound building in a slum area), may offer to sell to the authority (i.e. require them to buy) his interest in it. The order to demolish it must give a period of two months for the building to be vacated, and the offer may be made within that time. There is no need to prove that the property is "incapable of reasonably beneficial use": normal compensation principles apply.

# Chapter 25

# Compensation Under the Town and Country Planning Acts III Blight Notices

## 1. Compulsory Purchase Instigated by Land Owners: Blight Notices

Planning proposals which will eventually involve the compulsory acquisition of land may very well depreciate the value of a property or even make it virtually unsaleable.

It is true that such depreciation will be ignored – under Section 9 of the Land Compensation Act 1961 – in assessing the compensation when the property is actually acquired. But this will not help the owner in the meantime if he wants to sell his property in the open market.

For instance, the owner-occupier of a house affected by such proposals may, for personal reasons, be obliged to move elsewhere and may suffer considerable hardship if he has to sell at a very much reduced price.

In partial mitigation of such cases, a strictly limited class of owner-occupier is given power to compel the acquisition of interests in land provided that the claimant in each case can prove that he has made reasonable efforts to sell his interest, but has been unable to do so except at a price substantially lower than he might reasonably have expected to obtain but for the threat of compulsory acquisition inherent in the proposals. The power is derived from Sections 149–171 of the Town and Country Planning Act 1990.[1]

### (i) "Blighted Land"

The first requirement which the claimant must meet (1990 Act, Section 149) is to show that the land in question comes within one of the paragraphs in Schedule 13 to the 1990 Act and is therefore

---

[1] Part VI, Chapter II, of the Act.

"blighted land". These comprise (in effect) a list of the kinds of situation in which official proposals have reached a stage that points to the eventual acquisition of the claimant's land by some public authority, referred to as the "appropriate authority".[2] They are as follows, namely that the claimant's land:

(1) is shown in a "structure plan" (already approved, or submitted to the Secretary of State) as land which may be required for the functions of a government department, local authority, statutory undertaking or public telecommunications operator; or for an "action area" in the plan; or

(2) is shown as being required for the purpose of such functions by a "local plan" or defined in such a plan as the site of proposed development for any such functions; or

(3) is shown in a unitary development plan as required similarly to (1) above; or

(4) is shown in a unitary development plan as required similarly to (2) above; or

(5) is included in a plan approved by the local planning authority (exercising powers under Part III of the 1990 Act) as required for a government department, local authority or statutory undertaker's functions; or

(6) is to be safeguarded for development for functions as in (5) above; or

(7) is included in an order published in draft designating the site of a New Town; or

(8) is included in such an order which has been approved; or

(9) is included in an urban development area similarly to (7) and (8) above; or

(10) is included in a "clearance area"; or

(11) is surrounded by or adjoins a "clearance area" and the local authority decide to acquire it; or

(12) is included in a "renewal area" under Part VII of the Local Government and Housing Act 1989; or

(13) is indicated in a development plan as required for a new, or altered highway; or

(14) is liable to be acquired for a highway proposed to be constructed, improved or altered as shown in an order or scheme under Part II of the Highways Acts 1959 or 1980

---

[2] Town and Country Planning Act 1990, Section 169.

relating to trunk roads, special road, or classified roads, with availability of compulsory purchase powers, whether already in force or still before the Secretary of State; or

(15) is shown on plans approved by a local highway authority as land comprised in the site of a highway to be constructed or altered; or

(16) is a site in respect of which the Secretary of State has given detailed written notification of his intention to provide a trunk road or special road; or

(17) is included in a plan approved by resolution of a local authority which proposes to acquire it under Section 246 of the Highways Act 1980 (for mitigating adverse effects of highways); or

(18) is indicated by the Secretary of State (in writing to the local planning authority) as proposed to be acquired by him under that Section for a trunk or special road; or

(19) is included within the minimum width prescribed for a new street and comprising all or part of a dwelling (or its curtilage) built or being built when that width was prescribed; or

(20) is subject to a local authority's published intention to acquire it for a "general improvement area" under Part VIII of the Housing Act 1985; or

(21) is liable to be compulsorily acquired under a special enactment (within "limits of deviation" specified, if any); or

(22) is subject to a compulsory purchase order (or order to acquire rights compulsorily) but not yet subject to a notice to treat; or

(23) is liable to acquisition by compulsory order under the Transport and Works Act 1992.

An "agricultural unit" may be only partially "blighted". The part "blighted" is the "affected area" and the remainder is the "unaffected area". In a case of this sort the owner-occupier can serve a blight notice which extends to the "unaffected area" as well as the "affected area" if he can show that the "unaffected area" as such is not capable of being reasonably farmed even with any other available land which the claimant can add to it (1990 Act, Section 158).

## (ii) "Qualifying Interests"

The second requirement which the claimant must meet, under Section 149(2) of the 1990 Act, is to show that he has a "qualifying interest" in the land. He must be either an owner-occupier, or the mortgagee[3] (with a power of sale currently exercisable), or the personal representative[4] of an owner-occupier, who is:

(a)  an individual "resident owner-occupier" of the whole or part of a dwelling-house (without any limit of rateable value); or

(b)  the "owner-occupier" (person or partnership) of the whole or part of any other hereditament of which the net annual value for rating does not exceed a certain prescribed limit (at present £18,000);[5] or

(c)  the "owner-occupier" (person or partnership) of the whole or part of an agricultural unit, e.g. a farm.

The term "owner-occupier" is defined in some detail in Section 168 of the 1990 Act. The claimant must have an "owner's interest" in the land, i.e. the freehold or a lease with at least three years to run, and must have been in physical occupation of the property for at least six months up to the date of service of the blight notice, or up to a date not more than 12 months before such service, provided that (in such a case) the property (unless it is a farm) was unoccupied in the intervening period. If an owner-occupier's mortgagee is entitled to serve a blight notice, the relevant period is extended by six months for his particular benefit.

## 2.  Blight Notice Procedure

The notice to purchase the owner-occupier's "qualifying interest" must, under Section 150 of the 1990 Act, relate to the whole of his interest in the land, be in the prescribed form,[6] and be served on the "appropriate authority" (that is, the authority which, it appears from the proposals, will be acquiring the land).

Within two months of the receipt of the claimant's notice, the "appropriate authority" may under Section 151 serve a counter notice of objection, in the prescribed form[7] on any of the following grounds:

---

[3] 1990 Act, Section 162. The mortgagee cannot serve a blight notice if his mortgagor has already served one which is still under consideration (and the converse is also true).

[4] 1990 Act, Section 161.

[5] Town & Country Planning (Blight Provisions) Order 1990 (SI 1990 No 465).

[6] Town & Country Planning General Regulations 1992 (SI 1992 No 1492).

[7] 1990 Act, Section 151(1).

(a)  that no part of the hereditament or agricultural unit is "blighted land";
(b)  that they do not intend to acquire any part of the hereditament, or of the "affected area" of the agricultural unit, affected by the proposals;
(c)  that they intend to acquire a part of the hereditament or "affected area" but do not intend to acquire any other part;
(d)  that, in the case of "blighted land" within paragraphs (1) (3) or (13) of Schedule 13 (above), but not paragraphs (14), (15) or (16), they do not intend to acquire any part of the hereditament or "affected area" within 15 years[8] from the date of the counter notice;
(e)  that the claimant is not currently entitled to an interest in any part of the hereditament or agricultural unit;
(f)  that the claimant's interest is not a "qualifying interest;
(g)  that he has not made "reasonable endeavours" to sell " (except when the land is within paragraphs (21) or (22) of Schedule 13); or that the price at which he could sell is not "substantially lower" than that which might reasonably have expected to get but for the planning proposals.[9]

Within two months of the counter notice, the claimant may, under Section 153 of the 1990 Act, require the authority's objection to be referred to the Lands Tribunal who may uphold the objection or declare that the claimant's original notice is a valid one. If, in case (c), the authority's objection is upheld, or is accepted by the claimant, they will, of course, have to purchase the part referred to in the counter notice as land they intend to acquire.

If the authority object on grounds (a), (e), (f) or (g) above, the claimant must satisfy the Lands Tribunal that their objection is "not well-founded". In other words, the burden of proof lies on the claimant to show that the land is "blighted land",[10] that his interest is a "qualifying interest", and that he has made reasonable but unsuccessful attempts to sell at a fair market price.

---

[8] They can specify a longer period.
[9] 1990 Act, Section 150(1).
[10] In *Bolton Corporation* v *Owen* [1962] 1 QB 470, the Court of Appeal held that statements in a development plan which "zone" the claimant's land for residential use are not sufficient to justify a finding that it is required for the "functions of a local authority", even if the likelihood is that it will be redeveloped for council housing and not private housing.

However, if they object on grounds (b), (c) or (d), it is for the authority to satisfy the Lands Tribunal that their objection is "well-founded". In other words the burden of proof lies on them to show that they have no intention to acquire the land (or part of the land) in question, within 15 years or at all. In *Duke of Wellington Social Club and Institute Ltd v Blyth Borough Council*,[11] the Lands Tribunal held that an objection that the authority do not intend to acquire can be held to be "not well-founded" if there would be hardship on the claimant as a result; but in *Mancini v Coventry City Council*[12] the Court of Appeal disapproved the assertion that hardship could be relevant to an authority's intention not to acquire.

Section 159 of the 1990 Act empowers the authority serving a counternotice in response to a blight notice relating to a farm which includes an "unaffected area" (i.e. part of the farm is not "blighted land") to challenge the assertion that the "unaffected area" is incapable of being reasonably farmed on its own (or with any other available land added to it). Indeed, the authority must challenge that assertion if they claim an intention to take part only of the "affected area" (i.e. of the "blighted" part of the farm). If in either circumstance the Lands Tribunal uphold the counternotice they must make clear to which (if any) land the blight notice will apply, and specify a date for the deemed notice to treat.

Where no counter notice is served, or where the claimant's notice is upheld by the Lands Tribunal, the appropriate authority will be deemed under Section 154 of the 1990 Act to be authorised to acquire the claimant's interest in the hereditament, or (in the case of an agricultural unit) in that part of the unit to which the proposals that gave rise to the claim relate. But if the authority succeed in an objection that they should only take part of a claimant's property, Section 166 of the Act preserves the right to compel them to take all or none in accordance with Section 8 of the Compulsory Purchase Act 1965, on proof that otherwise the property, if it is a "house, building or factory", will suffer "material detriment", or that if it is "a park or garden belonging to a house", the "amenity or convenience" of the house will be seriously affected.[13]

In general, notice to treat will be deemed to have been served at the expiration of two months from the service of the claimant's

---

[11] (1964) 15 P&CR 212.

[12] (1982) 44 P&CR 114.

[13] See Chapter 15.

notice or – where an objection has been made – on a date fixed by the Lands Tribunal (Section 154 (2), (3)).

The general basis of compensation, where an authority are obliged to purchase an owner-occupier's interest under the 1990 Act, is that prescribed by the Land Compensation Act 1961 (and by the Housing Act 1985 if the claimant's interest is subject to a compulsory purchase order in relation to a clearance area). Thus, in principle, the market value figure that the claimant has tried in vain to secure by his own efforts in a private sale becomes payable to him by the "appropriate authority".

Except where the authority have already entered and taken possession under their deemed notice to treat, a party who has served a blight notice may under Section 156 withdraw it at any time before compensation has been determined, or within six weeks of its determination. In this case any notice to treat deemed to have been served will be deemed to have been withdrawn.

# Chapter 26

# Compulsory Purchase Compensation I
## Compensation for Land Taken

### 1. Introductory

The basis of compensation for land compulsorily acquired was in early times and up to 1919 regarded as a matter of valuation alone and not law.

The statutes, private and public alike, prescribed the procedural steps to be taken but left the question of payment to be dealt with by valuers.

But this separation from law was illusory, because no valuers can compel agreement. In the absence of agreement on compensation, arbitration was essential, whether by privately appointed arbitrators or official arbitrators prescribed under the 1919 Act (and earlier statutes) or ultimately the Lands Tribunal (Lands Tribunal Act 1949).

Arbitration is, however, a process regulated by law, because an allegation that any arbitrator's decision is faulty is a matter which can be made the subject of litigation. This has led to judicial decisions giving rise to case law and then to statutes governing principles of assessment applicable to compensation.

Arbitration in its essential common law form is a question of contract; so that the characteristic proceeding for enforcing proper compliance with arbitrations is a common law civil action for enforcement of arbitration contracts known as an "action on the award". The judicial decisions just referred to could be actions of this kind; alternatively, cases could reach the courts by reason of an arbitrator stating a "special case" on one or more points of law. In recent years, however, the distinction between public law and private law has come to be more clearly drawn, and compulsory purchase cases are undoubtedly within the public law field. They mostly arise as appeals by way of "case stated" from the Lands Tribunal (see below) to the Court of Appeal; alternatively, there are judicial review cases, or statutory challenges (e.g. to compulsory purchase orders), in procedural or jurisdictional disputes where the fundamental question is whether or not the relevant authority has acted *ultra vires*.

It should be noted that in *Melwood Units Property Ltd* v *Commissioner of Main Roads*[1] (a case taken on appeal from Australia) the Privy Council stated emphatically that the wrongful application of valuation principles is not merely a valuation issue but an issue of *law*.

The first statute to prescribe rules of assessment was the Acquisition of Land (Assessment of Compensation) Act 1919. Section 2 of this Act prescribed six rules governing the compensation payable for interests in land acquired compulsorily by any government department, local or public authority or statutory undertaking. The general basis prescribed was the price the land might be expected to realise if sold in the open market by a willing seller – known in practice as "open market value".

The Act also replaced the various earlier procedures (arbitrations, and verdicts of juries) by a standardised arbitration process conducted by a panel of official arbitrators, who were in turn superseded by the Act of 1949 (with effect from 1 January 1950) by the Lands Tribunal. Arbitrators could (and can) "state a case" for decision by the Queen's Bench Divisional Court of the High Court; and this method of taking legal points for further adjudication was an alternative to "actions on the award" as a source of early case law. From 1950 appeal from the Lands Tribunal has been by way of "case stated" (on points of law only) direct to the Court of Appeal.

The substantive principles of assessment of compensation, as distinct from procedural matters, are to be found in the Land Compensation Act 1961, as amended. As from 1 August 1961, this Act incorporates and re-enacts the Acquisition of Land (Assessment of Compensation) Act 1919, together with additional compensation provisions enacted in the Town and Country Planning Act 1959. The Act applies to all cases where land is authorised to be acquired compulsorily, including acquisitions made by agreement but under compulsory powers (which in compensation terms count as compulsory acquisitions).

The Land Compensation Act 1973 contains a variety of matters and deals only incidentally with purchase price compensation.

---

[1] [1979] AC 426.

Special compensation rules attributable to it and to other recent Acts will be referred to as appropriate.

It makes no difference *in principle* if what is acquired is only part of an owner's interest and not the whole (e.g. part of the garden of a dwelling-house, taken for road widening; or part of a farm, taken for a motorway), or even if it is a new and lesser right not hitherto in existence (e.g. an easement or wayleave, or even a new leasehold) though the House of Lords, in *Sovmots Ltd v Secretary of State for the Environment,*[2] has held that clear statutory authority must always be shown to justify acquisitions in the form of "new rights". Such authority now exists for local authorities (*not* other public bodies) in general terms, under the Local Government (Miscellaneous Provisions) Act 1976.

## 2. General Principles of Compensation

The general principles, established by judicial decisions under the Lands Clauses Consolidation Act 1845 (though it was intended to deal with procedural rather than compensation matters) apply to all cases of compulsory purchase and acquisition by agreement under compulsory powers, unless expressly or impliedly excluded by Statute. They may be briefly summarised as follows:

(i)  Service of the acquiring body's notice to treat, fixes the property to be taken and the nature and extent of the owner's interest in it. This is because the notice is a semi-contractual document and a step towards the eventual conveyance to the acquiring body of a particular owner's leasehold or freehold (as the case may be). Indeed, the courts have said that notice to treat plus compensation (when settled) in fact constitutes a specifically enforceable legal contract.[3] Any freehold or leasehold in respect of which notice to treat is *not* served therefore remains unacquired. The effect of Section 20 of the Compulsory Purchase Act 1965, is that holders of yearly tenancies and lesser interests are not entitled to receive notices to treat. The assumption is that the acquiring authority acquire the reversion to such an interest and the interest is then terminated at common law (e.g. by notice to quit) though compensation will have to be paid if it is terminated sooner.

---

[2] [1977] 2 All ER 385.
[3] As stated by Upjohn J, in *Grice v Dudley Corporation* [1957] 2 All ER 673.

Equitable as well as legal freeholds and leaseholds are subject to acquisition (e.g. options and agreements for leases[4]) and compensation will have to be assessed accordingly.

It should be remembered that, where general vesting declarations are used, in place of notices to treat and conveyances of the normal kind, a notice to treat is deemed to have been served, so that what has been said applies in those cases equally.

(ii)   The old leading case *Penny* v *Penny* (1868) was assumed to have established a rule that the value of the property acquired must be assessed as at the date of notice to treat. But in *Birmingham Corporation* v *West Midland Baptist (Trust) Association*[5] the House of Lords held that this supposed "rule" was without foundation, and expressed the view that the appropriate date for assessing the value of the land is either:

(a)   the date when compensation is agreed or assessed; or

(b)   the date when possession is taken (if this is the earlier); or

(c)   the date when "equivalent reinstatement" can reasonably be started, if that mode of compensation is to be applied (as it was in the *West Midland Baptist* case itself).

Where the compensation is assessed by the Lands Tribunal the valuation date is the last day of the hearing if possession has not been taken by then.

(iii)   Compensation must be claimed "once and for all... for all damages which can be reasonably foreseen".[6] But this can be varied if the parties agree; and under the Land Compensation Act 1973, Sections 52 and 52A (inserted by the Planning and Compensation Act 1991), claimants are entitled to an advance payment of 90% of the compensation, as estimated (in default of agreement) by the acquiring authority, plus payment of statutory interest, at rates set by the Treasury, on amounts of compensation outstanding after entry on to the land. Further advance payments can be made if the acquiring authority are

---

[5] [1969] 3 All ER 172; [1969] RVR 484.

[6] *Chamberlain* v *West End of London and Crystal Palace Rly Co* [1863] 2 B&S 617, *per* Erle CJ.

satisfied that their earlier estimate was too low. On the anniversary of any payment, if the aggregate amount of the accrued interest on any unpaid outstanding balances exceeds £1,000, then the amount due is payable by the authority.

(iv) Compensation must be based on the value of the land in the hands of the owner, not its value to the acquiring body.[7] It is the former value which is "market value" because it is the vendor, not the purchaser, who is at liberty (in a sale by agreement) to market the land or not as he chooses. Any increase in value solely attributable to the scheme involving the acquisition is not a market factor and must be disregarded.[8]

(v) Covenants, easements, etc. already in existence and affecting the land, whether by way of benefit or of burden, must be considered in assessing compensation.[9] For instance, the property may enjoy the benefit of a covenant restricting building or other works on adjoining land, or it may itself be subject to the burden of such a covenant and be less valuable in consequence.

Similarly, it may be a "dominant tenement" with the benefit of an easement of way, support, light, etc. over adjoining property; or, on the other hand, it may be a "servient tenement" subject to such a right which benefits adjoining property. Also the possibility of the removal or modification of restrictive covenants under Section 84 of the Law of Property Act 1925, as amended, is a factor which might properly be taken into account.

(vi) Where a lessee has a contractual or statutory right to the renewal of his lease, that right will form part of the value of his leasehold interest; but the mere possibility of a lease being renewed is not a legal right existing at the date of notice to treat and cannot be the subject of compensation.[10]

---

[7] *Cedar Rapids Manufacturing and Power Co* v *Lacoste* [1914] AC 569; *Re Lucas and Chesterfield Gas and Water Board* [1909] 1 KB 16.

[8] *Pointe Gourde Quarrying and Transport Co Ltd* v *Sub-Intendent of Crown Lands* [1947] AC 565. In this case, however, the disallowed item of claim was not in fact attributable to land value at all, only to estimated savings to the Crown of transport costs, etc.

[9] *Corrie* v *MacDermott* [1914] AC 1056.

[10] See *Re Rowton Houses' Leases, Square Grip Reinforcement Co (London) Ltd* v *Rowton Houses* [1966] 3 All ER 996.

(vii)   An owner is entitled to compensation not only for the value of the land taken but also for all other loss he may suffer in consequence of its acquisition.[11] For example, the occupier of a private house compulsorily acquired will be put to the expense of moving to other premises and will suffer loss in connection with his fixtures. An occupier of trade premises will suffer similar losses and in addition may be able to claim for loss on sale of his stock or for injury to the goodwill of his business. It will be convenient to discuss the "disturbance" compensation payable under these heads in Chapter 28. But it is important at this point to recognise them as part of the compensation to be paid to the owner for the compulsory taking of his interest in the land.

(viii)  Where part only of an owner's land is taken, Section 7 of the Compulsory Purchase Act 1965, makes it quite clear that the owner is entitled not only to the value of land taken, but also to compensation for severance or injurious affection of other land previously held with the land taken which he retained after the acquisition.

(ix)    There are Acts, for example the Local Government (Miscellaneous Provisions) Act 1976, which expressly give specified bodies the power to acquire "new rights" (e.g. easements, leases, etc., not currently in existence) and the appropriate payment for those (in principle) will presumably be the prevailing market figure for the grant of such a right.

The underlying principle throughout is that a normal compulsory purchase is in truth a compulsory assignment of an existing freehold or leasehold right in land; and consequently the fundamental question which a valuer should ask himself is – what capital sum would a purchaser of this freehold or leasehold expect to pay to acquire it in the open market? In this chapter we concentrate upon the question of compensation for land actually taken, with that principle in mind. Before considering the matter in further detail, it is important to reconsider and distinguish the general principle referred to under (iv) above from one arising from *Pointe Gourde etc. Transport Co Ltd* v *Sub Intendent of Crown Lands* commonly referred to as the "Pointe Gourde Principle".

---

[11] *Horn* v *Sunderland Corporation* [1941] 2 KB 26; *Venables* v *Department of Agriculture for Scotland* [1932] SC 573.

## The "Pointe Gourde Principle"

In the *Pointe Gourde* case, land in Trinidad containing a quarry had been acquired for a naval base. The owners claimed, in addition to land value and disturbance, a sum apparently representing savings to the Government by reason of having stone from the quarry conveniently on hand to build the base installations instead of needing to transport it there from afar. This item was rejected, being (in the words of the Privy Council) "an increase in value which is entirely due to the scheme underlying the acquisition". The reported facts of the case would show the reader the difficulty of understanding how the disputed item could have been justified and how the amount could have been assessed.

The true justification of the decision is that all compensatable items had already been accounted for in the sums agreed by the parties and that the particular benefit accruing to the Government over building materials was of no concern to the claimant owners whatever. For the Privy Council to speak of the disallowed figure as "an increase in value" was most unfortunate, because it was in fact an increase in *claim*, which is a totally different thing. The error has given rise to a "Pointe Gourde principle", that tends to distract the attention of lawyers and valuers, to say nothing of courts and tribunals, from the "willing seller" rule, which is the true rule of market value. Where it produces results at variance with those of the "willing seller" rule, the "Pointe Gourde rule" is dubious; and where the results are the same it is redundant.

The leading cases in compulsory purchase which illustrate the problem show that the "Pointe Gourde rule" works irrationally and unpredictably. In *Wilson v Liverpool Corporation*[12] it deprived the claimant of some of the development value which on the facts clearly accrued to his land. Conversely, in *Jelson Ltd v Blaby District Council*,[13] it gave the claimant development value which, on the facts, his land did not possess (this happened because it was said that the "rule" requires valuers to "disregard a decrease", as well as an increase, if it is "entirely due to the scheme underlying the acquisition"). In *Trocette Property Co Ltd v Greater London Council*,[14] it gave the claimant's leasehold interest in a cinema substantial

---

[12] [1971] 1 WLR 302.
[13] [1977] 2 EGLR 14.
[14] (1974) 28 P&CR 408.

"marriage value" in respect of redevelopment prospects which did not in fact exist because no such redevelopment was permitted (whether it could be assumed to be permitted, under Sections 14–16 of the Land Compensation Act 1961, might profitably have been considered instead). Yet in *Birmingham City District Council* v *Morris & Jacombs Ltd*[15] the "rule" was not applied when valuing a piece of access land which could in fact have been used for additional housing development had this use not been prohibited. The contrast between this and the *Jelson* decision is remarkable. Moreover, the three last-mentioned cases all arose on acquisitions under purchase notices, where there can be no relevant "scheme" at all because the authority are acquiring against their will and therefore without any kind of functional purpose in view.

## 3. Market Value

*Six Basic Rules of Valuation*

Section 5 of the Land Compensation Act 1961 amended as to rule 3 by the Planning and Compensation Act 1991, Section 70, prescribes six rules for assessing compensation in respect of land. Five of these rules relate to the valuation of land and interests in land; the sixth preserves the owner's right to compensation for disturbance and other loss not relating to land value which is suffered in consequence of the land being taken from him.

## Rule 1. No allowance shall be made on account of the acquisition being compulsory.

In assessing compensation under the 1845 Act it had become generally recognised custom to add 10% to the estimated value of the land on account of the acquisition being compulsory. This addition was no doubt intended originally to cover the cost of reinvesting capital and other incidental expenses to which the owner might be put and in that sense may be regarded as part of the value of the land to him. There was nothing in the 1845 Act which expressly authorised such an allowance. It is now expressly excluded (although see Home Loss Payments and Farm Loss Payments in Chapter 28). (Percentage additions made in order to

---

[15] [1976] 2 EGLR 143.

attempt to quantify *development* value are another matter. The modern law concerning the inclusion of development value in compensation is considered later in this chapter).

**Rule 2. The value of land shall, subject as hereinafter provided, be taken to be the amount which the land if sold in the open market by a willing seller might be expected to realise.**

The purpose of this rule is to indicate the true basis of compensation as being open market value, which is what the courts meant in earlier cases by stressing "value to the owner". It takes for granted that there will be a "willing buyer", i.e. that there exists a demand for the land – presumably because of the compulsory nature of the acquisition. Yet whether in the open market there would be a willing buyer at a price acceptable to a willing seller is a question which may well cause problems in certain cases. (Some of these problems may be solved by applying Rule 5; see below).

Where interests in land are purely of an investment nature, there is probably no difference in any valuer's mind between value to owner and value in the open market. In other cases the rule should put it beyond doubt that there must be excluded from the valuation any element which would have no effect on the price obtainable for the property under normal conditions of sale and purchase, i.e. anything that distorts a proper calculation of "market value".

The meaning of the words "amount which the land if sold in the open market by a willing seller might be expected to realise" was fully examined by the Court of Appeal in the case of *Inland Revenue Commissioners* v *Clay & Buchanan*,[16] as follows:

(i) "In the open market" implies that the land is offered under conditions enabling every person desirous of purchasing to come in and make an offer, proper steps being taken to

---

[16] [1914] 3 KB 466. In that case the judgment of the court was directed to almost identical words used in the Finance (1909–1910) Act 1910, with regard to taxation. See also *Re Hayes' Will Trusts* [1971] 2 All ER 341, where Ungoed-Thomas J speaks of ". . . a range of price, in some circumstances wide, which competent valuers would recognise as the price which property would fetch if sold in the open market". (The judge was dealing with estate duty valuation.)

advertise the property and to let all likely purchasers know that it is in the market for sale.

(ii) "A willing seller" does not mean a person who will sell without reserve for any price he can obtain. It means a person who is selling as a free agent, as distinct from one who is forced to sell under compulsory powers.

(iii) "Might be expected to realise" refers to the expectations of properly qualified persons who are informed of all the particulars ascertainable about the property and its capabilities, the demand for it and likely buyers.

In assessing compensation, then, it must be assumed that the owner is offering the property for sale of his own free will, but is taking all reasonable measures to ensure a sale under the most favourable conditions. The compensation payable will be *that price which a properly qualified person, acquainted with all the essential facts relevant to the property and to the existing state of the market, would expect it to realise under such circumstances.*

An estimate of market value on this basis will take into account all the potentialities of the land, including not only its present use but also any more profitable use to which, subject to the requisite planning permission, it might be put in the future, i.e. prospective development value as well as existing use value, subject to planning control.

For example, if land at present used as agricultural land is reasonably likely (having regard to demand) to become available for building in the future, the prospective building value may properly be taken into account under Rule 2, provided that it is deferred for an appropriate number of years. Again, if buildings on a well-situated site have become obsolete or old-fashioned, so that the rental value of the property could be greatly increased by capital expenditure on improvements and alterations (again, having regard to demand), compensation may properly be based on the estimated improved rental value, provided that the cost of the necessary works is deducted from the valuation.

It should be emphasised, however, that in the case of land capable of further development, the price which it might be expected to realise under present-day conditions of strict planning control depends very largely on the kind of development for which planning permission has already been obtained, or is reasonably likely to be given having due regard to the provisions of the development plan for the area. But it is essential to bear in mind the words used by Lord Denning MR, in *Viscount Camrose v Basingstoke Corporation*:[17]

> Even though (land may) have planning permission, it does not follow that there would be a demand for it. It is not planning permission by itself which increases value. It is planning permission coupled with demand.

Thus, existing use value rests on demand, whereas development value rests on demand plus planning permission; in principle, both go to make up "market value" of land.

In practice, a prospective purchaser in the open market would probably obtain permission for his proposed development before deciding the price he was prepared to pay. This factor is normally absent in compulsory purchase cases, and the Land Compensation Act 1961, therefore, provides that, besides taking into account any existing planning consents, certain assumptions shall be made as to the kinds of development for which planning permission might reasonably have been expected to be granted but for the compulsory acquisition. These assumptions are considered in detail later in this chapter.

In general, compensation will be based on the value – as at the date of assessment or of entry on the land – of the interests existing at the date of the notice to treat. But in all cases to which the Acquisition of Land Act 1981 applies[18] the Lands Tribunal "shall not take into account any interest in land or any enhancement of the value of any interest in land, by reason of any building erected, work done or improvement or alteration made", if the Tribunal "is satisfied that the creation of the interest, the erection of the building, the doing of work, the making of the improvement or the alteration, as the case may be, was not reasonably necessary and was undertaken with a view to obtaining compensation or increased compensation".[19] This does not prevent recognition of a merger or surrender as between landlord and tenant, provided that

---

[17] [1966] 3 All ER 161.
[18] Section 4 (re-enacting a similar provision in earlier legislation).
[19] The provision applies not only to the land being acquired but also to "any other land with which the claimant is, or was at the time of the erection, doing or making of the building, works, improvement or alteration, directly or indirectly concerned . . .".

compensation claimed as a result is a reasonable market figure.[20]

In the case of leasehold interests there seems no doubt that their value should be determined by reference to the length of the unexpired lease at the date when the valuation is made. (As to security of tenure, see pages 607–8 *supra*).

**Rule 3 (as amended). The special suitability or adaptability of the land for any purpose shall not be taken into account if that purpose is a purpose to which it could be applied only in pursuance of statutory powers, or for which there is no market apart from the requirements of any authority possessing compulsory purchase powers.**

Under the Lands Clauses Acts, if land could be shown to be specially suited or adapted to a particular purpose so that anyone requiring it for that purpose might be expected to pay a higher price on that account, this "special adaptability" might be regarded as giving the land a special value in a certain limited market and thereby increasing its value in the hands of the owner.

Thus, in *Manchester Corporation* v *Countess Ossalinski*[21] the fact that land was specially suitable and adaptable for the construction of a reservoir for a town's supply was held to enhance its value to the owner; and, in later cases under the Lands Clauses Acts, it was held that such "special adaptability" might properly be considered even where the land could only be used for the purpose for which it was best suited by bodies armed with statutory powers. This is because there exists no rule that a "market" situation requires a minimum of (say) two prospective purchasers; the fact that, in a

---

[20] *Banham* v *London Borough of Hackney* (1970) 22 P&CR 922. But if a residential tenant is rehoused by a local authority, under Section 39 of the Land Compensation Act 1973, after the date of his landlord's notice to treat, the landlord's interest is to be valued without regard to the tenant's giving up possession (Section 50(2) of the 1973 Act). Conversely, a tenant whose leasehold is expropriated must not be given less compensation by reason of his being rehoused by, or by arrangement with, the acquiring authority (Section 50(1)).

[21] Tried in the Queen's Bench Division of the High Court on 3 and 4 April 1883, but not reported. However, extracts from the judgment are contained in the report of the decision in *Re Lucas and Chesterfield Gas and Water Board* [1909] 1 KB 16.

particular case, there is a sole purchaser, who happens to be armed with statutory powers, does not affect the principle as such.

Rule 3 excludes the consideration of "special suitability or adaptability" of the land, provided, however, that this relates to its use for a "purpose" and not merely to the position of any party to the transfer.[22] In relation to any "purpose", the fact that its application depends solely on compulsory purchase or the existence of statutory powers suffices to bring the rule into operation. For instance, the circumstances existing in *Manchester Corporation* v *Countess Ossalinski* could not now affect the compensation payable for land, if there were no likelihood of its being used for the purpose of a reservoir except under statutory powers.

It is clear that "purpose" in Rule 3 means some prospective physical use of the land itself and does not extend to a purpose connected with the use of the products of the land, e.g. minerals, nor to a factor such as the special attraction of a landlord's reversion to a "sitting tenant" (i.e. to the latter the purchase of the reversion means an enlargement of his interest; to anyone else it means merely the acquisition of an investment).[23] If, therefore, "special suitability" of the land for "any purpose" is not established, then Rule 3 does not come into play in any case. It follows that "ransom value" is payable as being genuine market value; this is endorsed by the House of Lords decision in *Hertfordshire County Council* v *Ozanne*[24] and the Court of Appeal decision in *Wards Construction (Medway) Ltd* v *Barclays Bank plc*.[25]

The same reasoning underlies the decision in *Stokes* v *Cambridge Corporation*,[26] in which the market value price payable for the claimant's land was held not to include additional value merely because the acquiring body owned other land which provided

---

[22] In *Inland Revenue Commissioners* v *Clay & Buchanan* [1914] 3 KB 466, it was held that the special price already offered and paid for a house by the adjoining owner who wanted the premises for the extension of his nursing home was properly taken into account in assessing market value for the purposes of the Finance (1909–10) Act 1910.

[23] See *Pointe Gourde Quarrying and Transport Co Ltd* v *Sub-Intendent of Crown Lands* [1947] AC 565; and *Lambe* v *Secretary of State for War* [1955] 2 QB 612.

[24] [1991] 1 All ER 769.

[25] [1994] 2 EGLR 32.

[26] (1961) 13 P&CR 77.

access to facilitate development, since the claimant's ownership of the land did not have such access but only access sufficient for existing use. Conversely, in *Earl Fitzwilliam's Wentworth Estates Co* v *British Railways Board*,[27] the claimant's land already enjoyed enhanced access which the acquiring body could not disregard merely because they already owned suitable access land.

**Rule 4. Where the value of the land is increased by reason of the use thereof or of any premises thereon in a manner which could be restrained by any Court, or is contrary to law, or is detrimental to the health of the inmates of the premises or to the public health, the amount of that increase shall not be taken into account.**

The general purpose of this rule seems clear.

In *Hughes* v *Doncaster Metropolitan Borough Council*,[28] the House of Lords held that the rule excludes value attributable to a use carried on in breach of planning control, but does not exclude such a use after it is no longer susceptible to enforcement proceedings because of lapse of time.

**Rule 5. Where land is, and but for the compulsory acquisition would continue to be, devoted to a purpose of such a nature that there is no general demand or market for land for that purpose, the compensation may, if the Lands Tribunal is satisfied that reinstatement in some other place is bona fide intended, be assessed on the basis of the reasonable cost of equivalent reinstatement.**

There are certain types of property which do not normally come up on the market and whose value cannot readily be assessed by ordinary methods of valuation, such as that of estimating the income or annual value and capitalising it. Such properties include churches, alms-houses, schools, hospitals, public buildings and certain classes of business premises where the business can only be carried on under special conditions.[29]

---

[27] (1967) 19 P&CR 588.
[28] [1991] 1 EGLR 31.
[29] See *Festiniog Rail Co* v *Central Electricity Generating Board* (1962) 13 P&CR 248.

Rule 5 gives statutory authority to assessing compensation on the basis of the cost of providing the owner, so far as reasonably possible, with an equally suitable site and equally suitable buildings elsewhere – as an alternative to assessment on the basis of market value – provided that:

(i) the land is devoted to a purpose for which there is no general demand or market for land;

(ii) compensation is limited to the "reasonable cost" of equivalent reinstatement; and

(iii) the Lands Tribunal is satisfied that reinstatement in some other place is bona fide intended.

"Equivalent reinstatement" would seem to imply putting the claimant in the same position, or in an equally advantageous position, as that which he occupied when his land was acquired.

In certain cases the only practical method of reinstatement may be the provision of a new site and new buildings. But where, for instance, claimants are using an old building which has been adapted to their purposes, the term "equivalent reinstatement" might cover the cost of acquiring another similar property, if that is possible, together with the expenses of any necessary adaptations.

In a leading decision of the House of Lords[30] the point at issue was whether the cost of reinstatement should be assessed as at the date of notice to treat (14 August 1947) – agreed at £50,025 – or as at the earliest date when the work might reasonably have been begun (30 April 1961) – agreed at £89,575. It was unanimously decided by the House of Lords (confirming the decision of the Court of Appeal and reversing a decision of the Lands Tribunal) that the later figure was the proper basis for compensation.

Bearing in mind that, subject to the "special suitability" concept in Rule 3, there is no reason in strict principle why one prospective purchaser should not constitute a "market", it will be apparent that the application of Rule 5 depends not on there being "no market"

---

[30] *West Midland Baptist (Trust) Association (Inc)* v *Birmingham City Corporation* (1969) 3 All ER 172; (1969) RVR 484. For the phrase "devoted to a purpose" see *Zoar Independent Church Trustees* v *Rochester Corporation* [1974] 3 All ER 5 where a church was devoted to a purpose even though the congregation had fallen to single figures.

but on there being "no *general* market".[31] What therefore matters is that the Lands Tribunal should be satisfied that the facts of the case require Rule 5 to be applied.

Equally important is the question of development value. The wording of Rule 5 takes account only of the existing use of the land in question. If the existing use of land in circumstances envisaged by Rule 5 points to a (restricted) market value figure of (say) £1,000 and the reasonable cost of equivalent reinstatement is assessed at £20,000, it is obviously reasonable for the owner to claim the latter figure. If, however, that land has genuine development value assessed at £30,000, it is reasonable to claim this figure instead; but if there is development value assessed only at £15,000, there seems no reason in principle why "equivalent reinstatement" should not be claimed in preference to it.

## Rule 6. The provisions of Rule 2 shall not affect the assessment of compensation for disturbance or any other matter not directly based on the value of the land.

This rule merely preserves an owner's right (established under the Lands Clauses Acts) to be compensated not only for the value of his land but for any other loss he suffers through the land being taken from him, chiefly "disturbance". This is discussed in Chapter 28.

---

[31] This distinction was endorsed by the House of Lords in *Harrison and Hetherington* v *Cumbria County Council* [1985] 2 EGLR 37, in which the claimants were held to be entitled to equivalent reinstatement compensation for the site of the old cattle market in the centre of Carlisle; the demand for land for cattle markets normally extends only to one such site at a time in any given town, and therefore that demand, though genuine, cannot be described as "general". In *Lambe* v *Secretary of State for War* [1955] 2 QB 612, the court accepted that "market value" to a sitting tenant may well be different from (and higher than) "market value" to an investment purchaser when a reversion is sold; so presumably the demand (if any) from a sitting tenant for the reversion to his tenancy is not "general", though of course the question of equivalent reinstatement is not likely to arise for the reversioner.

## 4. Additional Rules of Assessment

Sections 6 to 9 of the Land Compensation Act 1961 prescribe three additional rules to be applied in assessing market value for compensation purposes.

*(a) Ignore notional increases or decreases in value due solely to development under the acquiring body's scheme*
Section 6 of and Schedule 1 to the 1961 Act give statutory authority to a principle that certain increases and decreases in value must be ignored. These are increases and decreases which are not attributable to market factors but are purely hypothetical. It is made clear, by the use of the words "land authorised to be acquired other than the land to be valued" and "other land", that what is in question is a modification of the value assessed for the claimant which is made because of what has happened, or is thought likely to happen, *on neighbouring land also being acquired.*

In principle, there is, of course, no objection to modifying an assessment of value for reasons of this kind. Section 6 and Schedule 1 therefore exclude such modifications only in the following circumstances:

(i) the "other land" is being acquired in order to be developed for the same purposes as the claimant's land because it is included in the same compulsory purchase order or in the same area of comprehensive development or of a New Town; *and*

(ii) an increase or decrease in value of the claimant's land is being envisaged on the hypothesis that it is not being acquired for those purposes – contrary to the reality of the situation, since its value is being assessed precisely because it is in fact being so acquired; *and*

(iii) such an increase or decrease in value cannot be justified independently in the way that it would be justified if the development would be "likely to be carried out" in other circumstances (that is to say, if there had been a market demand for such development and a likelihood of its being permitted).

In other words, what is being ruled out is an unreal modification of value, dependent on the blatant contradiction that the land is assumed *not* to be acquired, when assessing the price for its acquisition.

Table A

| Type of Case | | Ignore increase or decrease in value due to the following, if they are unlikely to occur but for the compulsory acquisition or designation (as the case may be): |
|---|---|---|
| Case 1 | All cases where the purpose of the acquisition involves development of any of the land to be acquired. | Development for the purpose of the acquisition, of any land authorised to be acquired, other than the land to be valued. |
| Case 2 | Where any of the land to be taken is in an area designated in the current development plan as an area of comprehensive development. | Development of any other land in that area in accordance with the plan. |
| Case 3 and 3A | Where on the service of notice to treat any of the land to be taken is in an area designated as the site of a New Town under the New Towns Act 1965,[32] or an extension of a New Town, designated in an order taking effect after 13 December 1966.[33] | Development of any other land in that area in the course of the development of the area as a New Town or extension as the case may be. |

---

[32] See Part II of the First Schedule for special provisions applying where notice to treat is served on or after the date on which the new town development corporation ceases to act. The current general Act is the New Towns Act 1981.

[33] New Towns Act 1966.

Table A continued

| Case 4 | Where any part of the land to be taken forms part of an area designated in the current development plan as an area of town development. | Development of any other land in that area in the course of town development within the meaning of the Town Development Act 1952.[34] |
|---|---|---|
| Cases 4A and 4B | Where any part of the land to be taken forms part of an area designated as an urban development area,[35] or part of a housing action trust area.[36] | Development of any other land in that area in the course of development or redevelopment of the area as an urban development area or a housing action trust area. |

In assessing compensation in any of the cases referred to in Table A above, no account is to be taken of any increase or decrease in the value of the interest to be acquired due to the development of the kind noted against that particular case.

"Development", which includes the clearing of land, refers both to development which may already have taken place and also to proposed development – being in either case development which would not have been likely to be carried out except for the compulsory acquisition (or, in cases other than Case 1, if the areas mentioned had not been designated).

In Case 1 of Table A the rule would apply to the effects on value of the development, or the prospects of development of other land acquired under the same compulsory purchase order or Special Act. In the remaining cases, which deal with large-scale undertakings, the rule extends to actual or prospective development of other land within the defined area of the scheme.

---

[34] Repealed by the Local Government and Housing Act 1989, which means that no additional such areas will be designated.
[35] By an order made under the Local Government, Planning and Land Act 1980, Section 134.
[36] Established under Part III of the Housing Act 1988.

In *Davy* v *Leeds Corporation*[37] the claimant's land was acquired (at "site value", because it comprised houses unfit for habitation) as part of a clearance area. He claimed that the prospect of clearance and redevelopment of the adjoining land in the area would increase the value of his own land if it were not being acquired, and that Section 6 did not rule out this notional increase because it was not justifiable to say that the clearance "would not have been likely to be carried out except for the compulsory acquisition". This contention depended on the possibility that the clearance might have been brought about in circumstances other than compulsory purchase by the local housing authority. The House of Lords rejected this argument, on the ground that the facts made it inconceivable that the clearance would have come in any other way. Section 6, therefore, applied, and the argument for an increase in value failed.

The application of these rules to assessing the value of the claimant's land requires the valuer to decide what would have happened in the absence of the scheme for which the compulsory purchase is taking place and assuming that no authority would have been given compulsory powers for an alternative scheme.

So, for example, when valuing an area of land which is part of a new town, which is surrounded by houses built as part of the new town scheme, and which is itself to be developed with new houses, the valuer must decide what would have happened if there had been no new town scheme. Although he can assume that planning permission would be granted for residential development (see below), he cannot assume that the infrastructure to enable the development to go ahead exists (as it actually does), unless he can show that it would have been provided even if there had been no new town scheme. Since this will normally not be so then the valuation is of an area of land with planning permission for residential development but without the possibility of development taking place. That planning permission has no value.

It is obvious that any exercise which seeks to establish what would have happened in the absence of the scheme is fraught with difficulties. This was recognised by Lord Denning MR, in *Myers* v *Milton Keynes Development Corporation*[38] who stated that:

---

[37] [1965] 1 All ER 753.
[38] (1974) 230 EG 1275.

It was apparent, therefore, that the valuation in the present case has to be done in an imaginary state of affairs in which there was no scheme. The valuer must cast aside his knowledge of what had in fact happened in the last eight years due to the scheme. He must ignore the developments which would in all probability take place in the future ten years owing to the scheme. Instead, he must let his imagination take flight to the clouds. He must conjure up a land of make believe, where there had not been, nor would be, a brave new town.

*(b) Set-off increases in value of adjacent or contiguous land in the same ownership*

Where land is acquired compulsorily, it may well be that other adjoining land belonging to the same owner is increased in value by the carrying out of the acquiring body's undertaking on land taken. Neither the Lands Clauses Acts nor the Acquisition of Land (Assessment of Compensation) Act 1919 made any provision whereby the acquiring authority might benefit from this increase in value. But a number of special Acts, including the principal ones under which land is taken for road works, expressly provide that, in assessing compensation for land taken, any increase in the value of adjoining and contiguous lands of the same owner due to the acquiring body's scheme shall be deducted from the compensation payable.

The Land Compensation Act 1961 now applies this principle of "set-off" to all cases of compulsory acquisition, as follows.

The effect of Section 7 of the 1961 Act is that where, at the date of notice to treat, the owner has an interest in land contiguous or adjacent to the land acquired, any increase in the value of that interest in the land retained due to development under the acquiring body's scheme is to be deducted from the compensation payable for the interest in the land acquired.[39]

As under Section 6, "development" refers to either actual or prospective development under the acquiring body's scheme which would not be likely to be carried out but for the compulsory acquisition. In this case, however, the prospect of development on the land acquired must be considered, as well as development on

---

[39] In *Leicester City Council* v *Leicestershire County Council* [1995] 2 EGLR 169 it was held that the compensation could be reduced to nil if the increase in value equalled or exceeded the compensation otherwise payable.

Table B

| | Type of Case | Set off increase in value of contiguous or adjacent land of same owner due to thefollowing, if unlikely to occur but for the compulsory acquisition or designation (as the case may be: |
|---|---|---|
| Case 1 | All cases where the purpose of the acquisition involves development of any of the land to be acquired. | Development for the purposes of the acquisition, of any land authorised to be acquired including the land to be valued. |
| Case 2 | Where any of the land to be taken is in an area designated in the current development plan as an area of comprehensive development. | Development of the land taken, or any other land in the area designated, in accordance with the plan. |
| Cases 3 and 3A | Where on the service of notice to treat any of the land to be taken is in an area designated as the site of a New Town or extension of a New Town. | Development of the land taken, or any other land in the area designated. |
| Case 4 | Where any part of the land to be taken forms part of an area designated in the current development plan as an area of town development. | "Town development" on the land taken, or any other part of the area designated.[40] |
| Cases 4A and 4B | Where any part of the land to be taken forms part of an area designated as an urban development area, or part of a housing action trust area. | Development or redevelopment on the land taken, or any other land in the urban development area or housing action trust area, as the case may be. |

---

[40] See n. 34, *supra*.

other land taken under the same compulsory purchase order, or special Act, or included in the same area of development.

Table B, based on the First Schedule to the 1961 Act, may help to make the position clear.

The above provisions as to "set-off" do not apply to acquisitions under certain existing Acts – specified in Section 8(7) – which already provide for "set-off", nor to acquisition under any local enactment which contains a similar provision. In these cases the question of "set-off" will be governed by the express provisions of the particular Act. But any provision in a local enactment which restricts "set-off" to any increase in "the existing use value" of contiguous or adjacent land will cease to have effect.

Under the above provisions as to "set-off", the compensation payable to an owner may be reduced by the increase in the value of his interest in adjacent land. But there are also cases where the compensation for land taken will be increased on account of the injurious affection to other land held with it due to the acquiring body's scheme. In either case, if such adjacent land is *subsequently* acquired, the operation of Section 6, which requires increases or decreases in value due to the scheme to be ignored, might result in an owner either being paid a certain amount of compensation twice over or deprived of it on two occasions.

Section 8 therefore provides as follows:

(i)   If, either on a compulsory purchase or a sale by agreement, the purchase price is reduced by setting off (under Section 7 or any corresponding enactment) the increase in the value of adjacent or contiguous land due to the acquiring body's scheme, then if the same interest in such adjacent land is subsequently acquired that increase in value (which has served to reduce the compensation previously paid) will be taken into account in assessing compensation and not ignored as it otherwise would have been under Section 6.

(ii)  Similarly, if a diminution in the value of other land of the same owner due to the acquiring body's scheme has been added to the compensation payable for land taken, then if the same interest in that other land is subsequently acquired, that depreciation in value (for which compensation has already been paid) will be taken into account in assessing compensation and not ignored as it otherwise would have been under Section 6.

If in either of the above cases part only of the adjoining land is subsequently acquired, a proportionate part of the set-off or injurious affection will be taken into account.

Section 8 also ensures that where several adjacent pieces of land of the same owner are acquired at different dates – although perhaps under the same compulsory purchase order – an increase in value due to development under the acquiring body's scheme which has already been set off against compensation on one occasion shall not be set off again on the subsequent acquisition of a further part of the lands in question.

The underlying principle in Sections 6–8 can be perhaps best expressed by saying that Sections 7 and 8 require genuine market increases and decreases in value to be taken properly into account, whereas Section 6 requires that increases and decreases which are not genuine market calculations be disregarded. It should, however, be noted that "set-off" under Section 7 is, of its very nature, somewhat capricious in its incidence because the increase in value to which it relates can only be taken into account if the owner is losing other land by compulsory purchase, and not beyond the amount of the compensation payable in respect of that other land.

### (c)  Ignore loss of value due to prospect of acquisition

Section 51(3) of the Town and Country Planning Act 1947 provided that, in assessing compensation, no account should be taken of any depreciation in the value of the claimant's interest due to the fact that the land had been designated for compulsory purchase in the development plan. Section 9 of the Land Compensation Act 1961 extends this principle to depreciation in value due to any proposals involving the acquisition of the claimant's interest, whether the proposals are indicated in the development plan – by allocation or other particulars in the plan – or in some other way. For instance, no account should be taken of depreciation in the value of the land due to its inclusion in a compulsory purchase order which has been publicised under the Acquisition of Land Act 1981, or any comparable provisions.

### 5.  Development Value

*Assumptions as to Planning Permission*

Section 14 to 16 of the Land Compensation Act 1961, prescribe certain *assumptions* as to the grant of planning permission which are to be made in assessing the market value of the owner's interest in the land to be acquired.

These assumptions are to be made without prejudice to any planning permission already in existence at the date of the notice to treat – whether, for instance, given by the local planning authority or General Permitted Development Order. It may be that the land in question enjoys additional value by way of development value because of an actual planning permission to an extent that makes the "planning assumptions" superfluous; but if so, this is coincidental, and its occurrence is uncommon. The valuer must make such one or more of the prescribed assumptions as are applicable to the whole or part of the land to be acquired, and compensation will be related to whichever of these assumptions is most favourable from the point of view of value in the open market.

But the fact that, under the Act, land may be assumed to have the benefit of planning permission, e.g. for residential purposes, does not necessarily imply a demand for that land for those purposes. It is important always to remember the words of Lord Denning MR, referred to earlier in this chapter, about the elements necessary for development value: "It is not the planning permission by itself which increases value; it is planning permission coupled with demand."[41] In some cases normal demand – apart from the acquiring authority's scheme – might be so far distant as to warrant only a "hope" value for development.

Again, the fact that these assumptions are to be made does not imply that planning permission for other forms of development would necessarily be refused. So the possibility that prospective purchasers might be prepared to pay a higher price for land in the hope of being able to obtain permission for a certain form of development may be a factor in assessing market value, even though the permission in question is not covered by the assumptions to be made under the Act. But in deciding whether permission might reasonably have been expected to be granted – and might therefore influence the price a prospective purchaser might be expected to pay – regard is to be had to any contrary opinion expressed in any certificate of "appropriate alternative development" which may have been issued under Section 17 of the Act[42] (as amended).

---

[41] *Viscount Camrose* v *Basingstoke Corporation* [1966] 3 All ER 161.
[42] Such certificates are considered below.

In addition there is Part IV of the 1961 Act, revived with amendments by the Planning and Compensation Act 1991 and applicable to transactions completed after 24 September 1991, which provides for additional "prospective development value" to be paid by the acquiring authority to the expropriated owner if within 10 years after completion of the acquisition a later planning decision adds extra value to the land, provided that the later decision refers to different subsequent development.

This does not apply if the permission relates to a new town or urban development area, nor to compulsory acquisition of listed buildings for which "minimum compensation" is paid because of the owners' deliberate neglect.

The assumptions to be made under Sections 14–16 of the Land Compensation Act 1961, may be described broadly as (1) general assumptions and (2) assumptions related to current development plans.

### (1) General Assumptions as to Planning Permission

(i)  Where the acquiring authority's proposals involve development of the whole or a part of the land to be acquired, it shall be assumed that planning permission would be given for such development in accordance with the authority's proposals (Section 15(1) and (2)).

But where a planning permission for the proposed development is already in force at the date of the notice to treat, its terms will take the place of any assumed permission, unless it is personal only and does not enure for the benefit of all persons interested in the land.

As a general rule then, the possibility of carrying out the type of development for which the land is acquired may be reflected in the compensation payable. For instance, if the local authority require land for housing, it will be assumed that planning permission is available for the kind of housing development which the local authority propose to carry out.

But where the proposed development is for a purpose to which the land could only be applied in pursuance of statutory powers, the assumption of planning permission will not necessarily affect the compensation, since this element of value may well be excluded by Rule 3 of Section 5 of the Act.

(ii)  It shall be assumed that planning permission would be granted for any form of development specified in Schedule 3 to the Town and Country Planning Act 1990 (Section 15(3) and (4)). This, broadly, issues permission to rebuild a building which has been demolished and also to dividing a residential property into small residential uses, e.g. conversion of a house into flats.

But to avoid compensation being paid twice over, if at some point compensation was previously paid for loss of development value under the Town and Country Planning Act 1990 (or its predecessors) the same element of value must not be paid again.

(iii) It shall be assumed that planning permission would be granted for development of any class specified in a "certificate of appropriate alternative development" which may have been issued under Part III of the 1961 Act (Section 15(5)).

But this assumption is governed by the terms of the certificate which may indicate that planning permission would not be granted until some date in the future, or subject to certain conditions or both.

The cases in which such certificates may be issued and the procedure in connection with them are considered later in this chapter.

### (2) *Assumptions as to Planning Permission related to Current Development Plans*

Where at the date of notice to treat land to be acquired is comprised in a current development plan – whether the original plan or some amended form of it – certain assumptions as to planning permission are to be made according to how the land, or any part of it, is dealt with in the plan.

In some cases land may be "defined" in the plan as the site for some specific development, e.g. a new road, a public building, a public park. In other cases land may be "allocated" (or zoned) for some more general purpose, e.g. for agricultural, residential or industrial user. Or again, the land may be defined in the plan as an "area of comprehensive development" or "action area", i.e. one which, in the opinion of the local planning authority, should be developed or redeveloped as a whole.[43]

---

[43] The Town and County Planning Act 1990, Section 36(7), defines an "action area" as any part of a local planning authority's area "which they

In all the cases (i) to (iv) set out below the assumptions made as to planning permission must be subject to:

(a)  any conditions which might reasonably have been expected to be imposed on the grant of the planning permission in question; and

(b)  any indication in the development plan – whether on any map or in the written statement – that such permission would only be granted at some future time, thus suggesting that the benefit of the assumed permission should be deferred for an appropriate period.

Regard must be had "to any contrary opinion expressed" in a certificate of appropriate alternative development under Part III of the 1961 Act, discussed below (Planning and Compensation Act 1991, Section 70 and Schedule 15, inserting a new subsection 14 (3A) into the 1961 Act).

Cases (i)–(iii) which follow refer to land which is not in an area of comprehensive development.[44]

Case (i):  Land defined in the development plan as the site for development of a specific description (Section 16(1)).

In this case it is to be assumed that planning permission would be given for the particular development.

In many cases where the land is defined as the site of development for which there is no demand except by bodies armed with statutory powers, this assumption will have little or no effect on the market value of the land. In such cases the owner may be advised to apply for a certificate of "appropriate alternative development" under Part III of the Act (see below).

Case (ii):  Land allocated in the development plan for some primary use (Section 16(2)).

---

[43] *continued*

have selected for the commencement during a prescribed period of comprehensive treatment, by development, redevelopment or improvement or partly by one and partly by another method".

[44] "Area of comprehensive development" is to be construed as an "action area" for which a local plan is in force under Part II of the Town and Country Planning Act 1990. See preceding footnote.

In this case it is to be assumed that planning permission would be given for any development which:

(a)    is within the specified primary use; and

(b)    is of a kind for which planning permission might reasonably have been expected to be granted but for the compulsory acquisition.

For example, land may be zoned primarily for "residential purposes", a term which may include a number of different forms of development, e.g. houses or blocks of flats, with varying densities per hectare. Here, planning permission must be assumed for the type of residential development for which it might reasonably have been expected to be granted had the land not been acquired compulsorily.

In some cases there may be a reasonable possibility that permission would have been given for the development of the whole or part of the land for some purpose ancillary to the purpose for which the area is primarily zoned, e.g. a terrace of shops in an area zoned for residential use. In such a case it would seem that, although definite planning permission can only be assumed for the specified primary use, it would be permissible to consider any possible effect which the reasonable expectation of planning permission for the ancillary use might have had on the open market value.

Case (iii):  Land allocated in the development plan primarily for two or more specified uses (Section 16(3)).

In this case it is to be assumed that planning permission would be given for any development which:

(a)    is within the range of uses specified in the development plan;

(b)    is of a kind for which planning permission might reasonably have been expected to be granted but for the compulsory acquisition.

Case (iv):  Land defined in the development plan as being subject to comprehensive development (Section 16(4) and (5)).

As stated above, land subject to comprehensive development, in the opinion of the local planning authority, should be developed or redeveloped as a whole.[45] The development plan, in addition to defining the area in question, will also indicate the various uses to which it is proposed the different parts of the area should be put on redevelopment.

In this case, in assessing compensation for the land to be acquired, regard is to be had to the range of uses proposed for the area, but the proposed distribution of those areas under the development plan must be ignored, together with any development which at the date of notice to treat has already taken place in the area in accordance with the plan.

The open market value of the land in question must then be assessed on the assumption that planning permission would be granted for any form of development – within the range of uses proposed for the area – for which permission might reasonably have been expected to be granted if the area had not been defined as an area of comprehensive development.

## 6. Certificates of Appropriate Alternative Development

Where an interest in land is proposed to be acquired by an authority possessing compulsory powers, Part III of the Land Compensation Act 1961 allows the claimant to apply for a "Certificate of Appropriate Alternative Development". (Until repealed by the Planning and Compensation Act 1991, Section 65, that land, or part of it, could not consist or form part of an area defined in the current development plan as an area of comprehensive development or "action area" or shown in the current development plan as allocated primarily for residential, commercial or industrial uses, or a combination of any of those uses with other uses).

Section 17 of the Land Compensation Act 1961, which was amended by Section 47 of the Community Land Act 1975, enacts

---

[45] See notes 43 and 44, *supra*. An "area of comprehensive development" is now described as an "action area". Both kinds of area have been intended usually, though not invariably, for achieving town-centre redevelopment (or "urban renewal") by redistribution of sites to private developers or public authorities, for redevelopment in accordance with the requirements of the local planning authority as specified in the plan for the area.

that either the owner of any interest to be acquired, or the acquiring authority, may apply to the local planning authority for a certificate stating for what development (if any) permission would have been granted if the land had not been subject to compulsory purchase. When the certificate is issued, planning permission for such development as is indicated in the certificate will be assumed in assessing the market value of any interest in the land for compensation purposes, in accordance with Section 15(5) of the 1961 Act (referred to earlier in this chapter).

This provision is of most benefit in cases where insufficient guidance is given by the development plan as to the kind of development for which planning permission would have been given, or where the plan favours development of the land in question of a kind which is not likely to give rise to an appreciable amount of development value.

Typical cases where such a certificate might be applied for are:

(i) land defined in the development plan as the site for some purely public development, e.g. a sewage disposal works or a public open space;
(ii) "white area" land not allocated in the plan for any form of development, e.g. land in the green belt;
(iii) land in an area not covered by a current development plan.

The right to apply for a certificate arises where land is "proposed to be acquired", i.e.:

(a) where any required notice in connection with the acquisition is duly published or served; or
(b) where notice to treat is "deemed to have been served"; or
(c) where an offer in writing is made on behalf of an acquiring authority to negotiate for the purchase of the owner's interest in the land.

Provisions as to the issue of certificates and appeals therefrom are contained both in the 1961 Act and in the Land Compensation Development Order, 1974.[46] The following is a summary of the procedure.

The application for a certificate, accompanied by a plan or a map, may be made at any time before the date of any reference to the

---

[46] SI 1974 No 539 (made under Section 20 of the 1961 Act).

Lands Tribunal to determine the compensation in respect of the applicant's interest in the land. But it cannot be made after that date except with the written consent of the other party or by leave of the Tribunal.

The applicant must specify one or more classes of development which in his view would be appropriate for the land in question if it were not being acquired compulsorily. He must also state the date on which a copy of the application has been or will be served on the other party. Not earlier than 21 days from this latter date, but within two months from the receipt of the application, the local planning authority must issue a certificate to the applicant stating that, if it were not proposed that the land should be compulsorily acquired, in their opinion planning permission either:

(a)  would have been granted for development of one or more specified classes (which may or may not be classes specified in the application) and for any development for which the land is to be acquired, but would not have been granted for any other development, or

(b)  would not have been granted for any development other than the development (if any) which the acquiring authority propose to carry out.[47]

Where, in the opinion of the local planning authority, the permission referred to in (a) above would only have been granted subject to conditions or at some future time, or both subject to conditions and at some future time, the certificate must say so.

The local planning authority are not necessarily bound by the provisions of their development plan in determining what sort of development might reasonably have been permitted on the land in question.

If a certificate is issued for development other than that specified in the application, or contrary to written representations made by the owner of the interest or the acquiring authority, the local planning authority must give their reasons and also state the right of appeal to the Secretary of State.

The two main requirements of this procedure therefore are: (i) to assume the absence of compulsory purchase, and (ii) on the assumption to consider whether and what planning permission the

---

[47] This type of certificate tends to be called a "nil certificate".

local planning authority should reasonably be expected to grant. The High Court has held that in (i) the assumption is *not* that the compulsory purchase was cancelled at the compulsory purchase order stage, but that compulsory purchase was not embarked upon at all; and that in (ii) the relevant planning policies to take into account are those which prevailed at the time of the compulsory purchase order and not those which prevailed subsequently at the time of entry or assessment.[48]

When a certificate is issued to either owner or acquiring authority a copy must be served on the other interested party. A copy must also be sent to the local authority in whose area the land is situated.

Section 18 of the 1961 Act provides that either the owner or the acquiring authority may appeal against a certificate to the Secretary of State for the Environment who, after giving the parties and the local planning authority an opportunity to be heard, may either confirm the certificate, or vary it, or cancel it and issue a different certificate in its place.

Appeal may also be made to the Secretary of State if the local planning authority fail to issue a certificate within two months of an application to them to do so, although this time may be extended by agreement of the parties. In this case the appeal will proceed as though the local planning authority had issued a certificate to the effect that planning permission could not normally have been expected to be granted for any development other than that which the acquiring authority propose to carry out.

Appeals to the Secretary of State must be made within one month of the date of issue of the certificate, or the expiry of the above time limit of two months (or extended time), as the case may be.

Section 19 of the Act contains special rules for applying the procedure in respect of certificates of appropriate alternative development to cases where an owner is absent from the United Kingdom or cannot be found. Section 21 gives the right to challenge the legal validity of a decision given by the Secretary of State on appeal (within six weeks) by means of an application to the High Court, which may quash the decision if it is found to be *ultra vires* the 1961 Act or to involve some procedural error substantially prejudicing the applicant.

---

[48] *Fletcher Estates (Harlescott) Ltd* v *Secretary of State for the Environment* [1998] 3 EGLR 13, *per* Dyson J.

Since 25 September 1991, compensation payable to a claimant includes the reasonable costs incurred in connection with the issue of a certificate, including costs incurred in an appeal under Section 18 if any issues of the appeal are determined in the claimant's favour.[49]

## 7. Special Compensation Rules in Particular Cases

### (a) Reinvestment by Investors

The Land Compensation Act 1961, Section 10A (inserted by the Planning and Compensation Act 1991, Section 70), provides that the payment of compensation by the acquiring authority to a landowner with an interest in reversion and not in occupation – e.g. a landlord – shall include reimbursement for the costs of acquiring "an interest in other land in the United Kingdom". The costs must have been incurred within one year from the actual date of entry (as distinct from notice of entry) and will be analogous to disturbance compensation discussed in Chapter 28 below; though disturbance compensation relates to dispossession whereas Section 10A compensation relates to deprivation of investment property.

### (b) Statutory Undertakers' Land

Under the Town and Country Planning Acts 1947–54, the compensation for certain special types of property was assessed on the assumption that planning permission would be granted for development whereby the use of the land could be made to correspond with that prevailing on contiguous and adjacent land.

This comprised properties held by local authorities for general statutory purposes, operational land of statutory undertakers and land held and used for charitable purposes.

This "prevailing use basis", as it was called, was abolished by the Town and Country Planning Act 1959, and now, under Section 11 of the Land Compensation Act 1961, compensation in these cases will, in general, be assessed on the same basis as for any other type of property. But where land is acquired from statutory undertakers who themselves have acquired the land as operational land, i.e. for the purpose of their undertaking, compensation will be assessed in

---

[49] Planning and Compensation Act 1991, Section 65.

accordance with certain special rules contained in Section 280 of the Town and Country Planning Act 1990. Nothing in the 1961 Act shall apply to any purchase of the whole or part of a statutory undertaking under any enactment which prescribes the terms on which such a purchase is to be effected (Section 36 of the 1961 Act).

## (c) Outstanding Right to Compensation for Refusal, etc., of Planning Permission

There are some cases in which, under the Town and Country Planning Acts, compensation for adverse planning decisions became payable subject to an obligation to repay in the event of a later grant of planning permission (or other relaxation of restrictions). The liability to repay depends on "a compensation notice" having been registered as a land charge. Section 12 of the 1961 Act provides that the factor of liability for repayment of compensation is to be taken into account in the assessment of "market value" irrespective of whether a "compensation notice" was registered before or after the notice to treat or a claim for compensation in respect of the adverse planning decision has been made.

## (d) Special Provisions in Private Acts

Where compulsory purchase powers are derived from Private Acts there may be provisions for a particular basis of compensation unique to purchases under that Act. These Acts are normally passed to enable a particular scheme to be carried out and include powers to acquire the land required. Petitioners against the measure sometimes obtain concessions on compensation.

As an example, the Croydon Tramlink Act 1994 contains provisions for the compensation payable to the owners of houses who have a negative equity in the house (i.e. the outstanding mortgage debt is greater than the value of the house) to be on a more generous basis than would be the case where the compensation basis is the general basis described above.

It follows that advisers dealing with cases under Private Acts need to determine whether any special provisions exist.

## 8. Summary of General Basis of Compensation

It may be useful at this point to summarise briefly and in very general terms the basis of compensation for land taken prescribed by the 1961 Act.

Essentially it is the best price which the owner's interest might be expected to realise if voluntarily offered for sale in the open market. But in many cases it may be necessary to make two or more estimates of the value of the property, each based on different assumptions permitted by the Act in order to determine what that "best price" is and (in particular) whether in addition to the existing use of the land the assessment should also take account of the prospect of development (having regard to "planning permission coupled with demand", in the words of Lord Denning MR, quoted earlier in this chapter).

The possible bases for these estimates of value may be formulated from the valuation standpoint as follows:

(i)   The value of the property as it stands, but having regard to the possibility of any future development within the limits of Schedule 3 of the 1990 Act (as amended), and taking into account also the terms of any planning permission already existing at the date of notice to treat, but not yet fully implemented by development.

In the case of a very large number of properties, both in built-up and in rural areas, this will represent the highest price obtainable in the open market.

(ii)  The value of the land, as in (i), but with the assumption that planning permission would be given for the development which the acquiring body propose to carry out.

This may or may not give a higher value than (i) according to whether permission to carry out this kind of development would be of value (a) to purchasers generally or (b) only to bodies armed with statutory powers or one particular purchaser – in which case the element of value might be excluded under Section 5 Rule 3 of the Land Compensation Act 1961 if the development is for a purpose which can only be carried out by that body or purchaser.

(iii) The value of the land subject to such of the prescribed assumptions as to planning permission – based on the provisions of the current development plan – as are applicable to the particular case.

This does not necessarily rule out of consideration the possibility that one or more purchasers might be prepared to pay a higher price in the hope that permission might be obtainable for some other form of development.

(iv) The value of the land subject to the assumption that planning permission would be given for one or more classes of development specified in a "certificate of appropriate alternative development" – in cases where the development plan gives little or no guide to the kind of development which might be permitted.

In all the above cases the valuation will exclude any increase or decrease in the value of the land in so far as it is attributable to actual or prospective development under the acquiring body's scheme, either on the land taken or on other land, and also any decrease in value due to the threat of acquisition.

Where the owner has an interest in other land held with that taken, additional compensation for severance and injurious affection may be payable under the provisions of the Compulsory Purchase Act 1965.

Where the owner has an interest in land adjacent or contiguous to that taken, any increase in the value of such land due to actual or prospective development under the acquiring body's scheme, either on the land taken or on other land, is to be deducted from the compensation.

Where development properties are acquired under the compensation provisions of the Land Compensation Act 1961, the problem is mainly one of the choice of basis most favourable to the claimant. The following examples illustrate the choices available:

*Example 26–1*

Open land under grass acquired for housing and allocated as such in development plan, programmed 1–5 years.

| | | |
|---|---|---|
| Housing value | £600,000 | per ha |
| Agricultural value | £5,000 | per ha |
| *Compensation* | £600,000 | per ha |

*Note*: The value for purpose acquired and for allocation in the development plan are the same.

*Example 26–2*

Suppose the same land is acquired for allotments.

| | | |
|---|---|---|
| Housing value | £600,000 | per ha |
| Agricultural value | £5,000 | per ha |
| Allotment value | £8,000 | per ha |
| *Compensation* | £600,000 | per ha |

*Note*: The value for allocation in the development plan gives the highest figure.

*Example 26–3*

As for Example 26–1 but programming 6–20 years. Acquired for school purposes.

*Compensation* – £600,000 per ha deferred for length of time market would allow. For example, this could be three years if the programme is likely to be accelerated. It could alternatively be 15 years if (say) no drainage is available until then.

*Note*: Here again the value is that for the allocation in the development plan since the value for school purposes is unlikely to be as great.

*Example 26–4*

Open land under grass acquired for housing. Land is shown in a "white area" on the development plan.

Values as in previous examples.
*Compensation* – Housing value, say, £600,000 per ha
(Value for purpose acquired).

*Example 26–5*

Open land under grass allocated in development plan for school purposes and acquired for use as such. The land is in the middle of a residential area in the development plan. Values as in previous examples.

*Compensation*. In this case the claimant should apply to the planning authority for a certificate of appropriate alternative development, asking for a certificate to be granted for residential development.

If granted –      Compensation £600,000 per ha
(Value for certified purpose).

If refused –      "School value" or existing use value.
Compensation, say, agricultural value £5,000 per ha

*Example 26–6*

Land in a "white area" is to be acquired as a recreation ground. There is some possibility that the development plan might be changed to a residential allocation on review. The agricultural value is £5,000 per ha. Recreation ground value £12,000 per ha. A speculator would give £50,000 per ha in the hope that the allocation will be changed.

*Compensation – £50,000 per ha.*

In practice, in the absence of direct evidence of recent comparable sales in the same area it is difficult to prove "hope value".

*Example 26–7*

A freehold building used until recently as a private house is to be acquired compulsorily: it is in an area of comprehensive development. It is now vacant. The development plan shows that the area is to be re-developed for industrial and shopping purposes. If converted into a shop, the premises would be worth £20,000 p.a. – the cost of conversion would be £80,000.

     The site area is 400 m². Industrial land is worth £100 per m² (£1,000,000 per ha).

(i) Value with permission to convert into a shop:

| | | |
|---|---|---|
| Net rental value | £20,000 p.a. | |
| YP | 12 | |
| | £240,000 | |
| *Less* Costs, etc., say | £100,000 | (to include fees – profit, etc.) |
| Value for conversion | £140,000 | |

OR

(ii) Value as industrial land
400 m² at £100 m²      £40,000

OR

(iii) Value as a house, say   £120,000

*Compensation*. In this case, basis (i) would be chosen by the claimant. Here the claimant is entitled, in any case, to the existing use value of £120,000. He could only claim £140,000 as compensation if he could show that planning permission for conversion to a shop might reasonably have been expected to be granted if the area had not been defined as an area of comprehensive development.

*Example 26–8*

The garden of a freehold house is to be acquired compulsorily. It will be used as a road which will give access to a site of 20 ha which is to be developed as a housing estate. The value of the 20 ha site with the access road is £20 million.

*Compensation*. The value of the land as a garden is negligible. The value as a road is very high since it unlocks the value of the housing site. It has ransom value.

The compensation payable depends on the facts of the case. If this is the only possible access to the rear land, then the starting point would be for the claimant to seek 50% of the development value unlocked – around £10 million. However, if an alternative access is possible then something less than 50% would be payable.

There have been several cases of ransom value considered by the Lands Tribunal. The sums awarded vary from 15% to 50% of the development value, whilst in some cases a fixed figure has been awarded. The Tribunal stresses that each case turns on its own particular facts and it is not possible to argue for any specific percentage because that was awarded in another case.

There is a widely held view that, in ransom cases, the decision in *Stokes* v *Cambridge Corporation* (1961),[50] when the Tribunal awarded one third of development value, applies. This is totally fallacious.

## 9. Arbitration of Compensation Disputes

If disputes over the assessment of compensation cannot be resolved by agreement then they must be decided by arbitration, either privately or (failing that) by reference to the Lands Tribunal under the Lands Tribunal Act 1949. The same is true of disputed apportionments of rent under any lease of land where only part of

---

[50] (1961) 13 P&CR 435.

the land is being acquired (Land Compensation Act 1961, Section 1). Procedures for such references are governed by the Lands Tribunal Rules (SI 1975 No 299, as amended).

Of particular importance is the formal "sealed offer" procedure, under the 1961 Act Section 4 (1), (3). Either side may offer unconditionally to pay or accept a specified sum. If the offer is not accepted it may be made a "sealed offer" submitted *but not disclosed* to the Lands Tribunal. If after the Tribunal's award the "sealed offer" turns out to be as favourable to the side that rejected it, or more favourable, then the side that made the offer *must* (unless there are special reasons to the contrary) be awarded costs against the rejecting side, in so far as these are costs incurred after the "sealed offer" was made. "Sealed offers" resemble "payments into Court" in civil actions for damages, or "Calderbank letters" (i.e. "without prejudice" except as to costs) in rent review arbitrations, i.e. the basic principle is the same.

But this does not alter the general rule that "the cost of litigation should fall on him who caused it", as stated by the Lord President Hope in a leading case in Scotland, *Emslie & Simpson Ltd* v *Aberdeen City District Council*,[51] in the Court of Session. In a civil action this means that normally the loser pays the winner's taxed costs; in the Lands Tribunal the acquiring authority is regarded as having "caused the proceedings". In the *Aberdeen* case there was an estimate of compensation, and the claimants obtained an advance payment of 90% of the estimate. They then referred their claim to the Lands Tribunal for Scotland. At the hearing of this reference the claim figure was very high and the authority's offer figure was very low; but neither side made a "sealed offer". The award was *higher* than the offer but *lower* than the claimed sum. It was also lower than the advance payment, and the claimants had to repay the difference. But the authority having made a low offer had thereby "caused the proceedings" and therefore still had to pay the claimants' costs.

## 10. Valuation

In general, the approach to valuing an interest subject to compulsory purchase is no different from that adopted in

---

[51] [1994] 1 EGLR 33.

preparing a normal open market valuation. There are, of course, the special rules described in this chapter to be taken into account, but these do not affect the method of valuation.

In most cases the valuation will be on an existing use basis when the direct comparison or the investment methods are likely to be adopted.

In preparing a valuation to be presented to the Lands Tribunal, the quality of evidence is assessed in the same way as for any arbitration. The best evidence is open market transactions. However, such transactions are made in the 'scheme world': if the scheme has an adverse effect the value will be reduced, and increased if the property benefits from the scheme. If such evidence is 'tainted' in this way then an adjustment may be required to correct it to value in the no scheme world.

Where open market evidence is scarce, the valuer may seek to draw upon other settlements of claims as providing comparable evidence. This is a valid approach but open market evidence will always be regarded as better evidence.

The Lands Tribunal always welcomes open market evidence in support of a valuation. Traditionally, evidence of such transactions is used to determine the market value of the subject premises. If 10 similar properties have sold for around £x then the compensation will be £x.

Over recent years consideration has been given to definitions of value, particularly in the preparation of the Red Book (see Chapter 20). The Red Book does not apply to compulsory purchase. However, the definition of market value in Rule 2 is close to estimated realisation price (ERP), which is forward looking, rather than market value in the Red Book which assumes completion of a sale on the valuation date. If the market is rising, the claimant will expect a value above that of completed market transactions following the logic of ERP. The converse applies in a falling market. The authors are not, however, aware of any case where this approach has been promoted.

One method of valuation which the Lands Tribunal has been reluctant to accept is the residual method to arrive at the market value of land with development value (see Chapter 16). Although the use of the residual method is widespread, in practice it is true that it can be manipulated to enhance or diminish a land value. However, the introduction of the *Practice Statement and Guidance Notes for Surveyors Acting as Expert Witnesses* published by the RICS in 1997, which explicitly requires a valuer when presenting evidence as an expert witness to submit opinions honestly held with a view to assisting the Tribunal rather than supporting his client, may help to overcome the Tribunal's reluctance.

If evidence is scarce or non-existent, the valuer may present *ipse dixit* evidence derived from experience and knowledge and understanding of values. The Tribunal has expressed a cautionary view on such an approach. In *Marson (HMIT) v Hasler*[52] a valuer submitted a valuation based on his man and boy experience. The Tribunal commented that "It had every respect for able and practical surveyors who belong to the (man and boy) 'school', but the fact should be recognised that when a member of this 'school' finds himself unable to agree values with an equally able and practical member of the 'analytical school' then, on a reference to the Lands Tribunal, the latter surveyor is apt to have the easier passage".

## 11. Interest on Compensation

If the acquiring authority take possession of the land ("entry") before compensation is assessed and paid, the vendor is entitled to interest on the capital sum which constitutes the amount agreed or awarded.[53] This applies not only to compensation under the "willing seller" principle but also to "equivalent reinstatement", as held by the Court of Appeal in *Halstead v Manchester City Council*.[54]

---

[52] [1975] 1 EGLR 157.

[53] Land Compensation Act 1961, Section 32. The Treasury prescribes the rate of interest to be paid. Subject to the provisions of the Land Compensation Act 1973, Sections 52 and 52A, regarding advance payments (discussed in Section 2(iii) of this chapter) payment is, inevitably, retrospective.

[54] [1998] 1 All ER 33. The "willing seller" and "equivalent reinstatement" principles are discussed in Section 3 of this chapter.

# Chapter 27

# Compulsory Purchase Compensation II
# Compensation for Injurious Affection

## 1. General Principles

The principles governing the right to compensation for injurious affection were considered in Chapter 15 where a distinction was drawn between:

(i)   claims for injurious affection to (i.e., depreciation of) land owned by the claimant in consequence of the acquisition, under statutory powers, of adjacent land some or all of which was formerly "held with" it; and

(ii)  claims for injurious affection to land owned by the claimant in consequence of the exercise of statutory powers on adjacent land none of which was formerly "held with" it.

The right to compensation in both cases derives either from the Lands Clauses Consolidation Act 1845, or – in the majority of cases nowadays – from the corresponding provisions of the Compulsory Purchase Act 1965. The relevant Sections are 63 and 68 of the 1845 Act (for (i) and (ii) respectively) and 7 and 10 of the 1965 Act (similarly).

The following diagram may help to make this fundamental distinction clear:

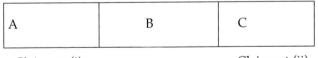

| A | B | C |

Claimant (i)                              Claimant (ii)

Land "B" is being acquired under statutory powers. Claimant (i), from whom it is being acquired, is entitled to claim compensation not only for its market value on acquisition but also for any injurious affection under Section 7 of the 1965 Act to his other land "A" which, until the acquisition, has been "held with" it. Claimant (ii) is only entitled to claim compensation for any injurious affection to his land "C" under Section 10 of the 1965 Act because none of land "B" has been "held with" it. Land "B" is being "severed" from land "A", but not from land "C".

Lands such as "A" and "B" are "held with" each other even if the claimant's rights in them are not the same (as in *Oppenheimer* v *Minister of Transport*[1] where the claimant's property right in the land taken was merely an option) and even if not actually contiguous, provided they are such that "the possession and control of each" gives "an enhanced value to the whole" (*Cowper Essex* v *Acton Local Board*[2]); and this is a valuation question of obvious practical importance.

The view is often expressed that compensation under this head is not governed by the provisions of the Land Compensation Act 1961. While this is undoubtedly true in regard to many items in the nature of disturbance which may arise where part only of the land is taken, it is submitted that, in so far as the claim for compensation is based on depreciation in the "value" of the land, it comes within the scope of Rule (2) of Section 5 of the Land Compensation Act 1961. Section 5 expressly applies in general terms to "Compensation in respect of any compulsory acquisition"; and the exception from its scope which is contained in Rule (6) is expressly confined to "disturbance or any other matter not directly based on the value of land" – not merely, it should be observed, the value of the land acquired.

It is therefore assumed in this chapter that, in all cases of compulsory acquisition pursuant to a notice to treat served after 29 October 1958, compensation for injurious affection – in that it is based on the value of the land – will represent the depreciation in the value of that land in the open market.

It would appear, however, that the assumptions as to planning permission to be made under Sections 14–16 of the 1961 Act, in the case of lands taken, do not apply to assessments of market value in the case of lands injuriously affected.[3]

---

[1] [1942] 3 All ER 485.

[2] (1889) 14 App Cas 153.

[3] Sections 14-17 apply to the "relevant interest" and "relevant land". Section 39(2) defines these, "in relation to a compulsory acquisition in pursuance of a notice to treat", as (respectively) "the interest acquired in pursuance of that notice" and "land in which the relevant interest subsists", and no mention is made of any land retained by the claimant.

## 2. Compensation where Part Only of the Land is Taken

This is "severance". Cases of this type frequently occur, particularly in connection with the construction of new roads or the widening of existing roads or town-centre redevelopment. An example occurred in *Ravenseft Properties Ltd* v *Hillingdon London Borough Council*:[4] though in that case the acquiring authority were eventually compelled to buy the whole of the property in dispute instead of severing it as they had intended by acquiring part of the garden.

It is necessary to consider the principal losses likely to be suffered in connection with the land formerly "held with" the land taken and the bases on which claims for compensation should be prepared.

Compensation where part of a property is taken falls under three main heads: land taken; severance and other injurious affection to land held therewith; other incidental losses resulting from the compulsory taking. Other matters which may have to be considered are accommodation works, and apportionment of rents.

As a matter of practice the last two items are the first to be addressed since they affect the amount of compensation payable. The accommodation works, if any, are reflected in the injurious affection to the land retained whilst the apportionment of rent has a direct bearing on both the value of the land taken and also the land retained.

### (a)  Accommodation Works

In the case of railway undertakings the acquiring body was obliged under Section 68 of the Railways Clauses Act 1845, to provide certain "accommodation works" for the benefit of owners of land adjoining the undertaking. These included bridges and other means of communication between severed portions of land, the adequate fencing of the works, means of drainage through or by the side of the railway and provision of watering places for cattle.

---

[3] *(continued)*
But see *Abbey Homesteads Group Ltd* v *Secretary of State for Transport* [1982] 2 EGLR 198 where the member of the Lands Tribunal held that in determining the planning position in relation to the land retained one could have regard to the planning background to a Section 17 Certificate in respect of the relevant land.

[4] (1968) 20 P&CR 483.

These works tended to reduce the effect of severance and other injury likely to be caused by the construction of the railway and were, of course, taken into account in assessing compensation.

In the case of other undertakings there is no such obligation on the acquiring body to provide accommodation works, or on the owner to accept in lieu of compensation. But in practice it is frequently agreed between the parties as a matter of joint convenience that the acquiring body will carry out certain "accommodation works" or "works of reinstatement", and that compensation shall be assessed on the basis of these works being provided. For instance, where a new road is constructed across agricultural or accommodation land, the acquiring body will not only fence along the boundaries of the road, but will also provide new gates where convenient. If part of the front garden of a house is taken under a street widening scheme, the acquiring body will probably agree to provide a new boundary wall and gates, to plant a quickset hedge to screen the house, to make good the connection to the house drains and similar works. If the use of the works will be noisy, such as a motorway, the authority will probably agree to double glazing or secondary glazing.

All these are items for which compensation would otherwise have to be included in the claim. Some acquiring authorities compare the cost of accommodation works with the consequent reduction in the compensation payable and, if the cost is significantly greater than the savings to them, will decline to carry out the works. For example, if a farm is severed by a motorway, the provision of a bridge or tunnel will allow the two parts of the farm to continue to function as a single entity. The injurious affection will thereby be reduced but bridges or tunnels can be costly to provide. If a bridge costs £80,000 and the reduction in the injurious affection is around £80,000 or more then it makes sense to provide a bridge; but if the reduction in the injurious affection is, say, only £5,000, then there seems little point in providing a bridge since it clearly does not have any significant impact on the use of the farm whether it is severed in two parts or not. It is usually convenient for both parties that the acquiring body should do the accommodation works while the work on the scheme is still in progress. Details of the works agreed upon will therefore be included in any settlement of compensation by agreement or, in case of dispute, may be referred to the Lands Tribunal and considered by them in awarding compensation.

It is also important in road cases to come to an understanding as to the maintenance of slopes, banks and retaining walls, and as to the owner's right of access to the road. The acquiring body will usually undertake responsibility for slopes, banks and retaining walls. Normally any retaining wall built to support a highway, or wall built to prevent soil from higher land falling on the highway, will be part of the highway. The owner will usually stipulate for access to the new road across any intervening slopes or banks and the right to lay sewers, and also for public services such as gas and water mains, electric cables, etc., to be permitted in the soil of the road.

### (b) Apportionment of Rent (Compulsory Purchase Act 1965, Section 19)

Where part of land subject to a lease is taken it may be necessary to apportion the rent between the part taken and that which is left. Naturally the lessee should not be required to continue to pay the same rent, particularly if the portion taken is considerable; but it is often difficult for the lessor and lessee to agree on the amount of the deduction to be made. The lessee can only demand a fair apportionment of the rent paid, not of the annual value of the property. Also, he cannot include in the reduction of rent any figure representing reduced value to the rest of the property. That loss should be met by compensation from the acquiring body to the lessor or lessee as appropriate.

In view of these difficulties it is a common practice, particularly where small portions of urban properties are taken for road widening purposes, to agree on a nominal apportionment of, say, one pound a year on the land taken. Under this arrangement the lessee will continue to pay the rent reserved under his lease for the balance of his term, but on the other hand will receive full compensation for the loss of rental value from the acquiring body; while the freeholder will be compensated for the injury to his reversion. Where the portion of land taken is significant it is common for the rent payable to be apportioned in the ratio of the rental value of the part retained to the rental value of the whole, ignoring the CPO scheme so as to exclude any effect of injurious affection caused by the scheme. This can lead to problems where a rent review takes place shortly before the valuation date as the parties will be likely to reflect the reduced value of the land retained in agreeing the new rent.

Where a freehold property is subject to a building lease at a ground rent and sub-lease at the full annual value or rack rent, apportionment might be agreed on a nominal basis of 1p per annum on the part taken as regards the rent under the ground lease and £1 per annum for the part taken as regards the rent under the sub-lease.

Any dispute as to apportionment of rent under a lease may be referred to the Lands Tribunal. But in that case the apportionment must be a true apportionment as distinct from a nominal one. The costs of apportionment are not covered by the words "costs of all conveyances" in Section 23 of the Compulsory Purchase Act 1965 and so are not automatically borne by the acquiring body (*Ex parte Buck* (1863)).

## (c) Land Taken

The area of land to be taken will usually be indicated on the plan accompanying the notice to treat. In the case of a road this will necessarily include the land required not only for the road itself but also for the slopes of any cuttings or embankments.

The compensation for the land taken will be assessed in accordance with the principles prescribed by the Land Compensation Act 1961 (as amended). In the case of agricultural land its "value" will probably be estimated by reference to prices paid for small areas of land of the same type in the district; but this is, of course, a point of valuation rather than of law, and in law it is justifiable in principle just as valuation by reference to comparable properties is justifiable generally.

In addition to the value of the land itself the owner is entitled to compensation for the loss of any buildings or fixtures or timber on the land, whether or not these items are included in the overall purchase price or are subject to a separate valuation. Obviously the same item of value must not be compensated for twice over. The tenant or owner-occupier of agricultural land may also have a claim for "tenant-right", which will be discussed in more detail under the heading of "other incidental losses". The relevant principles have been discussed above, in Chapter 26.

## (d) Severance and other Injurious Affection

It is important to remember that "severance" compensation is one variety of "injurious affection" compensation. That is to say an owner suffers compensatable loss (referred to in Section 7 of the 1965 Act as "damage") because his "other land" which he retains is depreciated whether by severance or some other factor. The loss by "severance" is loss whereby the market values of the land taken and the land retained amount to less than the total market value of the land before it was severed. A classic case is *Holt v Gas Light & Coke Co*[5] in which a small portion of certain land used as a rifle-range was compulsorily purchased. The part taken was peripheral only, but so situated that the safety area behind the targets was no longer sufficient and the land retained could not in future be used as a range.

The main kinds of severance can perhaps be best understood diagrammatically:

Land "B" is acquired from the owner and severed from land A in the first example and from lands "A" and "C" in the second, which the owner retains. The essential "severance" is that of land "B" from the other land; the severance of land "A" from land "C" is a secondary matter (though none the less important). More complicated forms of severance may, of course, cause further fragmentation of an owner's land; but the basic principles of compensation are not affected by this.

A typical modern example is where part of a farm is separated from the rest of the farm, including the farmhouse and farm buildings, by the construction of a road across the property, or where the area of a farm is significantly reduced in size by the taking of part even though the part remaining is not split up.

Such severance is likely to result in increased working costs. For example, if a portion of arable land is cut off by severance of intervening land there will probably be an increase in the cost of all the normal operations of ploughing, sowing, reaping, carting, etc., as well as additional supervision necessitated by the separation of the land from the rest of the farm. If the severed portion is pasture land, extra labour and supervision may be required in driving cattle to pasture as well as possible risk of injury to cattle, as for

---

[5] (1872) 8 LR 7 QB 728.

instance in crossing a busy new road. These additional expenses are likely to involve some reduction in rental value of the land which, capitalised, will represent the injury due to severance.

The extent of loss from severance will naturally vary greatly according to the nature of the undertaking and other circumstances, for example, whether or not a new road or railway is situated on a level with the land remaining or is put on an embankment or viaduct or in a cutting or tunnel, and whether or not "accommodation works" such as a private bridge or "underpass" are carried out to enable the owner to cross the road or railway. It must not be forgotten that land ownership extends physically downwards and upwards from the land surface, so that viaducts and tunnels involve "severance" of land in principle just as surface works do; though it may be that the financial loss caused to an owner by having a tunnel made below his land surface will in some circumstances be nil (see *Pepys* v *London Transport Executive*).[6] Ownership of property fronting a highway commonly includes half the highway, subject to the dedication of the surface. So in *Norman* v *Department of Transport*[7] the ownership of the sub-soil of a highway which was indicated in a draft CPO entitled the owners of the property fronting the highway to compensation (in this case a blight notice).

In whichever sense it is "severed", the property as a whole may suffer depreciation in value by reason of the fact that it can no longer be occupied and enjoyed as one compact holding.

The valuer must be careful to include in the claim for compensation any injury likely to be caused to the rest of the property by the authorised user of the works. At one time this only applied to such user in so far as it occurred on the land acquired from the claimant, i.e. land held by him with the injuriously affected land until the severance - as was held by the Court of Appeal in *Edwards* v *Minister of Transport*.[8] But Section 44 of the Land Compensation Act 1973 abolished this restriction, and loss is compensatable in respect of the user of the works as a whole. The user must, however, be *authorised*. If it is not covered by the statutory authorisation conferred on the acquiring authority they will be liable instead to damages in a civil action.

---

[6] [1975] 1 All ER 748.
[7] [1996] 1 EGLR 190.
[8] [1964] 1 All ER 483.

Here "injurious affection" arises "otherwise" than by severance, and is akin to depreciation actionable at common law in the tort of nuisance. For example, the use, as a major road, of a strip of land across a residential estate may seriously depreciate the value of the mansion by reason of noise and fumes, loss of privacy, and spoiled views. Such factors (other than loss of privacy and views) are actionable nuisances if they occur to any significant degree. All of them (including loss of privacy and views and any other non-actionable loss) can and should be taken into account in claiming injurious affection compensation in cases where land retained is depreciated (leaving aside the actual fact of severance) in consequence of the carrying out of the acquiring body's project.

Here the leading authority is the decision of the House of Lords in *Buccleuch (Duke of)* v *Metropolitan Board of Works*,[9] when part of the garden of the claimant's house (in fact its entire riverside frontage) was compulsorily acquired for the construction of the Victoria Embankment at Westminster. The full depreciation of the claimant's property was held to be payable, on the basis of severance and injurious affection in combination, in so far as the insignificant purchase price or market value of the strip of garden actually taken was insufficient of itself to make up the total loss in land value which the claimant had suffered (i.e. on a "before and after" valuation).

In the case of agricultural land, the possibility of crops being damaged by fumes or dust from a new road, or the yield of milk from pasture land being reduced by disturbance of cattle due to heavy traffic on the road, may result in loss of rental value.

In addition to consideration of the effect on value of the retained land by the authorised use of the works, the actual carrying out of the works can also have an impact. The date of valuation is commonly the date when possession is taken of the land required for the scheme. This establishes the valuation date. At that time the valuation of the retained land needs to reflect the fact that work is to be carried out on neighbouring land which may have a significant impact on the valuation. For example, if some garden land of a house is taken to build a motorway, the valuation of the retained land is of a house which will be close to the site of a proposed motorway where work may be going on for several years

---

[9] (1872) LR 5 HL 418.

which will be noisy and unpleasant, following which there will be a motorway with heavy traffic. All of these factors should be reflected in the valuation.[10]

Every head of damage which can reasonably be anticipated should be included in the claim, since no further claim can be made later for damage which might have been foreseen at the time when the land was taken.

The claim for depreciation in value of the land retained will be based on its market value which will take into account the benefit of any actual planning permission already given. It may be permissible, in some cases, to consider the effect on value of the possibility that planning permission might have been given, having regard to the terms of the development plan and other circumstances. But no assumptions of planning permission can be made, as in the case of land taken, because the relevant provisions of the Land Compensation Act 1961 do not refer to land retained; thus nothing other than genuine "hope value" existing independently of the actual grant of planning permission could possibly be justified in any relevant valuation.

The Land Compensation Act 1973, Section 8(4A), requires that there be entered in the local land charges register details of: (a) the works for which the land taken is required, and (b) that part of the land which is not taken. This is to avoid duplication of compensation in the event of a possible claim under Part I of the Land Compensation Act 1973 (discussed below).

## (e)  Other Incidental Injury

In addition to the value of the land taken and injurious affection to other land held therewith, the owner is entitled to compensation for all other loss and expense which he may incur in consequence of the

---

[10] Such an approach is supported by, for example, *Cuthbert* v *Secretary of State for the Environment* [1979] 2 EGLR 183 correctly applying Section 44 of the Land Compensation Act 1973 that compensation for injurious affection "shall be assessed by reference to the whole of the works and not only the part situated on the land acquired" (from the claimant). Section 44 reversed *Edwards* v *Minister of Transport* (1964) – see n.8 above. In *Budgen* v *Secretary of State for Wales* [1985] 2 EGLR 203 the Member adopted the approach of "looking at the property before and after the acquisition and construction of the road". Whether this gives full effect to Section 44 is debatable.

compulsory acquisition. The generic term for these items of compensation is "disturbance", which is the subject of the next chapter. But some may in practice be closely connected with claims for severance and injurious affection compensation and can for the sake of convenience be touched on here. Such losses will vary with the circumstances of each case, so that it would be misleading to attempt to suggest an exhaustive list of possible items. A consideration of some of the principal heads of damage likely to arise in connection with the taking of a strip of land through an agricultural estate will indicate the general nature of items which may be included with this part of the claim. They may be briefly summarised as follows:

(i)   *Disturbance and Tenant-Right on the Land Taken*
      Either an owner-occupier or a tenant is entitled to compensation for any loss suffered through having to quit a portion of farm land at short notice. Such loss may arise from the forced sale of stock which can no longer be supported on the reduced acreage of the farm, or from the forced sale of agricultural implements. He may also claim for other similar consequential losses, for example, temporary grazing and storage. Improvements to the land made at a tenant's expense may have to be taken into account. Both the owner-occupiers and tenants may claim for "tenant right".[11]

      Disturbance is payable for "having to vacate the premises", i.e. being disturbed in the occupation of the land taken, as described in Chapter 28. Where a claimant suffers losses due to the scheme but these are derived from the effect of the scheme on business conducted on the retained land then no compensation for disturbance as such is payable. For example, a forecourt to a shop may be acquired. No business is conducted on the forecourt. However, the business may be severely depreciated as a result of the work carried out for the scheme for which the forecourt was acquired. Strictly no compensation for the loss of profit is payable.

---

[11] Additional compensation for reorganisation of the tenant's affairs, up to four times the annual rent of the part taken, may also be payable under the Agriculture (Miscellaneous Provisions) Act 1968, Section 12, (as amended by the Agricultural Holdings Act 1986, Schedule 14). The significance of this is best considered in connection with "disturbance" compensation (see Chapter 28).

(ii)  *Damage During Construction*

It is almost unavoidable that during construction a certain amount of damage should be caused to crops, etc., on land immediately adjoining the works. It is also likely that certain parts of the farm, particularly those portions which are to be severed, will be more difficult to work during this period, so that a claim for increased labour costs or total loss of rental value may be justified. If the works are being carried out close to the mansion or farmhouse, the impact on the occupier may be considerable and in extreme cases it may be necessary for him to obtain temporary accommodation elsewhere.

(iii)  *Reinstatement Works*

Owing to the construction of the undertaking on the land taken, the owner or occupier may be forced to carry out various works to the remaining land. For example, small awkward-shaped pieces of land may be left on either side of the works which it will be necessary to throw into adjoining fields by re-arranging fences, grubbing-up hedges, filling in ditches, etc.

*(f)  Assessment of Compensation*

It is common practice to assess a global figure for compensation for both the land taken and also for injurious affection. This is achieved by the "before and after" method. The "before" value is determined by assessing the market value of the whole interest including the land to be taken as if no scheme of acquisition existed. The "after" value is the market value of the interest retained reflecting the full impact of the scheme.

The difference between these figures represents the value of the land taken and the injurious affection to the land retained. It is necessary to determine a value of the land taken which, when deducted from the full difference, leaves the balance as compensation for injurious affection.

However, where it is possible to determine both items of claim directly then this approach should be adopted.

This is considered more fully later in Section 4 of this chapter.

*Example 27–1*

A Highway Authority is about to acquire a strip of land through a farm of 80 ha for the construction of a road 30 m wide, running roughly north and south and likely to carry a considerable volume of traffic. Where the road first enters the farm at the north, it will run for approximately 100 m on an embankment of a maximum height of 3 m. For the remainder of its course it will be on the level. The total area of land acquired, including the slopes of the embankment will be 3 ha.

The farm consists of 64 ha of pasture, 15 ha of arable, and the rest roads and buildings. It is occupied by the owner. Similar land in the neighbourhood but without buildings has sold recently for £6,000/ha.

The new road will sever 24 ha of the best land from the homestead and interfere with the water supply and land drainage of three fields.

The Highway Authority will fence along the whole length of the road, and at two points will provide underpasses giving access from the fields on one side of the road to those on the other. They will provide adequate land drainage along the sides of the road and to the slopes of the embankment and will make good any interference with the water supply.

Estimate the compensation payable to the owner-occupier.

*Valuation*

Land Taken:
3 ha at £6,000/ha                                                   £18,000

Injury Due to Severance
24 ha will be severed from the farmhouse and buildings and extra labour and risk will be involved in driving cattle to and from those fields.
Loss of value, say £600/ha on 24 ha                                 £14,400

Disturbance on land taken, including possible loss
on sale of stock, say                                               £2,000
Tenant-right on land taken, say                                     £1,000

Interference during construction of works:
(i) Depreciation of rental value on 24 ha for, say, 6
months at £250/ha                                                   £3,000

(ii) Total loss of rental value on 8 ha immediately
adjoining the works, say                                            £2,000

| Cost of re-arranging fences between fields and ploughing up small area of pasture | £2,000 |
|---|---|
| Re-arrangement of land drainage to three fields | £1,500 |
| Compensation | £43,900 |

Plus surveyor's fees and legal costs.

*Note*: The works of fencing, draining, etc., which the Highway Authority have undertaken to provide come under the heading of "accommodation works" or "works of reinstatement".


*Example 27–2*

A local authority is carrying out a town centre shopping scheme, the land for which is being acquired under a CPO. The scheme will take around two years to complete.

X Traders Limited own the leasehold interest in shop premises comprising a two-storey shop with a single storey rear extension and a rear yard with access from a rear service road from which goods are brought into the shop.

The CPO allows the local authority to acquire the single storey rear extension and the rear yard. The service arrangements for the new scheme allow for service access to the retained shops via a covered service way from nearby lorry unloading service bays.

X Traders Limited have a lease for 20 years from 8 years ago at a current rent of £30,000 p.a. The current rental value in the absence of the scheme is £36,000 p.a. of which the rental value of the retained shop is £33,000 p.a. However, the disruption caused by the scheme reduces the rental value of the retained premises to £29,000 p.a. but the rental value will rise to £35,000 p.a. when the scheme is completed and open for trading. There is a rent review in two years' time.

*Estimate the compensation payable to X Traders Limited and the freeholders.*

1. *Accommodation Works*
The local authority will reimburse the costs of alterations to the internal layout of the shop necessitated by the loss of the rear extension and the alteration to the service access and will carry out the external works required such as the provision of a covered way from the lorry off-loading point to the rear access door to be incorporated in the re-designed rear wall following removal of the single storey extension.

2. *Apportionment of Rent Payable*
The existing rent is £30,000 p.a. The rental value of the whole is £36,000 p.a. and of the retained shop is £33,000 p.a. The rental value of the land taken is therefore £3,000 p.a.

Hence:

$$30,000 \times \frac{33,000}{36,000} = £27,500$$

∴ the rent will be reduced to £27,500.

3. *Valuation of Land Taken and Injurious Affection*

### A. X Traders Limited

| *No Scheme* | | |
|---|---|---|
| FRV | £36,000 p.a. | |
| *Less* Rent Payable | £30,000 | |
| Profit Rent | £6,000 p.a. | |
| YP 2 yrs @ 7/3/IT | 1.21 | £7,260 |
| (FRV payable at next review) | | |
| *Subject to Scheme* | | |
| FRV | £29,000 p.a. | |
| *Less* Rent Payable | £27,500 | |
| Profit Rent | £1,500 p.a. | |
| YP 2 yrs @ 7/3/IT | 1.21 | £1,815 |
| Value of Land Taken and Injurious Affection | | £5,445 |
| Value of Land Taken (No scheme world) | | |
| FRV | £3,000 p.a. | |
| *Less* Rent Payable | £2,500 | |
| Profit Rent | £500 p.a. | |
| YP 2 yrs (as before) | 1.21 | £605 |
| ∴ Injurious Affection | | £4,840 |
| Value of Land Taken | | £605 |

*Note*: The profit rent of the land taken of £500 equals the balance of the whole profit rent apportioned in the same way as the existing rent, i.e.

$$£6{,}000 \times \frac{33{,}000}{36{,}000} = £5{,}500,$$

which is the profit rent for the land retained.

4. *Disturbance*
Loss of Profits and other costs (see next chapter).

## B. Freeholders

*Valuation for Land Taken and Injurious Affection*

*No Scheme*

| | | |
|---|---:|---:|
| Rent Received | £30,000 p.a. | |
| YP 2 yrs @ 6.5% | 1.82 | £54,600 |
| Reversion to FRV | £36,000 p.a. | |
| YP in perp def'd | | |
| 2 yrs @ 6.5% | 13.56 | £488,160 |
| | | £542,760 |

*Subject to Scheme*

| | | |
|---|---:|---:|
| Rent Received | £27,500 p.a. | |
| YP 2 yrs @ 6.5% | 1.82 | £50,050 |
| Reversion to FRV | £35,000 p.a. | |
| YP in perp def'd | | |
| 2 yrs @ 6.5% | 13.56 | £474,600 |
| | | £524,650 |

∴ Value of land taken and injurious affection

| | |
|---|---:|
| No Scheme | £542,760 |
| *Less* | |
| Subject to Scheme | £524,650 |
| | £18,110 |

Value of Land Taken
(No scheme world)

| | | |
|---|---:|---:|
| Rent Received | £2,500 p.a. | |
| YP 2 yrs @ 6.5% | 1.82 | £4,550 |
| Reversion to FRV | £3,000 p.a. | |
| YP in perp def'd | | |
| 2 yrs @ 6.5% | 13.56 | £40,680 |
| | | £45,230 |

| | |
|---|---:|
| Value of Land Taken and Injurious Affection | £18,110 |
| Value of Land Taken | £45,230 |
| ∴ Injurious Affection (Betterment) | £(27,120) |

*Note*: The enhancement in value due to the scheme on reversion (betterment) is incorporated so as to offset it against the injurious affection. In this case the value of the reversion of the land retained is increased from £33,000 p.a. to £35,000 p.a. = £2,000 p.a. × 13.56 YP = £27,120.

Thus the value of the land taken of £45,230 is reduced by the betterment of £27,120 to produce compensation payable of £18,110.

## 3. Compensation where No Part of the Land is Taken

The four rules which govern the right to compensation for injurious affection in this type of case in consequence of the decision of the House of Lords in *Metropolitan Board of Works* v *McCarthy* (1874), have already been mentioned in Chapter 15. It should be remembered that these cases are treated very differently from those arising where some land has been taken from the claimant, as discussed above. Whereas in the latter cases compensation is paid in full for any injurious affection that is proved to occur, in these cases where no land is taken such proof is quite insufficient if the four McCarthy rules cannot be satisfied in full.

The four rules can be briefly re-capitulated as follows:

(i)   the works causing the injurious affection must be authorised by statute;
(ii)  the harm must be of a kind which would be actionable in the civil courts but for that authorisation;
(iii) the loss to be compensated must be confined to land value (e.g. not business loss or other "disturbance");
(iv)  the cause of the harm must be the execution of the works and not their use (e.g. the building of a new road or railway, not the effects of the traffic on it when opened).

Typical examples of successful claims are where authorised development by the acquiring body deprives an adjoining owner of the benefit of some easement attached to his land (such as a private right of light, as in *Eagle* v *Charing Cross Railway Co*),[12] or a

---

[12] (1867) LR 2 CP 638. If it were not for the statutory authority (rule i) this would be an actionable private nuisance (rule ii). The same is true of physical damage caused by temporary operations on the adjoining land compulsorily acquired: see *Clift* v *Welsh Office* [1998] 4 All ER 852. Court of Appeal.

private right of way, or is in breach of some restrictive covenant to the benefit of which the adjoining owner is entitled, as in *Re Simeon and the Isle of Wight Rural District Council*.[13] But compensation for infringing a restrictive covenant must not exceed the diminution in value of the land to which the benefit of the covenant attaches: *Wrotham Park Settled Estates* v *Hertsmere Borough Council*.[14]

Since the greatest stumbling-block to claimants has in most cases in practice been caused by rule (iv) – though rule (iii) also in practice frustrates claims from time to time, as shown in *Argyle Motors (Birkenhead) Ltd* v *Birkenhead Corporation*[15] – Parliament enacted Part I of the Land Compensation Act 1973 to give owners (on and after 23 June 1973) a right to compensation for depreciation to their land caused by the use (as distinct from the execution) of public works. The use must have begun on or after 17 October 1969, and must give rise to depreciation by reason of certain prescribed "physical factors", namely "noise, vibration, smell, fumes, smoke and artificial lighting and the discharge on to the land in respect of which the claim is made of any solid or liquid substance". In practice it is noise which is the major cause of depreciation, although the other factors do contribute. In *Blower* v *Suffolk County Council*[16] the sole contributory factor was the artificial lighting where the depreciation due to light emanating from a new road led to compensation even though the amount of light which reached the claimant's home was minimal. The "responsible authority" which has promoted or carried out the works is liable, including the Crown (but no claim may be made in respect of "any aerodrome in the occupation of a government department"). The claimant must hold an "owner's interest" in the depreciated land, namely the freehold or a leasehold with three or more years to run. If it is a dwelling it must be his residence if his interest entitles him to occupy it; landlords of dwellings may also claim. If it is a farm he must occupy the whole even though his interest need only be in part. If it is any other kind of property his interest and occupation

---

[13] [1937] 3 All ER 149.

[14] [1993] 2 EGLR 15.

[15] [1974] 1 All ER 201, (1974) 229 EG 1589; and see *Wildtree Hotels Ltd* v *Harrow London Borough Council* [1998] 3 All ER 638 (Court of Appeal): no compensation for noise, dust and vibrations, as such (but see n 12 above: *Clift* v *Welsh Office*).

[16] [1994] 2 EGLR 204.

must relate to all or a "substantial part" of it and, in accordance with the Town and Country Planning (Blight Provisions) Order, SI 1990 No 465, the rateable value must not exceed £18,000. Claims for £50 or less are ignored.

The period for submitting a claim begins one year after the date when the use of the works began, called (the "relevant date"), subject to certain exceptions, and the normal six-year limitation period under the Limitation Act 1980 then starts to run. Where residential long leaseholders are claiming enfranchisement against their landlords under the Leasehold Reform Act 1967 or under the Leasehold Reform, Housing and Urban Development Act 1993, and the "relevant date" under Part I of the 1973 Act falls *before* they acquire their freeholds or extended leases as the case may be but *after* service of notice to claim, they are regarded as having an "owners interest" for the purposes of compensation under the 1973 Act (Section 12 and 12A of that Act).

The valuation date is the relevant date. Prospective development value outside the Third Schedule to the Town and Country Planning Act 1990 (as amended) is not compensatable. Corresponding benefit to other land of the claimant will be "set off" against the compensation (if any). There must be no duplication of any compensation otherwise payable, e.g. under Section 7 of the 1965 Act, or as damages in tort for nuisance. It was held in *Vickers v Dover District Council*,[17] that liability in nuisance is not excluded if the authority are acquiring under a power (i.e., permissively) and not a duty, so that Part I of the 1973 Act does not apply in such a case. Availability of grants for soundproofing must be taken into account.

The use giving rise to compensation must be taken to include reasonably foreseeable future intensification. Beyond this, subsequent alterations to works, or changes of use apart from mere intensification and apart from aerodromes or highways, will give an additional right to compensation. "Physical factors" caused by aircraft as such give no right to compensation except in respect of new or altered runways, "taxiways" or "aprons".

Once the legal right to compensation is established, it is then a question of determining the depreciation in the market value of the land due to the injury complained of. In *Hickmott* v *Dorset County*

---

[17] [1993] 1 EGLR 193.

*Council*[18] a claim made under Part I of the Land Compensation Act 1973 failed because the claimant could show nothing beyond an increased fear of danger to property from passing traffic in consequence of road widening.[19] It is clear from *Pepys v London Transport Executive*[20] that in order to establish depreciation in market value it is not sufficient to show merely that a sale at a higher figure fell through by reason of the authority's works while a sale at a lower figure went ahead subsequently.

Two methods are commonly adopted to determine the amount of injurious affection. One is to make two valuations of the property – (i) as it previously stood, and (ii) after the interference with the legal right in question. This "Before and After" method, as it is often called, can be used both in this type of case and also in cases where part of land has been taken.

The other is to determine the rate of depreciation to the value of the land retained due to the factors to be taken into account.

## 4. "Before and After" Method of Valuation

Besides the case where no part of the owner's land is taken, this method is also used in many cases of the taking of part of a property where, in practice, it is very often difficult if not impossible to separate the value of the land taken from the injurious affection likely to be caused to the rest of the land by the construction and use of the undertaking (see Example 28-2 above).

As applied to this latter type of case it will involve two valuations:

(i)   of the property in its present condition unaffected by statutory powers; and
(ii)  of the property as it will be after the part has been taken and in contemplation of the construction and use of the undertaking in its entirety.

The amount by which valuation (i) exceeds valuation (ii) will

---

[18] [1977] 2 EGLR 15.

[19] See also *Streak and Streak v Royal County of Berkshire* (1976) 32 P&CR 435: failure to attribute depreciation to proximity of new motorway. But the claimant in *Davies v Mid-Glamorgan County Council* [1979] 2 EGLR 158, succeeded in a claim based on depreciation attributed to successive extensions to an airfield near his land.

[20] [1975] 1 All ER 748.

represent the compensation payable both for the loss of the land taken and also for injurious affection to the remainder. In addition, the owner might be entitled to claim for injury during the carrying out of the works, and it would be necessary to come to an understanding as to the works of reinstatement which the acquiring body are prepared to provide.

An example is the taking of a strip of land forming part of the garden or grounds of a house. Here it is usually very difficult to assign a value to the strip of land taken without at the same time considering all the consequences of the taking. Suppose that the strip forms part of a garden of a fair-sized house and is to be used in the construction of a new road; the following are some of the questions which will naturally suggest themselves in assessing the fair compensation to the owner:

(i)   Will the land remaining be reasonably sufficient for a house of this size and type?
(ii)  Will the proposed road be on the level, on an embankment, or in a cutting?
(iii) Will it be visible from the house, and if so from what parts of it?
(iv)  What volume of traffic may be expected?
(v)   How close will the house be to the road and to what extent is it likely to suffer from loss of privacy, noise, fumes and other inconvenience both during construction and when in use?

It is obvious that the most realistic approach to the problem is that of comparing the value of the house as it now stands (i.e. with no road to be built), with its estimated value after a portion of the garden has been taken for the new road (as in the *Duke of Buccleuch's* case, above).

In theory, at any rate, there are two statutory obstacles to the use of this method:

(i)   the fact that the statutory assumptions as to planning permission apply to land taken, but not to land injuriously affected; and
(ii)  the fact that Section 4(2) of the Land Compensation Act 1961 requires that the owner's statement of claim shall distinguish the amount claimed under separate heads and show how the amount claimed under each head is calculated.

Even so, it is probable that the valuer's first approach to a problem of this kind will be along the lines of a "before and after" method and that, having in this way arrived at what he considers a reliable figure, he will then proceed to apportion it if necessary as between compensation for land taken and compensation for severance and injurious affection to the remaining land.

The Lands Tribunal accepted the method in *Budgen* v *Secretary of State for Wales*[21] and in *Landlink Two Ltd* v *Sevenoaks District Council*[22] but in *ADP&E Farmers* v *Department of Transport*[23] preferred that there should be separate valuations of land taken and land retained.

In the case of claims under Part I of the 1973 Act the injurious affection is often determined by the alternative direct approach and expressed as a percentage of the open market value of the house in the absence of the scheme. The problem of applying the before and after method to such cases is that the "after value" represents the value of the property taking account of both the execution and also the use of the scheme whereas only the use gives rise to compensation. For example, a house situated in open countryside has a value which reflects both the view and the tranquillity. If a new road is built nearby the diminution in value will result from the loss of view (not compensatable) and the loss of tranquillity (compensatable as a noise factor). This may be the case for a comparable where a house is sold post the scheme opening and the price is compared with those achieved for houses unaffected by the scheme.

Consequently, compensation is often expressed as a percentage of value. The percentage is derived from the total diminution in value and depends on the impact of the use factors rather than the proximity to the scheme, although in most cases, of course, the closer the property is to the scheme the more likely is the use to have a greater adverse impact. The percentages many range from a modest 5% to 30% or more. In practice, a pattern of percentages will emerge so that compensation becomes dependent on the level of comparable settlements – although the Lands Tribunal has made it clear in many cases that market evidence will prevail against evidence based on settlements.

---

[21] [1985] 2 EGLR 203.
[22] [1985] 1 EGLR 196.
[23] 1988] 1 EGLR 209.

*Example 27–3*

The factory illustrated overleaf consists of a brick and slated building on two floors, in a reasonable state of repair, and is held on ground lease having 41 years still to run, the ground rent being £15 p.a. The leaseholder-occupier is a manufacturer and the estimated rack rental is £13,500 p.a. (£30 per m$^2$ ). The total floor space is 450 m$^2$ and the apportionment of the ground rent on the land to be taken has been agreed at £3 p.a.

A Highway Authority are acquiring a strip of land 8 m wide on the west side of the main road for road-widening purposes and notice to treat was served on all interested parties a month ago. Following the loss of the land the total floor space will be reduced to 355m$^2$.

The forecourt is used for loading and unloading goods, the entrance doors being at present in front. It may be assumed that the Highway Authority will not permit lorries to stand on the new road for these purposes, and that the claim put forward by the owners of the two interests concerned, that the whole of the premises should be acquired, will not be upheld.

It is required to assess the compensation payable and to list the probable "accommodation works" to be agreed.


*Valuation*

*Compulsory Purchase Compensation – II*
*Note*: The statutory assumptions do not appear to affect the case, and the "Before and After" method of approach is therefore adopted.

The accommodation works need to be agreed first as these will be reflected in the After valuation.

The accommodation works would probably include:

(i)   Demolition of existing front wall, erection of new wall and making good.
(ii)  Making side entrances and rear yard fit for traffic, e.g. tarmac surface, etc.
(iii) Any re-arrangement of drainage necessary.
(iv)  Forming opening in side or rear wall and provision of entrance doors.
(v)   New front gates and boundary walls.

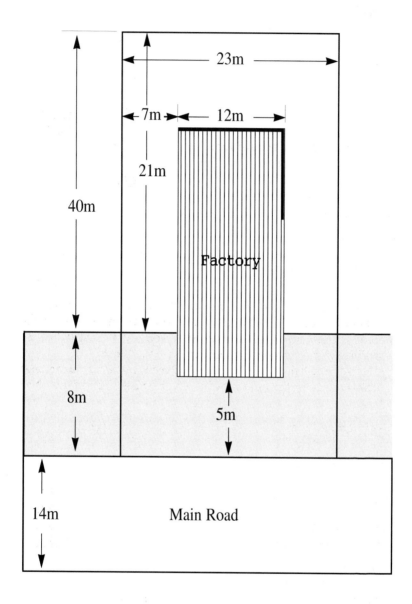

*Freeholder*
Value of interest *before* acquisition:

| | | | |
|---|---|---|---|
| Ground rent reserved | £15 p.a. | | |
| YP 41 yrs at 14% | 7.11 | | £106 |
| Reversion to | £13,500 p.a. | | |
| YP perp def'd 41 yrs | | | |
| @ 10% | 0.20 | | £2,700 |
| | | | £2,806 |

Value of interest *after* acquisition:

| | | | |
|---|---|---|---|
| Ground rent reserved | £12 p.a. | | |
| YP 41 yrs at 14% | 7.11 | £85 | |
| Reversion to 355 m² | | | |
| say, £22 per m² | £7,810 p.a. | | |
| YP perp def'd 41 yrs | | | |
| @ 11% | 0.13 | £1,015 | £1,100 |
| Compensation for land | | | |
| taken and injurious affection | | | £1,706 |

*Note*: The Years' Purchase and rental value per square foot used in the "after acquisition" valuation are lower than those used in the "before acquisition" valuation, as the property will then be less desirable, e.g. no forecourt loading and side entrance.

If it is required to divide this between the land taken and injurious affection the amount of the former could be obtained as follows:

| | | |
|---|---|---|
| Loss in ground rent | £3 p.a. | |
| YP 41 yrs @ 14% | 7.11 | £21 |
| Loss in rental value in | | |
| reversion 95 m² @ £30 per m² | £2,850 p.a. | |
| YP perp def'd 41 yrs @ 10% | 0.20 | 570 |
| | | £591 |
| Land taken, say | | |
| Injurious affection | | £1,115 |
| Compensation as above | | £1,706 |

*Leaseholder*
Value of interest before acquisition:

| | |
|---|---|
| Rental value | £13,500 p.a. |
| Ground rent | 15 |
| Profit rent | £13,485 |

| | | |
|---|---:|---:|
| YP 41 years @ 12 and 3% | | |
| (Tax at 33%) | <u>7.20</u> | £97,092 |
| Value of interest after acquisition: | | |
| Net rental value | £7,800 p.a. | |
| Ground rent | <u>12</u> | |
| Profit rent | £7,788 | |
| YP 41 years @ 13 and 3% | | |
| (Tax at 33%) | <u>6.70</u> | <u>£52,180</u> |
| Compensation for injurious affection | | |
| and land taken | | £44,912 |
| Cost of internal re-arrangement of factory | | |
| due to entrance doors being at side and | | |
| rear, say | | £20,000 |
| Temporary loss of net profits while | | |
| production impaired (see Chapter | | |
| 28), say | | <u>£15,000</u> |
| Total | | <u>£79,912</u> |

*Note*: If required, the sum of £44,912 mentioned could be apportioned between "land taken" and "injurious affection" in a similar manner to that done above in respect of the freehold interest.

*Example 27–4*

A sports ground with a total area of 3.2 ha is let by the freeholder to a club on annual tenancy at a rent of £8,000 p.a. The whole of the land is zoned residential at a density of 25 houses to the hectare in the Development Plan and it was anticipated that development would commence in two years' time when the land would be worth £1.25m per hectare.

The freeholder has just been served with a notice to treat in respect of 1.2 ha of the land which is required for a car park and for the erection of public conveniences.

The acquisition will not delay development of the land not being acquired but it is considered that its value for residential development will be only £1 million per hectare and that as a sports ground it will let for £3,000 p.a.

Assess the compensation payable to the freeholder for the land taken and injurious affection to the remainder.

*Valuation*

The most profitable assumption will be the value for the purpose for which the land is zoned and the claim will be made on this basis.

Value of interest before acquisition:

| Rental value as sports ground | £8,000 p.a. | | |
|---|---|---|---|
| YP 2 yrs @ 10% | 1.74 | £13,920 | |

| Reversion to building value: 3.2 ha @ £1.25m per hectare | £4,000,000 | | |
|---|---|---|---|
| PV £1 in 2 yrs @ 13% | 0.78 | £3,120,000 | £3,133,920 |

Value of interest after acquisition:

| Rental value as sports ground | £3,000 p.a. | | |
|---|---|---|---|
| YP 2 yrs @ 10% | 1.74 | £5,220 | |

| Reversion to building value 2.0 ha @ £1.0m per ha | £2,000,000 | | |
|---|---|---|---|
| PV £1 in 2 yrs @ 13% | 0.78 | £1,560,000 | £1,565,220 |

| Compensation for land taken and injurious affection | | | £1,568,700 |
|---|---|---|---|

1. Compensation for land taken

| Rental value as sports ground 1.2 ha @ £2,500 per ha | £3,000 p.a. | | |
|---|---|---|---|
| YP 2 yrs @ 10% | 1.74 | £5,220 | |

| Reversion to building value 1.2 ha @ £1.25m per ha | £1,500,000 | | |
|---|---|---|---|
| PV £1 in 2 yrs @ 13% | 0.78 | £1,170,000 | £1,175,220 |

| 2. Compensation for injurious affection | | | £393,480 |
|---|---|---|---|

*Note*: In calculating the compensation payable, allowance would have to be made in the valuations for any compensation payable by the landlord to the sports club as tenants under Part II of the Landlord and Tenant Act 1954.

*Chapter 28*

# Compulsory Purchase Compensation III Compensation for Disturbance

## 1. Disturbance Compensation and Disturbance Payments

Disturbance compensation is an element of compulsory purchase compensation, that is to say it is a sum added to the purchase price of land compulsorily acquired. It is not payable in respect of land retained by a claimant – a point re-affirmed in *Argyle Motors (Birkenhead) Ltd* v *Birkenhead Corporation*[1] – and is therefore a totally separate head of compensation from severance and injurious affection. It has been evolved as a principle of case law, without express authorisation in any statute (apart from Rule 6 of Section 5 of the Land Compensation Act 1961, which provides that the rules governing the assessment of land value "shall not affect the assessment of compensation for disturbance . . ."). Nevertheless, it has a statutory origin by implication, dating back to *Jubb* v *Hull Dock Co*,[2] in which a brewer obtained compensation for loss of business in addition to the purchase price of the compulsorily acquired brewery, the claim having been upheld on an interpretation of a private Act containing provisions similar to those of the Lands Clauses Consolidation Act 1845.

It is therefore in a sense "parasitic", because there must first be an acquisition price and, if there is, disturbance compensation can then be included in it. In *Inland Revenue Commissioners* v *Glasgow & South Western Railway Co*,[3] Lord Halsbury LC referred to acceptable claims for such well-known disturbance items as "damages for loss of business" and "compensation for the goodwill" but added: "in strictness the thing which is to be ascertained is the price to be paid for the land . . ."

In *Hughes* v *Doncaster Metropolitan Borough Council*,[4] Lord Bridge said:

---

[1] [1974] 1 All ER 201.
[2] (1846) 9 QB 443.
[3] (1887) 12 App Cas 315.
[4] [1991] 1 EGLR 31.

It is well-settled law that whatever compensation is payable to an owner on compulsory acquisition of his land in respect of disturbance is an element in assessing the value of the land to him, not a distinct and independent head of compensation.

It may be that the acquisition price is nominal, particularly in the case of leasehold interests, so that no actual sum is paid for the interest. This does not preclude a claim for disturbance.

The kinds of loss which are compensatable as "disturbance" are not exhaustively listed, and perhaps the only discernible principle is that they can constitute anything *other than* land value (because that is already compensatable as market value of land taken, or depreciation of land retained). The claimant must, however, "mitigate" his loss by taking care not to incur, and claim for, unreasonable and unnecessary expenditure. In *K&B Metals* v *Birmingham City Council*,[5] the claimant firm had bought equipment which could not be used after a draft discontinuance order was confirmed; the Lands Tribunal held the claim for the ensuing loss was reasonable (it would not have been if the purchase was *subsequent* to the confirmation of the order).

Removal costs are the most obvious example of disturbance; but it will be seen that there are many other examples of loss particularly in regard to business premises.

The connection between this loss and the compulsory acquisition of the land in respect of which it is suffered is causal and not chronological.

Costs incurred in *advance* of dispossession brought about by the compulsory acquisition are thus claimable provided that they are in no rational sense premature but satisfy the essential requirements of being (i) reasonably attributable to the dispossession and (ii) not too remote. The Court of Appeal so held in *Prasad* v *Wolverhampton Borough Council*,[6] as did the Court of Session in Scotland in *Aberdeen City District Council* v *Sim*.[7] These decisions upholding claims for anticipating disturbance items have been approved and followed by the Privy Council in an appeal from Hong Kong, *Director of Buildings and Lands* v *Shun Fung Ironworks Ltd*[8] the relevant

---

[5] [1976] 2 EGLR 180, a case concerning a discontinuance order though the principle is equally applicable to compulsory purchase.
[6] [1983] 1 EGLR 10.
[7] [1982] 2 EGLR 22.
[8] [1995] 1 All ER 846.

principles being the same in all three jurisdictions. This case is further discussed below (Judicial Approach to Disturbance Compensation).

Disturbance *payments*, on the other hand, are authorised expressly by statute (Sections 37–8 of the Land Compensation Act 1973) in certain cases in which disturbance compensation is not payable because, although a compulsory acquisition has in fact taken place (or some comparable occurrence – see below), the claimant himself has not had any interest compulsorily acquired from him but instead has merely been dispossessed. These are situations in which a tenant holding under a short-term leasehold has not had that term renewed by the acquiring authority which has taken his landlord's place on acquiring the latter's reversion. The dispossession must have occurred on or after 17 October 1972, in consequence:

(i) of the acquisition of the reversion under compulsory powers; or
(ii) of an order or undertaking under the Housing Acts; or
(iii) of redevelopment by an authority holding the land for a purpose for which the reversion was acquired under compulsory powers.

Licensees cannot claim these payments as of right, though authorities have a discretion to pay them. Agricultural land is excluded.

Section 38(1) of the 1973 Act provides as follows:

The amount of a disturbance payment shall be equal to:

(a) the reasonable expenses of the person entitled to the payment in removing from the land from which he is displaced; and
(b) if he was carrying on a trade or business on that land, the loss he will sustain by reason of the disturbance of that trade or business consequent upon his having to quit the land.

If a business tenant would also be entitled to compensation for loss of security of tenure under Section 37 of the Landlord and Tenant Act 1954, he is entitled to choose whichever amount is the higher, the payment under that Section or the disturbance payment, but not both (1973 Act, Section 37(4)). It should be observed that this provision for a disturbance payment is worded in pretty broad terms; though whether the Lands Tribunal (which settles disputes as to amounts, in the first instance) and the courts will interpret it broadly remains to be seen. Regard, however, must be had to the probable length of time for which the land from which the claimant has been dispossessed would have been available for his use, and also to the availability of other land, when "estimating the loss" (Section 38(2)).

It is important to remember that disturbance payments (under the 1973 Act) are claimable by a tenant who has been dispossessed but not expropriated; whereas disturbance compensation is claimable by any tenant or freeholder who has been *both* dispossessed *and* expropriated because it is included in (and treated as part of) his expropriation compensation. It follows from this that it is rare for a landlord to justify a claim for disturbance compensation, since he is not being dispossessed even though he may be expropriated. In *Lee* v *Minister of Transport*,[9] Davies LJ said: "Disturbance must, in my judgment, refer to the fact of having to vacate the premises". Thus, if premises are let by a freeholder to a periodic tenant, the former may be expropriated but not dispossessed while the latter (if served with a notice to quit and not notice to treat or notice of entry) will be dispossessed but not expropriated. The tenant would now qualify for a disturbance payment; but neither would be entitled to claim disturbance compensation.

An exception may, however, be recognised when both landlord and tenant are corporate bodies sufficiently closely connected with one another to be treated virtually as the same person, for example if they are related companies with the same directors and their separate identity is a mere matter of form. In *DHN Food Distributors Ltd* v *London Borough of Tower Hamlets*[10] the Court of Appeal held this to be the case for two companies with the same directors, of which one company owned the freehold of the acquired premises and the other occupied them as licensee; and the directors were awarded disturbance compensation. But in *Woolfson* v *Strathclyde Regional Council*,[11] the House of Lords insisted that normally where there are two companies they must expect to be treated separately,

[9] [1966] 1 QB 111.
[10] (1975) 30 P&CR 251.
[11] [1978] 2 EGLR 19.

since they are in law two separate "persons". Their Lordships distinguished the *DHN* case narrowly upon its own special facts. In *Woolfson* there were two companies, as in *DHN*, one owning the acquired premises and the other occupying them under a licence; ownership and membership of the two companies was not, however, identical, even though there was a very considerable overlap as regards both shareholders and directors. The *DHN* type of case will be rarely met.

A more recent exception, of express statutory origin, is that owners not in occupation, i.e. landlords and reversionary owners in general, can include in claims for the market value of their reversionary interests when expropriated, reimbursement of costs incurred in obtaining new premises in the United Kingdom (Land Compensation Act 1961, Section 10A, inserted by the Planning and Compensation Act 1991, Section 70). Such a claim must be made within a year of the entry on to the land by the acquiring authority. It is analogous to disturbance even though there is no physical dispossession but merely deprivation of an investment interest in landed property. As with physical disturbance, the claim is parasitic on the relevant value purchase price of the reversion: no purchase price, no reimbursement of expenses. There may be difficulties in practice if only an advance payment has been made within the 12-month period. Although the implication is that the incidental costs and expenses incurred in acquiring the new premises are for a property of similar value to the acquired premises the Act refers only to costs of "acquiring . . . an interest in other land in the United Kingdom".

If a periodic tenant or any tenant whose term does not exceed a year is in fact turned out (by notice of entry) before his tenancy is terminable at common law, then Section 20 of the Compulsory Purchase Act 1965, requires the acquiring authority to compensate him accordingly, since he is being expropriated. The market value of his interest may well be small, or nominal, but the right to compensation in principle carries with it the right to include disturbance items, which may well constitute the bulk of the claim in such a case.

## 2. Judicial Approach to Disturbance Compensation

Lord Nicholls, delivering the majority decision of the Privy Council in *Director of Buildings and Lands (Hong Kong)* v *Shun Fung Ironworks*

*Ltd*,[12] included several statements of basic principle about disturbance, all on the unstated assumption that the claimant is an owner-occupier. "To qualify for compensation a loss suffered . . . must satisfy the three conditions of being causally connected, not too remote, and not a loss which a reasonable person would have avoided." The third condition is usually expressed as a requirement that the claimant must "mitigate" his loss, which means that he must not unreasonably inflate the loss: it does not mean that he must make excessive efforts to assist the acquiring authority. It follows that, if reasonable, a claim to cover the cost of relocating a business can be acceptable notwithstanding that this may exceed the cost of total extinguishment of the business. "It all depends on how a reasonable businessman, using his own money, would behave in the circumstances." The same discount rates for capitalising lost profits should be used for relocation cases and extinguishment cases alike.

Relocation must not be confused with "equivalent reinstatement" which only applies when there is no general demand or market for the current use of the land.[13]

The duty to mitigate a loss arises when there is certainty that the compulsory acquisition will proceed, usually accepted as being when a notice to treat is served. So, if a notice to treat and notice of entry are served some while after an Order is confirmed such that relocation of a business is impossible in the time between the service of the notices and actual entry, it is not open to an authority to argue that the claimant should have anticipated the situation and looked for other premises following confirmation of the Order.

### 3.  Disturbance Compensation: Items which may be Claimed

*(a)  Residential Premises*

Where a dwelling-house is compulsorily acquired, the occupier, if he be a freeholder, or a lessee enjoying an appreciable profit rent, will be entitled to compensation based on the market value of his interest in the property, estimated in accordance with the statutory provisions already described. If he be a lessee or tenant holding under a contractual tenancy and paying approximately the full rack

---

[12] See n. 8, *supra.*
[13] Land Compensation Act 1961, Section 5, rule (v). See Chapter 26.

rental of the premises, the value of his interest in the land will be negligible, but he will be entitled to some measure of compensation for being forced to quit before the expiration of his term – similarly, when the interest is no greater than that of a yearly tenant and the tenant is required to give up possession before the expiration of his interest. No definite basis can be prescribed for these types of cases; but where the lease still has a number of years to run, a figure of one year's rent is often allowed for what may conveniently be called "extinction of lease". But if the acquiring body have compulsorily purchased the reversion they will become the tenant's new landlord, and they may well be content to await the termination of his tenancy contractually by effluxion of time or notice to quit, in which case there will be no purchase price in which disturbance can be included.

In addition to the value of his interest in the land, or a suitable allowance for the termination of that interest, and in theory part of it, the occupier will then also be entitled to disturbance compensation in respect of:

(i)   disposal of fixtures and other items which cannot be removed from the property being acquired other than at an unreasonable cost;
(ii)  cost of removal from the property being acquired to the new home; and
(iii) expenses incurred in acquiring the new property and other incidental expenses caused by having to leave the premises.

*Non-removable Items*

The occupier may remove any fixtures and other items to which he is entitled, or he may require the acquiring body to take them as part of the premises. Items which typically come within this category are carpets, curtains, cookers and other kitchen equipment ("white goods").

If the rule that a claimant must take all reasonable steps to mitigate his loss is strictly applied, it would seem that a claimant is entitled to claim the lesser of the value of these items on a sale to the authority or the costs of removing them to the new home including any costs of adaptation which may be required to fit them into their new surroundings or costs of plumbing them in or otherwise getting them into working order. In practice, acquiring authorities are usually happy to let the claimant decide whether he wishes to keep these items by transferring them to the new home or prefers to leave them.

Where items cannot be transferred, for example carpets or curtains which will not fit into the new property, claimants are not entitled to the cost of buying new carpets or curtains since they are treated as receiving value for money for their purchases.

In practice, the surveyors acting for the parties are able to agree the value of items not transferred. The basis of valuation is the price that a purchaser would pay in a normal open market transaction.

*Cost of Removal*
This item will include not only the actual cost of moving the owner's furniture to other premises but also any other expense incidental thereto, such as temporary storage of furniture. An authority normally requires three quotations to be obtained to show that the claimant is mitigating the cost. Storage will arise where there is a gap between giving possession and moving into a new home. Where the claimant is unable to find another home before entry by the authority he may claim the cost of renting furnished accommodation or living in an hotel until a new home is found. The problem of renting accommodation is that landlords normally offer tenancies of a minimum of six months which may seem to be excessive if a new home is found within that period.

Where a claimant hires a van and carries out the move himself or with the assistance of friends or family, he cannot claim the cost he would have incurred if he had employed a firm of furniture removers. On the other hand, the duty to mitigate the loss does not require the claimant to indulge in a self-help approach.

*Incidental Expenses*
In addition to cost of removal, there may be other incidental expenses incurred by an owner-occupier in consequence of being dispossessed of his premises. In *Harvey v Crawley Development Corporation*[14] Romer LJ summarised the principle governing the inclusion of such items in the claim as follows:

---

[14] [1957] 1 QB 485.

> The authorities . . . establish that any loss sustained by a
> dispossessed owner (at all events one who occupies his house)
> which flows from a compulsory acquisition may properly be
> regarded as the subject of compensation for disturbance
> provided, first, it is not too remote and secondly, that it is the
> natural and reasonable consequence of the dispossession of the
> owner.

In the case in question the claimant, who was required to sell and
give up possession of her house, was entitled to recover legal costs,
surveyor's fees and travelling expenses incurred in finding another
house to live in, together with similar expenses incurred to no
purpose in connection with the proposed purchase of a house on
which she received an unfavourable report from her surveyor. Such
fees are a genuine loss; whereas a purchase price paid for new
premises, large or small, represents not a loss but "value for
money".

Items within this category, often termed "Crawley Costs", will
depend upon the circumstances of each particular case and will
include such matters as value of stationery made redundant, costs
of notifying friends and organisations of the change of address,
costs of installing or transferring a telephone, loss on the forced sale
of school uniforms if children have to change schools, and so on. In
the case of a claimant who is disabled, the costs of adapting the new
property to his needs, or those of a member of his family, are
recoverable since they are not costs of adaptation which are
reflected in the price/value of the new house, in contrast to, say, the
costs of upgrading an outdated kitchen or bathroom.

Legal and other professional fees incurred after notice to treat in
connection with the preparation of the owner's claim are also
recoverable under rule 6 of Section 5 of the Land Compensation Act
1961. This practice was sanctioned by the Court of Appeal in
*London County Council* v *Tobin*.[15] But the same Court, in *Lee* v
*Minister of Transport*,[16] held that these items, though within Rule 6,
are not "disturbance" but are covered by the words "any other
matters not directly based on the value of the land". Thus,
solicitors' and surveyors' fees incurred in obtaining new premises
count as "disturbance" items; but solicitors' and surveyors' fees

---

[15] [1959] 1 All ER 649.
[16] [1966] 1 QB 111.

incurred in preparing the compulsory purchase claim itself count as "any other matter". As an exception to the general position, the legal fees incurred in conveying the interest to the authority are reimbursed under Section 23 of the Compulsory Purchase Act 1965.

### (b) Business and Professional Premises

Like the householder, the occupier of business premises will be entitled on compulsory purchase to compensation for:

(i)   the value of his interest in the land or for "extinction of lease";
(ii)  loss of or injury to fixtures; and
(iii) cost of removal.

In addition, a trader or business man may have a claim for loss on stock and trade disturbance. Unlike domestic claimants the occupier of business premises may be unable to find suitable alternative premises. In such cases, the claim for disturbance will be based on the loss suffered by the total extinguishment of the business rather than the costs and losses suffered by its removal to other premises. The claimant does not normally have a choice. If suitable alternative premises are available then the business should be transferred to them. If the claimant chooses not to move, the compensation will be assessed on the basis of a move – see, for example, *Rowley* v *Southampton Corporation*[17] and *Bailey* v *Derby Corporation*.[18] If the claimant does move to new premises but other premises were available which were suitable and would have resulted in lower compensation being payable then compensation is assessed as if the claimant had moved to the other premises – *Appleby & Ireland Ltd* v *Hampshire County Council*[19] and *Landrebe* v *Lambeth London Borough Council*.[20] An exception to this rule is where the claimant runs a business and he has reached the age of 60 at the date when he gives up possession and the premises from which he conducts the business have a rateable value not above a prescribed limit (currently £18,000). In such cases Section 46 of the Land Compensation Act 1973 gives the claimant the right to opt for total extinguishment so long as he gives undertakings not to dispose of

---

[17] (1959) 10 P&CR 172.
[18] [1965] 1 All ER 443.
[19] [1978] 2 EGLR 180.
[20] [1996] 36 RVR 112.

the goodwill or otherwise to resume business within a specified area or time.

*Compensation for Interest in Land*

As in the case of house property, if a lessee is enjoying a substantial profit rent he will be entitled to compensation for loss of his leasehold interest based on the ordinary principles of compulsory purchase valuation as explained above, in Chapter 26.

It is submitted that a claim of this kind is not affected by the provisions of the Landlord and Tenant Act 1954. For although under that Act the tenancy will, in most cases, automatically continue after the termination date, the landlord, by serving the necessary notice under the Act, can ensure that the new tenancy shall be at a rent at which (having regard to the other terms of the tenancy) the holding might reasonably be expected to be let in the open market. In any case, the Land Compensation Act 1973, Section 47, requires that both the business tenancy and the reversion to it be valued on the basis that the tenancy enjoys statutory protection.

Where the rent already paid under the lease is approximately the full rack rental value of the premises, the additional security of tenure given by the Landlord and Tenant Act 1954, would appear to strengthen the tenant's claim for compensation for "extinction of lease", although it must be remembered that this factor may well be reflected in the claim for "injury to goodwill" or "trade disturbance".

At all events, there is no justification for any item of loss being paid twice over, no matter what head of claim it may appear under. Whenever business premises are compulsorily purchased (as distinct from being terminated at common law) the purchase price, of the tenancy and of the reversion alike, is determined on the footing that the tenancy enjoys security of tenure under the Act of 1954 (unless, of course, the tenancy is "outside the Act" – see Chapter 18). The authority's right to obtain possession is ignored for this purpose; though if the landlord has in fact an independent ground to terminate the tenant's security of tenure, that must be taken into account. A similar principle applies to agricultural holdings (Land Compensation Act 1973, Sections 47 and 48).

In the case of a tenant from year to year, or for any lesser interest, whose right to compensation depends on Section 20 of the Compulsory Purchase Act 1965 (replacing Section 121 of the Lands Clauses Consolidation Act 1845) Section 39 of the Landlord and

Tenant Act 1954, as amended by Section 47 of the Land Compensation Act 1973 (also see Section 48 for agricultural holdings), expressly provides that the total compensation paid to him is not to be less than that which, in certain circumstances, he might have received under Section 37 of the 1954 Act if a new tenancy had been refused, i.e. an amount currently equal to one times the rateable value, or to two times the rateable value where the tenant or his predecessors in business have been in occupation of the holding for the past 14 years or more.

*Fixtures and Fittings, Plant and Machinery*
The occupier may insist on these being taken with the premises, particularly where the possibility of transferring and adapting them to other premises is somewhat problematical. He nevertheless has this choice open to him on the authority of *Gibson* v *Hammersmith and City Rail Co.*[21] In *Tamplin's Brewery Ltd* v *County Borough of Brighton*[22] the acquiring authority made new premises available to the claimants, and the outstanding question of compensation related to equipment. The claimants bought new equipment and relinquished the old. They then claimed the cost of the new equipment as an item of compensation, reduced only by an amount representing a saving in operating costs. The Lands Tribunal accepted this in principle, though with a reduction in the total sum. Yet it would seem that the claimants, in obtaining new equipment, got "value for money" in so doing; and this may cast doubt on the principle of the award in this case.

Normally the compensation will be the value of the existing fixtures to an incoming tenant for the business or, if sold by the claimant, the loss on forced sale being the difference between the value to an incoming tenant and the price achieved on a sale.

Value to an incoming tenant is the price that a purchaser of the business would pay for these items so that they represent the value to the business. The price is likely to be above that payable for similar items available elsewhere if only because they are in place and immediately available for use.

---

[21] (1863) 2 Drew & Son 603.
[22] (1971)P&CR 746.

The underlying principle in determining value within a disturbance claim is that it is value to the owner/claimant, not market value. In *Shevlin* v *Trafford Park Development Corporation*[23] the authority's valuer adopted the provisions of the RICS *Appraisal and Valuation Manual* in valuing the plant and machinery. This provided that:

> Where a valuation of plant and machinery is intended to reflect its work to an undertaking which is expected to continue in operation for the foreseeable future, the normal basis of valuation will be "value to the business" which is defined as the "value based upon the assumption that the plant and machinery will continue in its present existing use in the business of the company".

The Tribunal member took the view that "value to the business" is consistent with value to the owner and accepted the approach adopted by the authority's valuer which was to determine net current replacement cost at £92,965. However, it was accepted that the value to the claimant was above the replacement cost and a figure of £100,000, determined by "a robust approach" was adopted.

It is common for specialist plant and machinery valuers to be instructed to agree the values. The claimant's valuer will also be able to organise a disposal of the items. Where there is a significant amount of items an auction may even be held on the premises. The price achieved at auction, or on a sale by other means is deducted from the value to an incoming tenant. Since the price will be depressed because of the circumstances this is referred to as a forced sale price, and thus the claim is the loss on forced sale.

*Cost of Removal*
This will often be substantial, including in the case of industrial premises such items as the cost of moving machinery and its re-installation in new premises. Acquiring authorities commonly require two or three quotations to be obtained.

---

[23] [1998] 1 EGLR 115.

*Incidental Expenses*
These will include notification of change of address to customer
and suppliers, new sign writing on vehicles, value of redundant
stationery, statutory redundancy payments where staff cannot be
re-employed, as well as accrued holiday pay and statutory
compensation for short-term notice where applicable, travelling
expenses and management time in seeking new premises, and so
on. Each claim turns on its own particular facts and the acquiring
authority is liable to meet all reasonable expenses incurred which
are not too remote. The authority in effect takes the claimant as it
finds them. So, for example, where a claimant who owns an
affected business lives on the other side of the world, the fares and
other costs incurred in coming to the UK to deal with the removal
of the business may be reasonably incurred, even though the costs
are high. As in the case of private houses, there is authority for the
inclusion of legal and surveyors' fees incurred in connection with
the acquisition of new premises and also legal and other
professional fees such as accountants' or specialist valuers' fees
incurred in connection with the preparation of the claim. See for
example *LCC* v *Tobin*.[24] In the *Shevlin* case referred to above, the
Tribunal accepted accountants' fees for advice on the taxation
implications of the receipt of compensation.

*Loss on Stock*
This may consist either of depreciation in the process of removal to
other premises, or loss on forced sale in those cases where no other
premises are available, or where the trade likely to be done in new
premises will be of a somewhat different class, or where the stock
will not bear removal.

Loss on depreciation may be expressed as a percentage of the
value in the trader's books, but may be covered by the cost of
taking out an insurance policy before the move. Loss on forced sale
is often high since purchasers are aware of the circumstances
behind the sale – possibly 33–50% on retail prices. Where the
turnover of stock is high, as with, for example, retail grocers, a
valuation of the stock just prior to removal may be required. This
may require the payment of fees for a specialist stock valuer which
would be included in the claim.

---

[24] See n.15, *supra*.

*Trade Disturbance*

"Trade disturbance" is the term applied to the loss likely to result from a trader or business man being dispossessed of his premises. The capital value of the business, as distinct from that of the premises, is its "goodwill", and depends on the profits made in the course of trading. "Goodwill" is capital, and profits are income; but since the one is derived from the other it is important always to remember that the same loss is not compensatable twice over.

Compensation paid for loss of goodwill does not mean that the acquiring authority has purchased the goodwill. Any goodwill which exists following the displacement from the premises remains with the claimant. Acquiring authorities paying compensation for the loss of goodwill may seek undertakings from the claimant not to open a new business within a certain time or a certain distance whereby the claimant can resurrect the goodwill. Where claimants procure a claim for total extinguishment of their business as of right under Section 46 of the 1973 Act (see below), the Section provides for the acquiring authority to seek such undertakings as a condition for exercising the claimant's right under the Section.

It is also important to remember that the value of goodwill is not its open market value as such but the value to the owner.[25] Hence evidence from trade valuers as to the price for which a business would be sold is not necessarily the value to the owner and thus the compensation to which he is entitled.

It will be convenient to discuss the nature of this loss in relation to retail trades and then to see to what extent the same considerations apply to the wholesale trader, the manufacturer or the professional man.

The "goodwill" of a business has been defined as the probability of the business being carried on at a certain level of profit. In effect it is the value of the business itself as a capital asset, distinct from (a) the premises, (b) the stock-in-trade and other chattels, and (c) the profits and income from the business. Where a trader has been in business in certain premises for a number of years he will have acquired a circle of regular customers on whose patronage he can rely; he will also be able to count on a fairly steady volume of casual custom. As a result, by striking an average over three or four normal years he can make a fairly accurate estimate of his annual

---

[25] See, for example, *Sceneout Ltd* v *Central Manchester Development Corporation* [1995] 2 EGLR 179.

profits. It is the probability of those profits being maintained, or even increased, in the future which constitutes the "goodwill" of the business.

Although goodwill is a capital item assessable on the prospect of continuing profits in future years, the profits to be earned in the course of current trading are another matter. They are the subject of compensation in their own right (*Watson* v *Secretary of State for Air*)[26] if the compulsory acquisition prevents their being earned in fact because it cuts short the commercial activity which would, if continued, have produced them.

In *Bailey* v *Derby Corporation*,[27] the Master of the Rolls, Lord Denning, in regard to the claimant's demand for compensation in respect of the goodwill of his business, said: "All that is acquired is the land. The compensation is given for the value of the land, not for the value of the business." But the reason why the claim for goodwill failed in this case was that the claimant could have disposed of it on the open market had he tried. His excuse for not trying, namely age and ill-health, was not in fact valid, because age and ill-health need not prevent such a transaction being put in hand. It is to be noted that the claimant succeeded in obtaining £1,200 for loss of profits.

The sequel to this case was the enactment of Section 46 of the Land Compensation Act 1973. This allows claimants aged 60 or over (irrespective of health) to require compensation to include extinguishment of goodwill if they have not independently disposed of that asset. They must give an undertaking not to dispose of it nor to carry on the business themselves within a reasonable area and period to be specified, on pain of having to repay the amount in question. The premises must not have a rateable value in excess of a specified amount, currently £18,000. Special rules apply where a claim is made by a partnership or company.

The various factors which go to make goodwill can usually be divided into two classes:

(i)  those which, being of a personal nature, do not depend on the situation of the premises, e.g. the name and reputation of the firm or the personality of the proprietor;

---

[26] [1954] 2 All ER 582 – in this case, farming profits for the current year.
[27] [1965] 1 All ER 443.

(ii) those which are dependent on situation, e.g. the advantage derived from being on the main shopping street of a town or in a street or district which is the recognised centre for businesses of a particular type.

When business premises are taken under compulsory powers the tradesman's goodwill is not necessarily destroyed but it may be seriously damaged even if the business is transferred to other premises. How seriously depends upon whether the goodwill is mainly personal in character or largely dependent upon the situation of the premises. The test of whether goodwill can legitimately be said to have been partially or totally extinguished, thus justifying a claim for compensation, is to consider whether or not it could reasonably be sold on the open market and whether its marketability, if any, has been reduced because of the circumstances of the compulsory purchase.

The usual method of assessing this class of loss is by multiplying the average annual net profits by a figure of Years' Purchase which varies according to the extent of the injury likely to be suffered. In most cases the figure of net profits is based on the past three years' trading as shown by the trader's books. However, if the profits have been adversely affected by the scheme for a period then such profits would be ignored. For example, if the last year's accounts have been adversely affected, the previous three years "untainted" accounts might be adopted. The accounts are normally supplied to the valuer by the claimant's accountant. They are commonly accounts prepared for tax purposes rather than management accounts which show the true profits. Consequently adjustments need to be made to the accounts.

For example, the figures given will not usually have taken into account interest on the capital employed in the business, and it is argued that, as the trader could earn interest upon his capital by investing it elsewhere, the annual profit due to the business is really not the figure taken from the books, but that figure less interest on capital. The tenant's capital would comprise the value of the fixtures and fittings, stock, etc., and the necessary sum to be kept in hand for working expenses. The latter item would obviously be larger in the case of a credit business than in a cash business. The valuer needs, therefore, to examine the figure supplied to him as the average net profits and to make a deduction for interest on capital if that has not already been done. In practice it is usual to deduct the interest on the capital which is valued as at the valuation date. Clearly, since capital varies over the years it would be more accurate to take interest on the average value of the capital over the averaging period of the net profits, but this is normally an impossible exercise.

He must also ascertain what rent has been charged in the books. Usually the actual rent paid is charged in the accounts, and if the valuer estimates that the property is worth more than the rent paid, he must deduct the estimated profit rent. In the case of a freehold occupier, if no rent has been charged, the estimated annual rental used for arriving at the value of the property must be deducted. The reason for this is that, if the business ceased, the profit rent or rental value could continue to be obtained by letting the property.

The wages of assistants employed in a firm will naturally be deducted in finding the net profits. But it was held by the Lands Tribunal in *Perezic* v *Bristol Corporation*[28] that there was no evidence that it was customary to deduct a sum in respect of the remuneration of the working proprietor of a one-man business. This was extended to the remuneration for a husband and wife operating a business (*Zarraga* v *Newcastle-upon-Tyne Corporation*).[29] So any deduction made in the accounts for such remuneration will be added back. This also applies to any deductions for director's wages or pension payments where it is a family business owned and run by the director or directors. This is particularly common for businesses owned and run by a husband and wife.

In *Speyer AV* v *City of Westminster*,[30] the Tribunal expressed the view that whether such a deduction was appropriate or not must depend on the effect of the disturbance on the earning capacity of the claimant.

In *Appleby & Ireland Ltd* v *Hampshire County Council*[31] a considerable number of adjustments to the net profits shown in the accounts were made, including adjustments in respect of directors' reasonable remuneration (as against actual salaries paid), bad debts, and exceptional expenditure on research. Indeed, the accounts need to be examined and adjustments made to remove factors which distort the profit levels. For example, a large professional fee in one year only may arise because of a lease

---

[28] (1955) 5 P&CR 237.
[29] (1968) 205 EG 1163.
[30] (1958) 9 P&CR 478.
[31] [1978] 2 EGLR 180.

renewal or rent review. It would be more realistic if such costs were averaged over the period of the lease or the period to the next review as one-off, non-recurring items. Since interest on capital is deducted, any items for depreciation of plant, fixture, or leases are commonly added back, but in the *Shevlin* case referred to above, the Tribunal did not add back depreciation as the figure for plant and machinery adopted to determine interest on capital was the written down value allowing for depreciation.

Such of these deductions and adjustments as are applicable having been made, and true net profits having been found, the next step is to fix the multiplier which should be applied to this figure in order fairly to compensate the trader for injury to his goodwill.

Before considering the level of multipliers which may be appropriate in any particular case a general comment is appropriate. In practice, the multipliers range from 0.5 to 5. These are reflected in the many Lands Tribunal decisions on the value of goodwill but, as the Tribunal frequently stresses, each case turns on its facts and the multiplier adopted in one case cannot be used as a "comparable" for another. The problem in practice is that there are no true comparables. It may be possible to discover a multiplier from an actual market sale of a business but this is evidence of market value whereas what is sought is value to the owner. It is agreed that value to the owner may be equal to market value but it is often higher. It is common practice for a pattern of settlements to emerge which become comparables within the scheme. The multiplier in a disputed case is then derived from these comparables (as an example, see *Klein v London Underground Ltd*).[32] The multipliers referred to below are based on the general experience of the authors and those adopted in Lands Tribunal cases. They are intended to be indicative only of the levels that might be applied.

Where the taking of the premises will mean total extinction of trade the full value of the goodwill will be allowed as compensation. In the case of retail businesses, this normally ranges between two to five Years' Purchase of the net profits. The figure used will be influenced, amongst other things, by whether the profits appear to be increasing or decreasing.

Where the trade will not be totally lost but may be seriously injured by removal to other premises, a fair compensation may be

---

[32] [1996] 1 EGLR 249.

about half what would be paid for total extinction. But a much lower figure will be given where the trade is likely to be carried on almost as profitably as before.

A cash tobacconist adjoining the entrance of a busy railway station may lose the whole trade if the premises upon which the trade is carried on are taken from him and he can only relocate in a quite different location – in essence he will have to start up a new business. In such a case, up to the full value of the business may be appropriate compensation – possibly three to four Years' Purchase of the net profits.

A milkman, on the contrary, knowing all his customers by reason of delivery of milk daily at the houses, may face removal with little risk of loss of that portion of his trade, and a baker or family butcher or tailor may be much in the same category, although any of these may suffer some loss of casual custom, and also in respect of some items sold over the counter. In such cases as little as half-a-year's profits might be proper compensation.

It is obvious that, since the variety of trades is so great and the circumstances of each case may vary so widely, every claim of this sort must be judged on its merits. But the following are amongst the most important considerations likely to affect the figure of compensation:

(i)   *The nature of the trade and the extent to which it depends upon the position of the premises*
      For instance, a grocer or confectioner or draper whose business is done entirely over the counter will suffer considerable loss if he has to move to premises in an inferior trade position. Probably one and a half to two Years' Purchase of his net profits would not be excessive compensation. On the other hand, a credit tailor would be far less likely to be injured by removal since his customers might be expected to follow him, assuming that he can find premises convenient to them.

      Where the trade is partly cash and partly credit, the valuer may think it proper to separate the profits attributable to the two parts and to apply a higher figure of Years' Purchase to the former than to the latter.

(ii)  *The prospects of alternative accommodation*
      For instance, in many cases a shopkeeper turned out of his present premises will find it impossible to obtain a shop elsewhere in the town and a claim for total loss of goodwill

may be justified. Or, again, the fact that the only alternative accommodation available is in an inferior trading position may considerably affect the figure.

(iii) *The terms on which the premises are held*
The compensation payable to a trader holding on a lease with a number of years to run will probably be substantially the same as that of a claimant in similar circumstances who owns the freehold of his premises – particularly in view of the additional security of tenure enjoyed under the Landlord and Tenant Act 1954. But a shopkeeper who holds on a yearly tenancy only can scarcely expect as high a level of compensation. Where a lease contains a redevelopment clause or other power for repossession by the landlord, the effect on the YP will depend on the likelihood of the power being exercised.

(iv) *The pattern of profits*
The value of goodwill depends on the prospects for making profits in the future. The prospects can be extrapolated from recent evidence, normally, as previously related, the last three years accounts. If these show sharply rising profits then the prospects are bright, supporting a high YP, whilst falling profits have the opposite effect.

In addition to compensation for the value of the goodwill or the decrease in its value, a trader may have lost profits in the period running up to closure or transfer of the business. For example, a retailer selling women's clothes would aim to dispose of as much stock as possible if the business is to close. Profits would be below normal levels in the period as a result of both price cutting and also not having new season designs in stock. Thus, a claim for temporary loss of profits would be made in addition to the value of goodwill. This is also the sort of case where the final year's profits would be ignored (or adjusted) in determining the average net profits.

The difficulty of determining temporary loss of profits lies in establishing what the profits would otherwise have been. One approach is to extrapolate the profits from previous years if there is an identifiable trend. Another is to establish the pattern of profits from other premises to produce an index which can be applied to the subject premises. For example, if the retailer selling ladies' clothes is a multiple retailer, the profit patterns from similar outlets can assist in assessing the likely profits that would have been earned in the subject premises. The difference between these and the actual profits equals the temporary loss.

The question of compensation payable to wholesale traders and manufacturers involves similar considerations to those already discussed. It will usually be found, however, that the probable injury to trade is much less than in the case of retail business, since profits are not so dependent upon position. Provided other suitable premises are available, the claim is usually one for so many months' temporary loss of profits during the time it will take to establish the business elsewhere, rather than any permanent injury to goodwill (e.g. see *Appleby & Ireland Ltd* v *Hampshire County Council*).[33] In practice, alternative premises will need to be taken before the loss of the existing property. This will result in a period when the trader is incurring two payments of rent, perhaps rates, insurance and similar outgoings. The additional payments will be compensated as part of the temporary loss of profits, as they reduce profits. They are referred to as double overheads.

In this type of case the sum claimed for temporary loss of income has been regarded as not subject to income tax because compulsory purchase compensation, regardless of its particular components is supposed to be "one sum" in the nature of a "price to be paid for the land" (*Inland Revenue Commissioners* v *Glasgow & South Western Railway Co*[34] – a decision of the House of Lords). It was therefore held in *West Suffolk County Council* v *Rought Ltd*,[35] that the Lands Tribunal, in assessing the compensation payable for temporary loss of profits, should have deducted their estimate of the additional taxation which the claimant company would have had to bear if it had actually earned the amount which the interruption to its business prevented it from earning, because if earned in fact that amount would have been liable to tax as income. On this basis acquiring authorities for some time paid compensation to companies net of tax. However, changes in the tax status of compensation payments flowed from provisions in the Finance Acts 1965 and 1969 and the Income and Corporation Taxes Act

---

[33] See n. 19, *supra*.

[34] (1887) 12 App Cas 315 (the words of Lord Halsbury LC).

[35] [1957] AC 403 – a decision of the House of Lords following the House's decision in *British Transport Commission* v *Gourley* [1956] AC 185, which applied this principle to payments of damages in tort.

1970, and the current position, exemplified in the decision of the Court of Appeal in *Stoke-on-Trent City Council* v *Wood Mitchell & Co Ltd*,[36] is that temporary loss of profits, loss on stock and such revenue items as removal expenses and interest, are paid gross by the acquiring authority and then taxed as receipts in the hands of the claimant company.

In the case of professional men and women, such as solicitors or surveyors, the goodwill of the business is mainly personal and is not very likely to be injured by removal. But the special circumstances of the case must be considered, particularly the question of alternative accommodation. For instance, a firm of solicitors in a big city usually requires premises in the recognised legal quarter and in close touch with the courts, although this consideration might be of less importance to an old-established firm than to a comparatively new one.

In some cases of trade disturbance a basis somewhat similar to that of reinstatement is used in making the claim.

Thus, a claimant forced to vacate his factory, shop, etc., may show that he is obliged to move to other premises, which are the only ones available, and will base his claim for trade disturbance on the cost of adapting the new premises plus the reduced value of his goodwill in those premises.

The latter figure may be due to reduction in profits on account of increased running costs (other than rent) in the new premises. For instance, heating and lighting may cost more, transport costs may be higher, or it may be necessary to employ more labour owing to the more difficult layout of a production line.

Or, again, the new position may not be so convenient for some customers as was the old. Or it may involve some change in the nature of the business which tends to make the profits rather less secure. For instance, in *LCC* v *Tobin*[37] the Court of Appeal approved a figure for trade disturbance although it appeared that the profits in the new premises were likely to be higher than in the old. The compensation was based on the difference between the value of the profits in the old premises multiplied by 3 YP and those in the new premises multiplied by 1.5 YP, the lower figure being justified by changes in the general circumstances of the business in the new premises.

---

[36] [1978] 2 EGLR 21.
[37] See n.15, *supra*.

Notwithstanding the emphasis given to the generally accepted approach of applying a multiplier to average profits, there will be cases where the profits are small or where the business has been in recession and shows small profits over recent years or even losses following earlier years of prosperity. A strict application of the accepted approach may produce a small amount for goodwill or even nil. Depending on the facts of the case this may be recognised as an inadequate figure. For example, the years before acquisition were a bad period but the prospects for recovery are good. At its extreme, a loss-making business may nonetheless have a valuable goodwill. As a departure from the norm the valuers may agree that it is more appropriate to adopt a different approach – such as discounting projected future profits[38] – or simply plumping for a spot figure or a high multiplier with no evidence to support it, termed as 'adopting a robust approach' by the Lands Tribunal.[39] The aim is to arrive at what is seen to be and accepted as a fair figure of compensation.

It is common for the disturbance claim to be prepared some time after the claimant has moved from the acquired premises. If so, information can be obtained from the trading figures in the alternative premises which provide strong evidence as to the loss of profits suffered by the claimant in the period following the move. Regard will be had to such evidence applying the *Bwllfa* principle of the weight to be given to evidence of events after the valuation date.[40]

To justify a claim based on reinstatement of the business in other premises a claimant must show that he has acted reasonably and taken only such premises as a prudent person would take in order to safeguard his interests and mitigate his loss. In *Appleby & Ireland Ltd* v *Hampshire County Council*[41] the Lands Tribunal held that the claimants should have moved to premises in the town where they were when acquired and compensation was assessed on the notional move rather than costs and losses incurred in the actual move to another town.

If there are no premises to which he can remove, except at a cost much out of proportion to what he is giving up, he may be justified

---

[38] *Reed Employment Ltd* v *London Transport Executive* [1978] 1 EGLR 166.
[39] *W Clibbett Ltd* v *Avon County Council* [1976] 1 EGLR 171.
[40] *Bwllfa & Merthyr Dare Steam Collieries Ltd* v *Pontypridd Waterworks Co* [1903] AC 426.
[41] See n.19, *supra*.

in taking those premises; but he may in certain circumstances, particularly in the case of a trader, be compelled to allow a set-off from the cost of so reinstating himself in respect of the improved position he enjoys.

An acquiring body may sometimes offer alternative accommodation with a view to meeting a claim for trade disturbance.

Thus, if a retail trader claims that his business will suffer complete loss by the taking of the premises in which it is carried on, it is open to the acquiring authority to indicate premises which can be acquired for the reinstatement of the trade, or even to offer to build premises on land belonging to themselves, available for the purpose, and to transfer the premises to the claimant, in reduction of his claim. Arrangements of this kind are usually a matter of agreement between the parties. A claimant cannot be forced to accept alternative accommodation, although he may find it difficult to substantiate a heavy claim for trade disturbance in cases where it is offered.

In *Lindon Print Ltd* v *West Midlands County Council*[42] the Lands Tribunal accepted that the onus of proof is on the acquiring authority to show that the claimant has not mitigated his loss.

### Adverse Taxation Consequences

One consequence of a compulsory purchase is that the claimant owner is forced to sell at a time of the authority's choosing and not his own. The purpose of the compensation provisions previously discussed may achieve the goal of providing adequate compensation. However, the subsequent taxation of the compensation by the Inland Revenue or other tax authorities may result in inadequate sums being retained.

In most cases this will not be so. A residential owner-occupier will be exempt from capital gains tax, if the main residence exemption applies as it would on a market disposal. Both investors and occupiers of business premises will benefit from rollover relief (see Chapter 22).

However, there will be cases where the claimant will suffer unfair tax consequences. This was recognised in *Alfred Golightly & Sons Ltd* v *Durham County Council*.[43]

---

[42] [1987] 2 EGLR 200.
[43] [1981] 2 EGLR 190.

In that case the claimant incurred a development land tax liability which was greater than would have been incurred if the interest had not been purchased compulsorily. The acquiring authority had to pay increased compensation under rule 6 to offset this increased tax burden.

The same principle can apply to other cases. For example, a claimant may be the proprietor of a business. He is (say) aged 49. As a result of compulsory purchase the business is extinguished. He receives compensation for the value of his interest in the property and the goodwill. He will be liable to capital gain tax on this compensation.

In the absence of the scheme he can establish that his business plan was to continue the business until he reaches 50, and then to sell the business some time afterwards to suit himself when he would take advantage of the retirement relief from capital gains tax, perhaps paying no capital gains tax.

He is therefore faced with a tax liability as a consequence of the compulsory purchase. He can claim a tax indemnity from the acquiring authority who will be responsible for the tax payable on the compensation as further compensation.

### Example 28–1

Market Traders Ltd is a company owned by Mr and Mrs Trader. The company is run by Mr Trader whilst Mrs Trader acts as company secretary.

The company owns a head leasehold interest in a warehouse. The lease is for 80 years from 1945 at a fixed rent of £200 p.a. The current rental value is £10,000 p.a. The lease is being acquired compulsorily and the business will be extinguished. The profit and loss account for the past year is set out below. The balance sheet shows the tangible fixed assets apart from land and buildings to be:

| | |
|---|---|
| Motor vehicles | £58,749 |
| Plant and machinery | £62,460 |
| Fixtures and fittings | £12,370 |

They have been valued by a trade valuer at £180,000.

**Profit and Loss Account for Past Year**

| | | |
|---|---:|---:|
| Turnover | | 1,189,284 |
| *Cost of sales* | | |
| Opening stock | 139,869 | |
| Purchases | 944,632 | |
| Salaries and wages | 59,197 | |
| Closing stock | (115,440) | |
| | | 1,028,258 |
| Gross profit | | 161,026 |
| *Less Overheads* | | |
| Motoring expenses | 7,219 | |
| Telephone charges | 704 | |
| Printing, postage and stationery | 2,056 | |
| Staff refreshments | 558 | |
| Directors' bonus | 27,000 | |
| Directors' remuneration | 54,670 | |
| Directors' national insurance | 8,330 | |
| Directors' pension costs | 1,355 | |
| Heating and lighting | 4,837 | |
| Repairs and renewals | 15,771 | |
| Insurances | 4,397 | |
| Rent and rates | 11,581 | |
| Sundry expenses | 221 | |
| Cleaning | 4,448 | |
| Bank charges | 2,560 | |
| Bad debt provision | 170 | |
| Auditors' remuneration | 3,750 | |
| Accountancy fees | 962 | |
| Depreciation | 14,767 | |
| | | 165,356 |
| Net trading loss for the year | | (4,330) |

| | | |
|---|---:|---:|
| *Step 1* Assess the Adjusted Net Profit | | £ |
| Annual Profit (Loss) from accounts | | (4,330) |
| *Add* Directors' Renumeration | | |
| (husband and wife are sole proprietors) | | 91,355 |
| | | 87,025 |
| *Add* Depreciation | | 14,767 |
| | | 101,792 |

| | | |
|---|---:|---:|
| *Deduct* Profit Rent (10,000 – 200) | | <u>9,800</u> |
| Net Profit | | 91,992 |
| *Deduct* | | |
| Interest on Capital | | |
| Current Value of Vehicles, P&M, F&F | £180,000 | |
| Cash in hand say, 2 weeks' purchases, say | <u>£40,000</u> | |
| | £220,000 | |
| Interest @ 8% | <u>0.08</u> | <u>17,600</u> |
| Adjusted Net Profit | | £74,392 |

*Step 2* Prepare Claim for Disturbance
The Net Profits need to be assessed for each of the previous three years, as in Step 1. Assume the net profits are:

| | | |
|---|---|---:|
| Year 1 | Net Profit | £62,740 |
| Year 2 | Net Profit | £74,812 |
| Year 3 | Net Profit | £91,992 |

*Claim for Disturbance*
1. Goodwill

| | | |
|---|---:|---:|
| Net Profits last 3 years | £62,740 | |
| | £74,812 | |
| | <u>£91,992</u> | |
| | £229,544 | |
| | <u>÷ 3</u> | |
| Average Net Profits | £76,515 | |
| *Less* Interest on capital (as above) | <u>£17,600</u> | |
| Adjusted Net Profits | £58,915 | |
| Profits have risen by 50% over 3 years, say YP | <u>4</u> | £235,660 |

2. Loss on Forced Sale

| | | |
|---|---:|---:|
| Value to incoming tenant of Vehicles, F&F, P&M | £180,000 | |
| Price realised at auction (assume) | <u>£73,900</u> | |
| Loss on Forced Sale | £106,100 | |
| Stock (assume 40% of £115,440) | <u>£46,176</u> | £152,276 |

3. Redundancy Payments
Payments to staff calculated
on statutory basis (assume)   £37,000
*Add* Accrued holiday pay
(assume)                         £7,400                       £44,400

4. Miscellaneous
(Depends on circumstances of each case).

5. Fees
(a) Reasonable legal fees on conveyance of lease
(b) Surveyors' fees under Ryde's Scale

## 4. Disturbance in relation to Development Value

In *Horn* v *Sunderland Corporation*,[44] Lord Greene, MR, said that rule 6

> does not confer a right to claim compensation for disturbance.
> It merely leaves unaffected the right which the owner would,
> before the Act of 1919, have had in a proper case to claim that
> the compensation to be paid for the land should be increased on
> the ground that he has been disturbed.

In the same case this right was described as "the right to receive a
money payment not less than the loss imposed on him in the public
interest, but on the other hand no greater".

It is true that, in practice, the value of the land itself will be
assessed in accordance with rule 2 and another figure for disturbance may be arrived at under rule 6. But these two figures are, in
fact, merely the elements which go to build up the single total
figure of price or compensation to which the owner is fairly entitled
in all the circumstances of the case and which should represent the
loss he suffers in consequence of the land being taken from him.

It follows that an owner cannot claim a figure for disturbance
which is inconsistent with the basis adopted for the assessment of
the value of the land under rule 2.

For instance, in *Mizzen Bros* v *Mitcham UDC* (1929)[45] it was held
that claimants were not entitled to combine in the same claim a

---

[44] [1941] 2 KB 26; 1 All ER 480.
[45] Estates Gazette Digest, 1929, p. 258; reprinted *Estates Gazette*, 25 January,
1941, p. 102.

valuation of the land on the basis of an immediate sale for building purposes and a claim for disturbance and consequential damage upon the footing of interference with a continuing market garden business since they could not realise the building value of the land in the open market unless they were themselves prepared to abandon their market garden business. The existing use value of the land as a market garden was about £12,000, and the disturbance items totalled £4,640. These items when aggregated still fell short of the development value of the land for building, which was £17,280, but they could not justify any additional claim on top of that sum by way of disturbance, even though they were in fact being disturbed.

On the other hand, where a valuation on the basis of the present use of the land, plus compensation for disturbance, may exceed a valuation of the land based on a new and more profitable user, the owner has been held entitled to claim the former figure, though not both.

The facts in *Horn v Sunderland Corporation* were that land having prospective building value and containing deposits of sand, limestone and gravel was compulsorily acquired for housing purposes. The owner occupied the land as farm land, chiefly for the rearing of pedigree horses.

The owner claimed the market value of the land as a building estate ripe for immediate development and also a substantial sum in respect of the disturbance of his farming business.

The official arbitrator awarded a sum of £22,700 in respect of the value of the land as building land, but disallowed any compensation in respect of disturbance of the claimant's business on the grounds that the sum assessed could not be realised in the open market unless vacant possession were given to the purchaser for the purpose of building development.

It was held by a majority of the Court of Appeal that the arbitrator's award was right in law provided that the sum of £22,700 equalled or exceeded:

(i)   the value of the land as farm land; plus
(ii)  whatever value should be attributable to the minerals if the land were treated as farm land; plus
(iii) the loss by disturbance of the farming business.

If, however, the aggregate of items (i), (ii) and (iii) exceeded the figure of £22,700 the claimant was entitled to be paid the excess as part of the compensation for the loss of his land.

Claims for disturbance must relate to losses which are the direct result of the compulsory taking of the land and which are not remote or purely speculative in character. Loss of profits in connection with a business carried on the premises and which will be directly injured by the dispossession of the owner is a permissible subject of claim. But where a speculator claimed, in addition to the market value of building land, the profits which he hoped to make from the erection of houses on the land, the latter item was disallowed.[46] It was quite distinct from the development value of the land, which was a legitimate item of claim on the basis just described. The point at issue was succinctly put by Lord Moulton in a decision of the Privy Council when he said that "no man would pay for the land, in addition to its market value, the capitalised value of the savings and additional profits which he would hope to make by the use of it".[47] Again, where an owner of business premises compulsorily acquired was also the principal shareholder in the company which occupied the premises on a short-term tenancy, a claim in respect of the depreciation in the value of her shares which might result if the acquiring body gave the company notice to quit was held to be "too remote" for compensation.[48]

In *Palatine Graphic Arts Co Ltd v Liverpool City Council*,[49] the Court of Appeal decided that a regional development grant paid to claimants should not be deducted from their compulsory purchase compensation.

## 5. Home Loss and Farm Loss Payments

The Land Compensation Act 1973, makes additional provision for compensation broadly within the scope of disturbance, namely "home loss" and "farm loss" payments. These do not depend on the distinction between dispossession with expropriation and dispossession without expropriation, and therefore cut across the

---

[46] *Collins v Feltham UDC* [1937] 4 All ER 189; and see *D McEwing & Sons Ltd v Renfrew County Council* (1959) 11 P&CR 306.

[47] *Pastoral Finance Association Ltd v The Minister (New South Wales)* [1914] AC 1083.

[48] *Roberts v Coventry Corporation* [1947] 1 All ER 308 (the words of Lord Goddard CJ).

[49] [1986] 1 EGLR 19.

dividing line which separates disturbance compensation from disturbance payments.

Home loss payments may be claimed under Sections 29–33 of the 1973 Act, as amended by Section 68 of the Planning and Compensation Act 1991, by a freeholder, leaseholder, statutory tenant or employee, in respect of a dwelling substantially occupied by him (in that capacity) as his main residence for 12 months prior to displacement in circumstances similar to those giving rise to a disturbance payment as described earlier in this chapter. (Caravan-dwellers also can claim a home loss payment on proof of 12 months' residence within a caravan site). The claim must be made within six months of the displacement. If the claimant has been resident for 12 months but not in a capacity described above (i.e. he has been a mere licensee for part of the time), the period can be made up by adding the period of occupancy of his predecessors. The payment consists of a set figure, which is 10% of the value of the interest in the dwelling, subject to a maximum of £15,000 and a minimum of £1,500. A home loss payment has been upheld in respect of displacement from a "pre-fab" which was due for removal and replacement by a more substantial dwelling.[50]

Farm loss payments may be claimed under Sections 34–6 of the 1973 Act, as amended by Section 70 of the Planning and Compensation Act 1991, by freeholders or leaseholders, including yearly tenants, whose interest is compulsorily purchased with the result that they are displaced from their farms ("agricultural units") or part only of their farms (0.5 hectares or more) against their will. A claimant must begin to farm another "agricultural unit" in Great Britain not more than three years after being displaced, and at some time between the date when the acquiring authority were authorised to acquire his original farm and the date when he begins to farm the new one he must acquire a freehold or a tenancy in the latter. The amount of the payment is calculated as the average annual profit from the original farm during the three years immediately before the displacement, or during such shorter period as the claimant has been in occupation, deducting (i) an amount representing a reasonable rent (regardless of the actual rent) if the claimant is a tenant responsible for outgoings, and (ii) any items already included in a claim for disturbance compensation. If the new farm is not so valuable as the old, the

---

[50] *R* v *Corby District Council, ex parte McLean* [1975] 2 EGLR 40.

payment must be scaled down accordingly. This is done by reference to the existing use value of each unit (ignoring development value, that is to say) in accordance with market prices at the date when the new unit is begun to be farmed, as regards that property, and at the date of displacement, as regards the original property. The main dwelling on each property, if there is one, must also be excluded from the calculation; but the "willing seller" rule and attendant rules in Section 5 of the Land Compensation Act 1961 are to be applied. It should also be noted that if the purchase price of the original property includes development value the amount of the development value will be deducted from the farm loss payment. Claims must be made within one year after the farming of the new unit begins, and disputes over the amount claimed are to be settled by the Lands Tribunal. Interest is payable, at whatever rate is currently prescribed for compulsory purchase compensation, from the date of the claim, as also are sums for reasonable legal and valuation costs.

One most important complication in regard to farm loss payments is that no such payment can be claimed by a farmer already entitled to "additional compensation to tenant for disturbance" under Section 12 of the Agriculture (Miscellaneous Provisions) Act 1968 (as amended by Schedule 14 to the Agricultural Holdings Act 1986, applying Section 60(2)(b) of that Act). This is so regardless of whether the payment under the 1968 Act is the full amount, namely four times the yearly rent, or a mere fraction of that figure no matter how small. Section 12 of the 1968 Act provides that, in the event of a compulsory purchase, the full amount is payable in principle, but that where a term of two years or more is expropriated there is a "ceiling" on the compensation that is fixed at the amount of compensation that would be paid to a yearly tenant subject to compulsory purchase of the property. This means that a farmer with a short-to-medium length agricultural tenancy which is compulsorily purchased might be entitled to a compulsory purchase payment (including disturbance, which will normally be a major item in it) such that the "ceiling" is approached but not actually reached, which in turn means that a small fraction of the Section 12 payment under the 1968 Act is still to be added in order to reach that "ceiling". In turn, this means, under the 1973 Act, that if a Section 12 payment (however small) is due, no "farm loss payment" can be claimed. But the ceiling does not apply to the compulsory purchase compensation as such, which may reach or exceed it. If so a "farm loss payment" is claimable under the rules mentioned above.[51] The interrelation of these various compensation provisions is clearly a difficult and complicated matter and must be considered with great care in each particular case.[52]

---

[51] See also the references to agricultural holdings on pages 607–608, *supra*.
[52] For a further consideration of compulsory purchase compensation, including disturbance, see Chapter 9 of *Valuation: Principles into Practice* (4th Edition) Editor W.H. Rees (Estates Gazette).

# Index